MULTISTATE SYSTEM RELIABILITY WITH DEPENDENCIES

MULTISTATE SYSTEM RELIABILITY WITH DEPENDENCIES

AGNIESZKA BLOKUS

ELSEVIER

ACADEMIC PRESS

An imprint of Elsevier

Academic Press is an imprint of Elsevier
125 London Wall, London EC2Y 5AS, United Kingdom
525 B Street, Suite 1650, San Diego, CA 92101, United States
50 Hampshire Street, 5th Floor, Cambridge, MA 02139, United States
The Boulevard, Langford Lane, Kidlington, Oxford OX5 1GB, United Kingdom

Library of Congress Cataloging-in-Publication Data
A catalog record for this book is available from the Library of Congress

British Library Cataloguing-in-Publication Data
A catalogue record for this book is available from the British Library

ISBN 978-0-12-821260-8

For information on all Academic Press publications
visit our website at https://www.elsevier.com/books-and-journals

Publisher: Matthew Deans
Acquisitions Editor: Brian Guerin
Editorial Project Manager: Rafael G. Trombaco
Production Project Manager: Sojan P. Pazhayattil
Cover Designer: Victoria Pearson

Typeset by SPi Global, India

Contents

Acknowledgment

I would like to express my heartfelt thanks and appreciation to Professor Krzysztof Kołowrocki for his substantive support and encouragement that had invaluable and great impact on this book.

CHAPTER ONE

Introduction

In the reliability and safety analysis, the independence of system components is often assumed that could limit its applicability in many practical situations. The problem of system component dependence applies to both two-state systems and multistate systems. In two-state system reliability analysis, when considering the dependencies of component failures, it is assumed that a system component failure may cause the corresponding load to be distributed over the remaining components (Birolini, 1994; Blokus-Roszkowska, 2007a; Daniels, 1945; Harlow and Phoenix, 1978, 1982; Pradhan et al., 2010; Smith, 1982, 1983). Similar problem is analyzed by Song et al. (2014) assuming that failures of one component can impact all components, potentially causing them to fail more often.

The multistate system was introduced in the mid-1970s (Barlow and Proschan, 1975; Barlow and Wu, 1978; El-Neweihi et al., 1978; Murchland, 1975; Ross, 1979). El-Neweihi et al. (1978) developed the concept of a multistate coherent structure as a generalization of binary coherent structure. Different approaches to describe multistate systems and to estimate their reliability are found in (Ebrahimi, 1984; Fardis and Cornell, 1981; Griffith, 1980; Hudson and Kapur, 1982, 1983; Kołowrocki, 2003b, 2014; Murchland, 1975; Natvig, 2007; Ohio and Nishida, 1984; Ushakov, 1986; Zio, 2009). Liu et al. (2008) use fuzzy theory to assess reliability and performance for a system consisting of a single multistate element. The authors represent the intensity of the multistate elements' transition between each state and the performance rates by fuzzy values. The book of the authors Lisnianski and Levitin (2003) contains a comprehensive overview and presentation of methods of multistate systems reliability analysis. Further extension of the book's results on reliability assessment and optimization with applications is contained in the book (Lisnianski et al., 2010). This second book presents stochastic process methods for reliability analysis of multistate systems with multistate components, including Markov, Markov reward, and semi-Markov models, as well as universal generating function (UGF) method. Moreover, fuzzy multistate systems and their reliability evaluation are introduced in Lisnianski et al. (2010) in case performance and failure

Multistate System Reliability with Dependencies
https://doi.org/10.1016/B978-0-12-821260-8.00001-4

rates of system components cannot be exactly determined. Ding and Lisnianski (2008) propose fuzzy universal generating functions for multistate system reliability analysis, extended from the universal generating function method introduced by Ushakov (1986) and described in (Levitin, 2005). This method is also used for the analysis of multistate systems with dependent elements (Levitin, 2004). Considering failure dependency, in the literature the issue of multiple dependent competing failure processes for components and systems is often raised. Competing failure processes can cause either a component or a system failure. Rafiee et al. (2014) propose particular random shock patterns for changing degradation rate of such systems. Song et al. (2014) consider system of degrading components that experience s-dependent competing failure processes. Wang and Pham (2012) also analyze s-dependent competing failure processes resulting from degradation and random shocks, however using time-varying copulas. Wang et al. (2011) consider catastrophic (binary state) failure, degradation (continuous processes), and failure due to shocks (impulse processes). Chen et al. (2015) analyze failure mechanism dependence, including acceleration, inhibition, accumulation, and competition and propose reliability evaluation of system incorporating these dependence failure processes. Multistate systems with dependent components and subsystems often occur in reality, and their reliability and safety analysis is justified and really met in practice. Hence, this type multistate system is also widely discussed in the literature (Abouammoh and Al-Kadi, 1991; Amari and Misra, 1997; Aven, 1985, 1993; Brausch, 1987; Brunelle and Kapur, 1999; Butler, 1982; Griffith, 1980; Kołowrocki, 2000; Natvig, 1984; Natvig and Streller, 1984; Yingkui and Jing, 2012). Another practically important approach to multistate system reliability analysis considers the assumption about component degradation (departures from the reliability state subsets) instead of component failures (Blokus-Roszkowska, 2007a; Kołowrocki, 2008a,b, 2014; Xue, 1985; Xue and Yang, 1995a,b). In real technical systems, components often degrade with time by going to states corresponding to different performance levels. Degradation of components, or subsystems in case of complex systems, causes the decreasing of system reliability and its tasks performance.

This book is focused on the reliability analysis of multistate aging systems using the last approach with its development by the assumption of component dependency. Many multistate methods are difficult to apply practically with enough good accuracy because using them causes that it is necessary to consider a large number of transitions between the states what results that receiving exact solutions is not possible and evaluation of the approximate solution is often not precise. The approach proposed in this book allows to eliminate this problem and get the exact values of basic system reliability indicators, independent of the number components and the number of their reliability states. It means that large numbers of system components and their reliability states do not restrict the possibility of the proposed approach use and receiving exact solutions. The main idea of the approach to the multistate aging systems reliability analysis presented in this book is to select a model of load sharing that best describes the impact of changes in the reliability state of components on the reliability states of other components. This idea, developed in the book, is a generalization of two-state system reliability analysis, where equal load-sharing (ELS) and local load-sharing (LLS) rules are commonly used (Daniels, 1945; Harlow and Phoenix, 1978, 1982; Phoenix and Smith, 1983; Pradhan et al., 2010; Smith, 1982, 1983). The equal load-sharing (ELS) model, in which the load on the failed two-state component is transferred uniformly among the remaining components, has been studied early by Daniels (1945) and Smith (1982, 1983), and later by Pradhan et al. (2010). Local load-sharing (LLS) rule, in which the load on the failed two-state component is transferred to other components proportionally to their distance from the failed component, was introduced by Harlow and Phoenix (1978, 1982), and further analyzed by Phoenix and Smith (1983). Various other models of failure dependency of system components are found, and different approaches to this problem are applied, both analytical and numerical (Jain and Gupta, 2012; Kostandyan and Sørensen, 2014; Singh and Gupta, 2012; Singh et al., 2008). Some concepts of load-sharing models, Daniels model, and extensions of the Daniels model are described by Crowder et al. (1991), referring to the components as fibers and the system as a bundle as fibers. Load-sharing and dependency models in the system reliability modeling are also given by Misra (1992); however, they apply to two-state systems. Kostandyan and Sørensen (2014) estimate reliability of a system by structural reliability approach and the failure mechanism is based on a fracture mechanics model. In a multistate LLS dependency model, the mean values of component lifetimes in the reliability state subsets vary depending on the distance from the component that was just changed to its reliability state subset, or more generally, depending on component exposure to the reliability state change of the component that has left the reliability state subset. Whereas, in an ELS dependency model, the mean values of the component lifetimes in the reliability state subsets change with equal dependence on the number of components that have left their reliability state subset. Those multistate approaches to the reliability analysis of systems composed of dependent components have been presented by Blokus-Roszkowska and Kołowrocki assuming an ELS model (Blokus-Roszkowska, 2007b; Blokus-Roszkowska and Kołowrocki, 2014, 2015a) and in the case of an LLS dependency model (Blokus-Roszkowska and Kołowrocki, 2015b, 2017).

Samaniego (2007) also points out that dependencies among components and their lifetimes can occur due to stress, in reliability analysis of systems. However, the author provides a tool for qualifying reliability and identifying influence of component failures on system failures and its lifetime, assuming that the components are independent and have the same lifetime distributions. The signature system is this tool in structural reliability that allows evaluating features of the system's design. Samaniego (2007), presenting the system signatures in theory and applications, focuses on comparative reliability analysis. The book "Structural Reliability Analysis and Prediction" (Melchers and Beck, 2018) presents a broad overview of theory and techniques for structural reliability analysis and prediction of structural safety and serviceability. Measures of structural reliability are defined, and traditional methods as well as integration and simulation methods are included. Theory of structural reliability assessment is described paying attention to uncertainties in reliability assessment of complex systems especially during its prediction. Authors assumed time-invariant random loading in reliability analysis of structural systems; nevertheless, one chapter is devoted to load and load effect modeling. Stochastic processes are applied to modeling load processes concerning loads variability in time and location, and other dynamic effects. To illustrate the process of load modeling and its probabilistic description, wind loading, wave loading, and floor loading are discussed in Melchers and Beck (2018).

Different approaches to failure dependency, more exactly to reliability analysis of system subject to dependent failures processes, are proposed (Jiang et al., 2015; Li and Pham, 2005; Song et al., 2014). Song et al. (2014) developed reliability models considering different dependent patterns of shock effects on multiple failure processes for all components in a series system. Particularly, the authors consider multiple-component series system subject to hard failure, which is a failure that occurs suddenly when transmitted system shocks are large enough to cause any component failure, and soft failure, characterized by systematic and continuous degradation, which occurs when any component deteriorates to a certain threshold. In the reliability analysis of multistate systems composed of several subsystems, the impact of subsystem component degradation on other components degradation in this subsystem and the impact of those changes that have occurred within the subsystem to the reliability of other subsystems are taken into account (Blokus-Roszkowska and Kolowrocki, 2017). In such systems, various models of dependency among subsystem components and models of dependency among the subsystems are jointly applied. In this scope, the book proposes the mixed load-sharing (MLS) dependency model among components and subsystems, naturally assuming that the lifetimes of subsystem components in the reliability state subset decrease according to the LLS rule and the lifetimes of subsystems decrease according to the ELS rule. The MLS dependency model is applied in the book to reliability analysis of multistate parallel-series and "*m* out of *l*"-series systems.

In reliability analysis of a series system, it is automatically assumed that the system fails after any of its components failure. However, in reality, considering the aging system of multistate components, the deterioration of one's condition does not necessarily mean the system failure, although it may affect its lifetime. In that case, the influence of component departures from the reliability state subsets on the reliability state of remaining components is proposed. This approach to reliability analysis of aging-dependent multistate series system, initially introduced by Blokus-Roszkowska and Kołowrocki (2015c) and further developed in (Blokus and Kołowrocki, 2019), is the main source of innovative results of this book that are extended on the reliability analysis of systems with other more complex structures.

The main innovativeness of reliability models used in this book consists in combination of the multistate approach to reliability analysis of aging systems, introduced by Xue and Yang (Xue, 1985; Xue and Yang, 1995a,b), and widely developed in (Kołowrocki, 2008b, 2014; Kołowrocki and Soszyńska-Budny, 2011), with the dependence of their components degradation (Blokus-Roszkowska, 2007a; Blokus-Roszkowska and Kołowrocki, 2014, 2015a). Such analytical approach earlier was not presented in the literature and additionally this approach allows for finding new general solution presented in the book. The results, presented in the book, allow significantly to increase the accuracy of reliability analysis of real technical systems as the aging and components dependency are their obvious and natural features.

The book consists of seven parts, including this Introduction, Chapters 2 and 3 presenting results of reliability analysis of multistate systems, Chapter 4 presenting results of availability analysis of multistate systems, Chapter 5 related to safety of networks with cascading effects, Summary, and Bibliography.

In this chapter, the book introduction and its contents are given.

In Chapter 2, basic assumptions and definitions related to reliability analysis of multistate aging systems are introduced. The reliability functions of a multistate system and its components are defined. The system risk function and other basic reliability characteristics of a system and its components are also defined. Particularly, definitions of the mean lifetimes and the standard deviations of a system and its components in the reliability state subsets, and the mean lifetimes in particular reliability states are given. As a special case, Section 2.2 shows how multistate approach becomes the two-state system reliability analysis. It presents the form of assumptions in the two-state approach and definitions of the component reliability function and the reliability function of a two-state system. To illustrate a two-state and multistate approach to the reliability analysis, basic reliability characteristics of an exemplary system are determined. Discussing the results obtained in both cases, attention was paid to the benefits of a multistate approach allowing a more precise and closer to reality analysis of system reliability. In Section 2.3, basic reliability structures of multistate systems are introduced and their reliability functions are determined. Namely, the multistate series, parallel, "*m* out of *n*," series-parallel, parallel-series, series-"*m* out of *k*" and "*m* out of *l*"-series systems with aging components are defined, and their reliability functions are determined.

In Chapter 3, multistate systems with dependent components and dependent subsystems are analyzed and their basic reliability characteristics are determined. Based on the dependency among components and subsystems, the structure of the system, the specificity of its operation and the specificity of its components, three dependency models in a multistate system are considered: the LLS dependency model, the ELS dependency model, and the MLS dependency model. The reliability analysis of systems with those three dependency models is performed. Finally, to illustrate the way of theoretical result applications, the reliability analysis of an exemplary system and a shipyard rope elevator is presented. Chapter 3 is organized as follows.

Section 3.1 introduces the LLS dependency model. The approach to reliability analysis of multistate series systems with components following the LLS dependency rule is described. The reliability functions of such systems are determined in a general case and in particular case under the fixed components' reliability functions. Namely, for a series system with components having piecewise exponential distribution, Rayleigh distribution and Erlang distribution its reliability functions are determined. Additionally, for a piecewise exponential, aging series system, following the LLS dependency rule, its mean lifetime in the reliability state subset, the standard deviation of the system lifetime, and the intensities of system's departures from the reliability state subsets are given. Next, the LLS dependency model is extended for multistate series-parallel and series-"*m* out of *k*" systems, assuming the dependence according to the LLS rule among components in their series subsystems. For those systems, there are determined reliability functions in general case when their components have any reliability functions and in particular case when components have piecewise exponential reliability functions. As a special case, a homogeneous and regular series-parallel system and a homogeneous and regular series-"*m* out of *k*" system are considered as well.

Section 3.2 describes the ELS dependency model of components for multistate parallel and "*m* out of *n*" systems. Considering the components' lifetimes in the reliability state subsets for multistate systems following the ELS dependency rule, it is assumed that the influence of components departures from the reliability state subsets on other system components may differ depending on the system and components' features, expressed by a component stress proportionality correction coefficient. Under such assumptions the system reliability function, the mean lifetime, and the standard deviation of the system lifetime in the reliability state subset for multistate piecewise exponential parallel and "*m* out of *n*" systems are determined. Next, the ELS dependency model is extended for multistate parallel-series and "*m* out of *l*"-series systems, assuming the dependence according to the ELS rule among components in their subsystems. For those systems, it is assumed that after departure of some components of a subsystem from the reliability state subset, the lifetimes of remaining components in this subsystem decrease equally depending on the number of these components that have left the reliability state subset and on the component stress proportionality correction coefficient.

Section 3.3, linking the LLS and ELS dependency models, the MLS model for multistate parallel-series, and "*m* out of *l*"-series systems is proposed. It is justified to suppose that both transitions between reliability states of a subsystem's components may have an effect on the other components in this subsystem and that changes of the subsystem states may have an impact on the components of the remaining subsystems, causing a decrease of their lifetimes in the reliability state subsets. For the MLS dependency model, the reliability functions of aging parallel-series and "*m* out of *l*"-series systems are determined, under the assumption that their components follow piecewise exponential reliability functions. Moreover, the results are transferred to a special case of a homogeneous and regular parallel-series system and a homogeneous and regular "*m* out of *l*"-series system.

In Section 3.4, the obtained theoretical results of Sections 3.1–3.3 are illustrated by their application to the reliability evaluation of an exemplary system in Section 3.4.1. However, potential applications of load-sharing models are found in many areas, including textile engineering and material durability testing (Carlson and Kardomateas, 1996; Daniels, 1945; Harlow and Phoenix, 1978, 1982; Phoenix and Smith, 1983; Pradhan et al., 2010), technical systems reliability analysis (Blokus-Roszkowska, 2016; Blokus-Roszkowska and Kołowrocki, 2014, 2015a; Kołowrocki, 2003a), software reliability, civil and structural engineering (Amari et al., 2008; Kvam and Pena, 2005), safety assessment of power plant (Amari et al., 2008) and other. In Section 3.4.2, the MLS model for a multistate regular "*m* out of *l*"-series system is applied to reliability evaluation of the shipyard rope elevator, in case its components have piecewise exponential reliability functions assuming dependence of departures rates from the subsets of their reliability states. A discussion of the results obtained for different models of dependency and without considering dependence is presented.

Chapter 4 is devoted to maintenance analysis of multistate aging systems with dependent components under imperfect repair. The maintenance strategy for multistate systems and particularly for systems subject to dependent competing failure processes is considered in (Peng et al., 2010). The authors analyze reliability of complex systems that experience multiple dependent failure mechanisms due to degradation and/or shock loads. Liu et al. (2013) analyze redundancy and imperfect maintenance strategy for multistate systems assuming that even if the element is replaced by a new one, it cannot be restored to a virtually new condition. Imperfect maintenance of a deteriorating system is widely discussed in (Castro, 2009; Høyland and Rausand, 2004; Marquez, 2007; Pham and Wang, 1996; Wang and Pham, 2011). A maintenance strategy for a system that experience degradation and sudden shocks is proposed in (Caballé et al., 2015). The authors assume that a sudden shock provokes the total breakdown of a system. The maintenance strategy for a system subject to internal degradation and sudden shock is proposed, and under this maintenance strategy the expected cost rate function is determined (Caballé et al., 2015).

Many real technical systems are aging systems with multistate components degrading with time. In such systems, due to aging, their components degrade ranging from the state of full reliability to complete failure state

(Kołowrocki, 2003a; Kołowrocki and Soszyńska, 2006; Natvig and Morch, 2003; Ramirez-Marqueza and Coit, 2007; Szymkowiak, 2018, 2019). Considering maintenance and renewal of multistate systems, the replacement of a system component by a new one is often assumed. For a multistate aging system, it means that the component after repair is in the best "as good as new" reliability state, which in fact is not always possible. Moreover, even the perfect repair, due to conditions resulting from the state of the remaining components, may cause the component not to behave as new. In addition, its lifetimes in the reliability state subsets can be shorter than for a component in a new system (Blokus-Roszkowska, 2019). In this book, it is assumed that a multistate system is repaired after exceeding its critical reliability state. Taking into account this assumption under imperfect repair, the renewal of the system means that the system is restored to the critical reliability state or better, one of the previous degraded state, but not necessarily to the best state.

In Chapter 4 of this book, combining the results of the reliability analysis of multistate systems with dependent components and the results of the classical renewal theory (Badoux, 1987; Gertsbakh, 2000; Zio, 2007; Zio and Compare, 2013), the availability analysis of aging-dependent systems under imperfect repair (Levitin and Lisnianski, 2000; Muhammad et al., 2010; Nourelfath et al., 2012; Pham et al., 1997; Soro et al., 2010) is proposed. A repairable system with negligible renovation time is considered as well as in case when the renewal time is non-negligible and cannot be omitted. The system can be repaired after exceeding its critical reliability state, assuming that after some time or after a fixed number of repairs the system cannot be restored to the state of full reliability. Chapter 4 is organized as follows.

In Section 4.1, basic assumptions and definitions related to renewal stream and renewal process are presented. In next Sections 4.2 and 4.3, basic characteristics of a repairable system with negligible renovation time and in case when the renewal time is non-negligible are determined, under given in Section 4.1 assumptions of imperfect repair. Namely, the expected values of the times until the successive times that the reliability critical state is exceeded, and the expected values of the numbers of times that the reliability critical state is exceeded up to the fixed time point, are given. The procedure for determining the expected value of the number of times the system exceeds the reliability critical state up to the fixed time point are proposed for a system with negligible renewal time and in case the renewal time is significant and cannot be omitted. For a repairable system with non-negligible renewal time, the expected values of time until the successive system renewals, the expected values of the number of system renewals until the fixed moment, and the availability coefficient of the system at the fixed moment are determined.

Moreover, the obtained results of system availability analysis, assuming imperfect repair, are combined with the results of system reliability analysis obtained in Chapter 3 and applied to multistate systems with dependent components. In Section 4.4, the availability analysis of a multistate series system with components following the LLS dependency rule is performed and the availability of a series-parallel system and a series-"*m* out of *k*" system with dependent according to the LLS rule components of their series subsystems is discussed. In Section 4.5, a parallel system following the ELS dependency rule, an "*m* out of *n*" system following the ELS dependency rule, a parallel-series system with dependent according to the ELS rule components of its parallel subsystems, and an "*m* out of *l*"-series system with dependent according to the ELS rule components of its "*m* out of *l*" subsystems are considered. In Section 4.6, the availability analysis of multistate parallel-series and "*m* out of *l*"-series systems, assuming the MLS dependency model, is discussed as well.

The obtained theoretical results are applied in Section 4.7 to the availability evaluation of an exemplary transport navigation system. The results are compared for a single master navigation system and a navigation system with a back-up system. As a second application, the shipyard rope elevator in case when its components are independent and in case of the MLS dependency model, under assumption of imperfect repair, is analyzed.

Finally in Section 4.8, maintenance costs for a renewable system with negligible and non-negligible renewal durations are analyzed. Cost of repair as well as costs related to maintaining the continuity of the system and performing tasks are included. The procedure for determining the total cost of maintenance up to the fixed time point for fixed critical state, in both cases for a system with negligible renewal time and for a renewable system with non-negligible renewal time, is proposed. To minimize the maintenance costs for a renewable system, the procedure comparing the costs of repair and operation of the new system during time until the fixed moment, for different configuration of the number of repairs to various reliability states, is proposed. That procedure can help to select the most beneficial repair procedure that minimize the maintenance costs under assumption of imperfect repair.

In Chapter 5, the results of Chapter 3 are extended by considering networks with cascading effects. Instead of reliability, the safety analysis of critical infrastructure (CI) networks composed of assets and subnetworks is proposed.

Critical infrastructures are usually interconnected and mutually dependent in various and complex ways, creating critical infrastructure network. They are interacting directly and indirectly at various levels of their complexity and operating activity (Kjølle et al., 2012; Kotzanikolaou et al., 2013; Nieuwenhuijs et al., 2008; Ouyang, 2014; Rinaldi et al., 2001; Zio and Sansavini, 2011). Identifying and modeling dependencies depend on the level of analysis. The selected level of analysis can vary from micro- to macro-level (De Porcellinis et al., 2009; Han and Zio, 2019; Holden et al., 2013). A holistic approach as in Lauge et al. (2015) can be considered or a reductionist approach in which elementary components are identified and their behavior is described. For example, Svedsen and Wolthunsen (2007) focus on the components of critical infrastructure networks and they demonstrate several types of multidependency structures. Analysis performed in Chapter 5 of this book also focuses on the component level. Describing cascading effects in infrastructure networks, both the dependencies among subnetworks of this network and among their assets are considered. In that case, after changing the safety state subset by some of assets in the subnetwork to the worse safety state subset, the lifetimes of remaining assets in this subnetwork in the safety state subsets decrease. Models of dependency and behavior of components can differ depending on the structural and material properties of the network, operational conditions and many other factors.

In Section 5.1, basic assumptions related to safety analysis of multistate and aging CI networks, composed of assets, are introduced. The safety function of a multistate CI network and its assets are defined. Next, the safety functions of a multistate series CI network, composed of independent subnetworks and in case when its subnetworks are dependent according to the LLS rule, are determined. Moreover, the safety functions of multistate "*m* out of *k*" and parallel CI networks, composed of independent subnetworks and in case when their subnetworks are dependent according to the ELS rule, are determined. Special attention is paid to a multistate series CI network composed of series subnetworks. The safety of this CI network is analyzed in three cases. Namely, in case when subnetworks and their assets are independent, when subnetworks are dependent according to the LLS rule and consist of independent assets, and the last case when subnetworks are independent and assets in the subnetworks follow the LLS dependency rule. Additionally, the safety function of a network composed of "*m* out of *l*" subnetworks linked in series is determined. It is analyzed when the subnetworks and their assets are independent, when the assets of subnetworks are dependent according to the ELS rule, and when the subnetworks follow the LLS rule and their assets follow the ELS rule.

In Section 5.2, proposed theoretical results are applied to the safety analysis of an exemplary academic network and an exemplary distribution network. Since components of transmission and distribution networks require constant maintenance and degrading causes their properties deterioration over time, multistate approach to the safety analysis of distribution systems seems to be reasonable. In critical and overload states, the component degradation can proceed much faster. The multistate safety analysis of the distribution network is performed regarding its assets and subnetworks interdependencies. The transition between safety states of subnetworks or their assets can have impact on safety characteristics of remaining subnetworks and assets and as a consequence on safety of the whole system. Further, such approach to analysis of interconnections and interdependencies can help to capture the critical points and critical operations that can affect the whole network functioning. In both examples, the safety analysis results for the networks with dependent components are compared with the results obtained under assumption of component independence.

The book is completed with Summary, where the obtained theoretical results are summarized and the possibility of other theoretical developments and practical applications of the proposed reliability, availability, and safety analysis of multistate aging systems and infrastructures with dependent components and subsystems are suggested.

References

Abouammoh, A., Al-Kadi, M., 1991. Component relevancy in multi-state reliability models. IEEE Trans. Reliab. 40, 370–375.

Amari, S.V., Misra, R.B., 1997. Comment on: dynamic reliability analysis of coherent multistate systems. IEEE Trans. Reliab. 46, 460–461.

Amari, S.V., Misra, K.B., Pham, H., 2008. Tampered failure rate load-sharing systems: status and perspectives. In: Misra, K.B. (Ed.), Handbook on Performability Engineering. Springer-Verlag, London, pp. 291–308.

Aven, T., 1985. Reliability evaluation of multistate systems with multistate components. IEEE Trans. Reliab. 34, 473–479.

Aven, T., 1993. On performance measures for multistate monotone systems. Reliab. Eng. Syst. Saf. 41, 259–266.

Badoux, R.A.J., 1987. Availability and maintainability. In: Colombo, A.G., Keller, A.Z. (Eds.), Reliability Modeling and Applications. Kluwer Academic Publishers Group, Dordrecht, pp. 99–124.

Barlow, R.E., Proschan, F., 1975. Statistical Theory of Reliability and Life Testing. Probability Models. Holt Rinehart and Winston, Inc., New York.

Barlow, S.E., Wu, A.S., 1978. Coherent systems with multi-state components. Math. Oper. Res. 3 (11), 275–281.

Birolini, A., 1994. Quality and Reliability of Technical Systems. Theory – Practice – Management. Springer Verlag, Berlin, Heidelberg.

Blokus, A., Kołowrocki, K., 2019. Reliability and maintenance strategy for systems with aging-dependent components. Qual. Reliab. Eng. Int. 35, 2709–2731. https://doi.org/10.1002/qre.2552.

Blokus-Roszkowska, A., 2007a. Reliability Analysis of Homogenous Large Systems With Component Dependent Failures. PhD Thesis, Gdynia Maritime University – Polish Academy of Sciences, System Research Institute, Warsaw (in Polish).

Blokus-Roszkowska, A., 2007b. On component failures' dependency influence on system's lifetime. Int. J. Reliab. Qual. Saf. Eng. 14 (6), 1–19. Special Issue: System Reliability and Safety.

Blokus-Roszkowska, A., 2016. Reliability analysis of the bulk cargo loading system including dependent components. In: Simos, T., Tsitouras, C. (Eds.), Proceedings of the International Conference of Numerical Analysis and Applied Mathematics 2015 (ICNAAM 2015). AIP Publishing, pp. 440002-1–440002-4. AIP Conf. Proc. 1738.

Blokus-Roszkowska, A., 2019. Availability analysis of transport navigation system under imperfect repair. In: Zamojski, W., Mazurkiewicz, J., Sugier, J., Walkowiak, T., Kacprzyk, J. (Eds.), Contemporary Complex Systems and Their Dependability, Proceedings of the Thirteenth International Conference on Dependability and Complex Systems DepCoS-RELCOMEX. Springer International Publishing AG, part of Springer Nature 2019, pp. 35–45.

Blokus-Roszkowska, A., Kołowrocki, K., 2014. Reliability analysis of complex shipyard transportation system with dependent components. J. Pol. Saf. Reliab. Assoc. Summer Saf. Reliab. Semin. 5 (1), 21–31.

Blokus-Roszkowska, A., Kołowrocki, K., 2015a. Reliability analysis of ship-rope transporter with dependent components. In: Nowakowski, et al. (Eds.), Safety and Reliability: Methodology and Applications – Proceedings of the European Safety and Reliability Conference, ESREL 2014. Taylor & Francis Group, London, pp. 255–263.

Blokus-Roszkowska, A., Kołowrocki, K., 2015b. Reliability analysis of conveyor belt with dependent components. In: Podofillini, et al. (Eds.), Safety and Reliability of Complex Engineered Systems – Proceedings of the European Safety and Reliability Conference, ESREL 2015. Taylor & Francis Group, London, pp. 1127–1136.

Blokus-Roszkowska, A., Kołowrocki, K., 2015c. Reliability analysis of multistate series systems with dependent components. J. Pol. Saf. Reliab. Assoc. Summer Saf. Reliab. Semin. 6 (1), 31–36.

Blokus-Roszkowska, A., Kolowrocki, K., 2017. Modelling safety of multistate systems with dependent components and subsystems. J. Pol. Saf. Reliab. Assoc. Summer Saf. Reliab. Semin. 8 (3), 23–41.

Brausch, A., 1987. Calculation of critical importance for multi-state components. IEEE Trans. Reliab. 36, 247–249.

Brunelle, R.D., Kapur, K.C., 1999. Review and classification of reliability measures for multistate and continuum models. IIE Trans. 31, 1171–1180.

Butler, D., 1982. Bounding the reliability of multi-state systems. Oper. Res. 30, 530–544.

Caballé, N.C., Castro, I.T., Pérez, C.J., Lanza-Gutiérrez, J.M., 2015. A condition-based maintenance of a dependent degradation-threshold-shock model in a system with multiple degradation processes. Reliab. Eng. Syst. Saf. 134, 98–109.

Carlson, R.L., Kardomateas, G.A., 1996. An Introduction to Fatigue in Metals and Composites, first ed. Chapman and Hall, London.

Castro, I., 2009. A model of imperfect preventive maintenance with dependent failure modes. Eur. J. Oper. Res. 196 (1), 217–224.

Chen, Y., Yang, L., Ye, C., Kang, R., 2015. Failure mechanism dependence and reliability evaluation of non-repairable system. Reliab. Eng. Syst. Saf. 138, 273–283.
Crowder, M.J., Kimber, A.C., Smith, R.L., Sweeting, T.J., 1991–2017. Statistical Analysis of Reliability Data, first ed. Chapman & Hall/CRC. eBook.
Daniels, H.E., 1945. The statistical theory of the strength of bundles of threads I. Proc. Roy. Soc. Ser. A 183, 404–435.
De Porcellinis, S., Oliva, G., Panzieri, S., Setola, R., 2009. A holistic-reductionistic approach for modeling interdependencies. In: Palmer, C., Shenoi, S. (Eds.), Critical Infrastructure Protection III, 3rd IFIP Int. Conf. on Critical Infrastructure Protection (CIP-2009). Springer, USA, pp. 215–227.
Ding, Y., Lisnianski, A., 2008. Fuzzy universal generating functions for multi-state system reliability assessment. Fuzzy Sets Syst. 159, 307–324.
Ebrahimi, N., 1984. Multistate reliability models. Nav. Res. Logist. 31, 671–680.
El-Neweihi, E., Proschan, F., Sethuraman, J., 1978. Multistate coherent systems. J. Appl. Probab. 15 (12), 675–688.
Fardis, M.N., Cornell, C.A., 1981. Analysis of coherent multistate systems. IEEE Trans. Reliab. 30, 117–122.
Gertsbakh, I.B., 2000. Reliability Theory With Applications to Preventive Maintenance. Springer, Berlin.
Griffith, W., 1980. Multistate reliability models. J. Appl. Probab. 17, 735–744.
Han, F., Zio, E., 2019. A multi-perspective framework of analysis of critical infrastructures with respect to supply service, controllability and topology. Int. J. Crit. Infrastruct. Prot. 24, 1–13.
Harlow, D.G., Phoenix, S.L., 1978. The chain-of-bundles probability model for the strength of fibrous materials. J. Compos. Mater. 12, 195–214.
Harlow, D.G., Phoenix, S.L., 1982. Probability distribution for the strength of fibrous materials under local load sharing. Adv. Appl. Probab. 14, 68–94.
Holden, R., Val, D.V., Burkhard, R., Nodwell, S., 2013. A network flow model for interdependent infrastructures at the local scale. Saf. Sci. 53 (3), 51–60.
Høyland, A., Rausand, M., 2004. System Reliability Theory: Models, Statistical Methods and Applications. John Wiley & Sons, Inc., Hoboken, NJ.
Hudson, J.C., Kapur, K.C., 1982. Reliability theory for multistate systems with multistate components. Microelectron. Reliab. 22, 1–7.
Hudson, J.C., Kapur, K.C., 1983. Reliability analysis of multistate systems with multistate components. Trans. Inst. Ind. Eng. 15, 127–135.
Jain, M., Gupta, R., 2012. Load sharing M-out of-N: G system with non-identical components subject to common cause failure. Int. J. Math. Oper. Res. 4 (5), 586–605.
Jiang, L., Feng, Q., Coit, D.W., 2015. Modeling zoned shock effects on stochastic degradation in dependent failure processes. IIE Trans. 47 (5), 460–470.
Kjølle, G.H., Utne, I.B., Gjerde, O., 2012. Risk analysis of critical infrastructures emphasizing electricity supply and interdependencies. Reliab. Eng. Syst. Saf. 105, 80–89.
Kołowrocki, K., 2000. On asymptotic approach to multi-state systems reliability evaluation. In: Limnios, N., Nikulin, M. (Eds.), Recent Advances in Reliability Theory: Methodology, Practice and Inference. Birkhauser, Boston, pp. 163–180 (Chapter 11).
Kołowrocki, K., 2003a. An asymptotic approach to reliability evaluation of large multi-state systems with applications to piping transportation systems. Int. J. Press. Vessel. Pip. 80, 59–73.
Kołowrocki, K., 2003b. Asymptotic approach to reliability analysis of large systems with degrading components. Int. J. Reliab. Qual. Saf. Eng. 10 (3), 249–288.
Kołowrocki, K., 2008a. Reliability and risk analysis of multi-state systems with degrading components. Int. J. Mater. Struct. Reliab. 6 (2), 213–228.

Kołowrocki, K., 2008b. Reliability of Large Systems, Section in Encyclopedia of Quantitative Risk Analysis and Assessment. vol. 4. John Wiley & Sons, pp. 1466–1471.
Kołowrocki, K., 2014. Reliability of Large and Complex Systems, second ed. Elsevier, London.
Kołowrocki, K., Soszyńska, J., 2006. Reliability and availability of complex systems. Qual. Reliab. Eng. Int. 22 (1), 79–99. J. Wiley & Sons Ltd.
Kołowrocki, K., Soszyńska-Budny, J., 2011. Reliability and Safety of Complex Technical Systems and Processes: Modeling – Identification – Prediction – Optimization, first ed. Springer-Verlag, London.
Kostandyan, E., Sørensen, J., 2014. Dependent systems reliability estimation by structural reliability approach. Int. J. Performability Eng. 10 (6), 605–614.
Kotzanikolaou, P., Theoharidou, M., Gritzalis, D., 2013. Assessing n-order dependencies between critical infrastructures. Int. J. Crit. Infrastruct. 9 (1–2), 93–110.
Kvam, P.H., Pena, E.A., 2005. Estimating load-sharing properties in a dynamic reliability system. J. Am. Stat. Assoc. 1 (100), 262–272.
Lauge, A., Hernantes, J., Sarriegi, J.M., 2015. Critical infrastructure dependencies: a holistic, dynamic and quantitative approach. Int. J. Crit. Infrastruct. Prot. 8, 16–23.
Levitin, G., 2004. A universal generating function approach for the analysis of multi-state systems with dependent elements. Reliab. Eng. Syst. Saf. 84, 285–292.
Levitin, G., 2005. The Universal Generating Function in Reliability Analysis and Optimization, Reliability Engineering, first ed. Springer, London. https://doi.org/10.1007/1-84628-245-4.
Levitin, G., Lisnianski, A., 2000. Optimisation of imperfect preventive maintenance for multistate systems. Reliab. Eng. Syst. Saf. 67, 193–203.
Li, W., Pham, H., 2005. Reliability modeling of multi-state degraded systems with multi-competing failures and random shocks. IEEE Trans. Reliab. 54 (2), 297–303.
Lisnianski, A., Levitin, G., 2003. Multi-State System Reliability. Assessment, Optimisation and Applications. World Scientific Publishing Co. Pte. Ltd., New Jersey, London, Singapore, Hong Kong.
Lisnianski, A., Frenkel, I., Ding, Y., 2010. Multi-State Systems Reliability Analysis and Optimization for Engineers and Industrial Managers. Springer, London.
Liu, Y., Huang, H.Z., Levitin, G., 2008. Reliability and performance assessment for fuzzy multi-state element. J. Risk Reliab. 222 (4), 675–686.
Liu, Y., Huang, H.Z., Wang, Z., Li, Y., Yang, Y., 2013. A joint redundancy and imperfect maintenance strategy optimization for multi-state systems. IEEE Trans. Reliab. 62 (2), 368–378.
Marquez, A.C., 2007. The Maintenance Management Framework. Models and Methods for Complex Systems Maintenance. Springer, Londres, Reino Unido.
Melchers, R.E., Beck, A.T., 2018. Structural Reliability Analysis and Prediction, third ed. John Wiley & Sons.
Misra, K.B., 1992. Reliability Analysis and Prediction. Elsevier.
Muhammad, M., Majid, M.A., Mokhtar, A.A., 2010. Reliability evaluation for a multi-state system subject to imperfect repair and maintenance. Int. J. Eng. Technol. 10 (01), 59–63.
Murchland, J., 1975. Fundamental concepts and relations for reliability analysis of multistate systems. Reliability and fault tree analysis. In: Barlow, R.E. (Ed.), Theoretical and Applied Aspects of System Reliability. SIAM, pp. 581–618.
Natvig, B., 1984. Multi-state coherent systems. In: Johnson, N.L., Kotz, S. (Eds.), Encyclopedia of Statistical Sciences, Vol. 5. Wiley and Sons, New York, pp. 732–735.
Natvig, B., 2007. Multi-state reliability theory. In: Ruggeri, F., Kenett, R., Faltin, F.W. (Eds.), Encyclopedia of Statistics in Quality and Reliability. Wiley, New York, pp. 1160–1164.

Natvig, B., Morch, H., 2003. An application of multistate reliability theory to an offshore gas pipeline network. Int. J. Reliab. Qual. Saf. Eng. 10 (4), 361–381.

Natvig, B., Streller, A., 1984. The steady-state behaviour of multistate monotone systems. J. Appl. Probab. 21, 826–835.

Nieuwenhuijs, A., Luiijf, E., Klaver, M., 2008. Modeling dependencies in critical infrastructures. In: Papa, M., Shenoi, S. (Eds.), Critical Infrastructure Protection II, IFIP Series 253. Springer, Boston, MA, pp. 205–213.

Nourelfath, M., Chatelet, E., Nahas, N., 2012. Joint redundancy and imperfect preventive maintenance optimization for series–parallel multi-state degraded systems. Reliab. Eng. Syst. Saf. 103, 51–60.

Ohio, F., Nishida, T., 1984. On multistate coherent systems. IEEE Trans. Reliab. 33, 284–287.

Ouyang, M., 2014. Review on modelling and simulation of interdependent critical infrastructure systems. Reliab. Eng. Syst. Saf. 121, 43–60.

Peng, H., Feng, Q., Coit, D.W., 2010. Reliability and maintenance modeling for systems subject to multiple dependent competing failure processes. IIE Trans. 43 (1), 12–22.

Pham, H., Wang, H.Z., 1996. Imperfect maintenance. Eur. J. Oper. Res. 94 (3), 425–438.

Pham, H., Suprasad, A., Misra, R.B., 1997. Availability and mean life time prediction of multistage degraded system with partial repairs. Reliab. Eng. Syst. Saf. 56, 169–173.

Phoenix, S.L., Smith, R.L., 1983. A comparison of probabilistic techniques for the strength of fibrous materials under local load sharing among fibres. Int. J. Solids Struct. 19 (6), 479–496.

Pradhan, S., Hansen, A., Chakrabarti, B.K., 2010. Failure processes in elastic fiber bundles. Rev. Mod. Phys. 82, 499–555.

Rafiee, K., Feng, Q., Coit, D.W., 2014. Reliability modeling for dependent competing failure processes with changing degradation rate. IIE Trans. 46 (5), 483–496.

Ramirez-Marqueza, J.E., Coit, D.W., 2007. Multi-state component criticality analysis for reliability improvement in multi-state systems. Reliab. Eng. Syst. Saf. 92, 1608–1619.

Rinaldi, S., Peerenboom, J., Kelly, T., 2001. Identifying, understanding and analyzing critical infrastructure interdependencies. IEEE Control. Syst. Mag. 21 (6), 11–25.

Ross, S., 1979. Multivalued state component systems. Ann. Probab. 7, 379–383.

Samaniego, F.J., 2007. System Signatures and Their Applications in Engineering Reliability. International Series in Operations Research & Management Science, 110 Springer, New York.

Singh, B., Gupta, P.K., 2012. Load-sharing system model and its application to the real data set. Math. Comput. Simul. 82 (9), 1615–1629.

Singh, B., Sharma, K.K., Kumar, A., 2008. A classical and Bayesian estimation of a *k*-components load sharing parallel system. Comput. Stat. Data Anal. 52, 5175–5185.

Smith, R.L., 1982. The asymptotic distribution of the strength of a series-parallel system with equal load sharing. Ann. Probab. 10, 137–171.

Smith, R.L., 1983. Limit theorems and approximations for the reliability of load sharing systems. Adv. Appl. Probab. 15, 304–330.

Song, S., Coit, D.W., Feng, Q., Peng, H., 2014. Reliability analysis for multi-component systems subject to multiple dependent competing failure processes. IEEE Trans. Reliab. 63 (1), 331–345.

Soro, I.W., Nourelfath, M., Ait-Kadi, D., 2010. Performance evaluation of multi-state degraded systems with minimal repairs and imperfect preventive maintenance. Reliab. Eng. Syst. Saf. 95 (2), 65–69.

Svedsen, N., Wolthunsen, S., 2007. Connectivity models of interdependency in mixed-type critical infrastructure networks. Inf. Secur. Tech. Rep. 1, 44–55.

Szymkowiak, M., 2018. Characterizations of distributions through aging intensity. IEEE Trans. Reliab. 67 (2), 446–458.

Szymkowiak, M., 2019. Lifetime Analysis by Aging Intensity Functions. Monograph in Series: Studies in Systems, Decision and Control, Springer.

Ushakov, I.A., 1986. Universal generating function. Sov. J. Comput. Syst. Sci. 24 (5), 118–129.

Wang, Y., Pham, H., 2011. A multi-objective optimization of imperfect preventive maintenance policy for dependent competing risk systems with hidden failure. IEEE Trans. Reliab. 60 (4), 770–781.

Wang, Y., Pham, H., 2012. Modeling the dependent competing risks with multiple degradation processes and random shock using time-varying copulas. IEEE Trans. Reliab. 64 (1), 13–22.

Wang, Z., Huang, H.Z., Li, Y., Xiao, N.C., 2011. An approach to reliability assessment under degradation and shock process. IEEE Trans. Reliab. 60 (4), 852–863.

Xue, J., 1985. On multi-state system analysis. IEEE Trans. Reliab. 34, 329–337.

Xue, J., Yang, K., 1995a. Dynamic reliability analysis of coherent multi-state systems. IEEE Trans. Reliab. 44 (4), 683–688.

Xue, J., Yang, K., 1995b. Symmetric relations in multi-state systems. IEEE Trans. Reliab. 44 (4), 689–693.

Yingkui, G., Jing, L., 2012. Multi-state system reliability: a new and systematic review. Procedia Eng. 29, 531–536.

Zio, E., 2007. An introduction to the basics of reliability and risk analysis. In: Series in Quality, Reliability and Engineering Statistics 13. World Scientific Publishing Co. Pte. Ltd., New Jersey – London – Singapore – Beijing – Shanghai – Hong Kong – Taipei – Chennai.

Zio, E., 2009. Computational Methods for Reliability and Risk Analysis. World Scientific Publishing, Co. Pte. Ltd., New Jersey – London – Singapore – Beijing – Shanghai – Hong Kong – Taipei – Chennai.

Zio, E., Compare, M., 2013. Evaluating maintenance policies by quantitative modeling and analysis. Reliab. Eng. Syst. Saf. 109, 53–65.

Zio, E., Sansavini, G., 2011. Modeling interdependent network systems for identifying cascade-safe operating margins. IEEE Trans. Reliab. 60 (1), 94–101.

CHAPTER TWO

Multistate aging systems

Basic assumptions and definitions related to reliability analysis of multistate and aging systems are introduced. The reliability functions of a multistate system and its components are defined. The system risk function and other basic reliability characteristics of a system and its components are also defined. Particularly, definitions of the mean lifetimes and the standard deviations of a system and its components in the reliability state subsets, and the mean lifetimes in particular reliability states, are given.

Moreover, basic reliability structures of multistate systems are introduced and their reliability functions are determined. Namely, the multistate series, parallel, "m out of n", series-parallel, parallel-series, series-"m out of k," and "m out of l"-series systems with aging components are defined, and their reliability functions are determined.

2.1 Multistate approach description

In the multistate reliability analysis to define systems with aging components we assume that (Blokus-Roszkowska and Kolowrocki, 2017; Kołowrocki, 2001, 2008, 2014; Kołowrocki and Soszyńska, 2006; Kołowrocki and Soszyńska-Budny, 2011; Xue, 1985; Xue and Yang, 1995a,b; Yu et al., 1994):

– n is the number of components in the system,
– E_i, $i=1,2, \ldots, n$, are components of a system,
– all components and a system under consideration have the reliability state set $\{0,1, \ldots, z\}$, $z \geq 1$,
– the reliability states are ordered, the state z is a state of full reliability and the state 0 is a reliability state of complete failure,
– $T_i(u)$, $i=1,2, \ldots, n$, are independent random variables representing the lifetimes of components E_i in the predefined state subset $\{u, u+1, \ldots, z\}$, $u=0,1, \ldots, z$, while they were in the reliability state z at the moment $t=0$,
– $\boldsymbol{T}(u)$ is a random variable representing the lifetime of a system in the reliability state subset $\{u, u+1, \ldots, z\}$, $u=0,1, \ldots, z$, while it was in the state z at the moment $t=0$,
– the system reliability state degrades with time t without repair,

https://doi.org/10.1016/B978-0-12-821260-8.00002-6

– $e_i(t)$ is a component reliability E_i state at the moment t, $t \in (-\infty, \infty)$, given that it was in the reliability state z at the moment $t=0$,
– $s(t)$ is a system reliability state at the moment t, $t \in (-\infty, \infty)$, given that it was in the reliability state z at the moment $t=0$.

These assumptions mean that the reliability states of the system with degrading components may be changed in time only from better to worse (Kołowrocki, 2014; Kołowrocki and Soszyńska-Budny, 2011). The exemplary realizations of the system reliability states changing are given in Figs. 2.1 and 2.2.

According to the assumptions mentioned, $\boldsymbol{T}(u)$ denotes the system lifetime in reliability state subset $\{u, u+1, \ldots, z\}$, $u=0,1, \ldots, z$. Thus, $\boldsymbol{T}(z)$ is

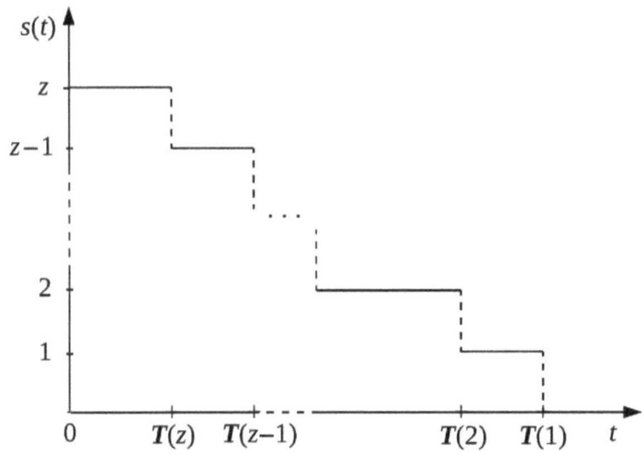

Fig. 2.1 The exemplary realization of reliability state changing in a multistate system.

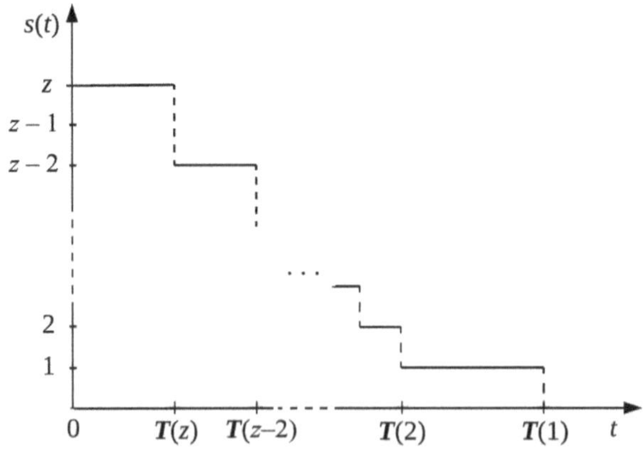

Fig. 2.2 The exemplary realization of reliability state changing in a multistate system.

the system lifetime in state z, $\boldsymbol{T}(z-1)$ is the time of system being in a not worse state than z-1, i.e., in z state or z-1 state, etc. Consequently, $\boldsymbol{T}(1)$ is the time of system being in a state not worse than state 1, i.e., in one of reliability states from the subset $\{1,2, \ldots, z\}$.

Consequently, the assumptions justify Definitions 2.1.1–2.1.3 of basic reliability notions given below (Kołowrocki, 2014; Kołowrocki and Soszyńska-Budny, 2011).

Definition 2.1.1. The reliability function of a multistate component E_i is a vector

$$R_i(t, \cdot) = [R_i(t,0), R_i(t,1), \ldots, R_i(t,z)], t \in (-\infty, \infty),$$
$$i = 1,2,\ldots,n, \tag{2.1.1}$$

where

$$R_i(t,u) = P(e_i(t) \geq u \mid e_i(0) = z) = P(T_i(u) > t), \quad u = 0,1,\ldots,z, \tag{2.1.2}$$

is the probability that a component E_i is in the reliability state subset $\{u, u+1, \ldots, z\}$, $u=0,1, \ldots, z$, at the moment t, $t \in (-\infty, \infty)$, while it was in the reliability state z at the moment $t=0$.

Under this definition (Kołowrocki, 2014; Kołowrocki and Soszyńska-Budny, 2011)

$$R_i(t,0) \geq R_i(t,1) \geq \cdots \geq R_i(t,z), t \in (-\infty, \infty), \quad i = 1,2,\ldots,n.$$

Further, if we denote by

$$p_i(t,u) = P(e_i(t) = u \mid e_i(0) = z), t \in (-\infty, \infty), \quad u = 0,1,\ldots,z,$$

the probability that the component E_i is in the reliability state u at the moment t, while it was in the state z at the moment $t=0$, then by (2.1.2)

$$R_i(t,0) = 1, R_i(t,z) = p_i(t,z), t \in (-\infty, \infty), \quad i = 1,2,\ldots,n, \tag{2.1.3}$$

and

$$p_i(t,u) = R_i(t,u) - R_i(t,u+1), u = 0,1,\ldots,z-1, t \in (-\infty, \infty),$$
$$i = 1,2,\ldots,n. \tag{2.1.4}$$

Further, from (2.1.3), we replace $R_i(t,0)$, existing in (2.1.1), by 1.

Moreover, if

$$R_i(t,u) = 1 \text{ for } t \leq 0, u = 1,2,\ldots,z, \quad i = 1,2,\ldots,n,$$

then

$$\mu_i(u) = \int_0^\infty R_i(t, u)dt, \ \ u = 1, 2, \dots, z, \ \ i = 1, 2, \dots, n, \tag{2.1.5}$$

is the mean lifetime of the component E_i in the state subset $\{u, u+1, \dots, z\}$,

$$\sigma_i(u) = \sqrt{n_i(u) - [\mu_i(u)]^2}, u = 1, 2, \dots, z, \ \ i = 1, 2, \dots, n, \tag{2.1.6}$$

where

$$n_i(u) = 2\int_0^\infty tR_i(t, u)dt, u = 1, 2, \dots, z, \ \ i = 1, 2, \dots, n, \tag{2.1.7}$$

is the standard deviation of the component E_i lifetime in the reliability state subset $\{u, u+1, \dots, z\}$ and

$$\bar{\mu}_i(u) = \int_0^\infty p_i(t, u)dt, \ \ u = 1, 2, \dots, z, \ \ i = 1, 2, \dots, n, \tag{2.1.8}$$

is the mean lifetime of the component E_i in the reliability state u, in the case when the integrals defined by (2.1.5), (2.1.7), and (2.1.8) are convergent.

Next, according to (2.1.3), (2.1.4), (2.1.5), and (2.1.8), we have

$$\bar{\mu}_i(u) = \mu_i(u) - \mu_i(u+1), \ \ u = 1, 2, \dots, z-1, \bar{\mu}_i(z) = \mu_i(z), \\ i = 1, 2, \dots, n. \tag{2.1.9}$$

Definition 2.1.2. The reliability function of a multistate system is a vector

$$\boldsymbol{R}(t, \cdot) = [\boldsymbol{R}(t, 0), \boldsymbol{R}(t, 1), \dots, \boldsymbol{R}(t, z)], \ \ t \in (-\infty, \infty), \tag{2.1.10}$$

where

$$\boldsymbol{R}(t, u) = P(s(t) \geq u | s(0) = z) = P(\boldsymbol{T}(u) > t), \ \ u = 0, 1, \dots, z, \tag{2.1.11}$$

is the probability that a system is in the reliability state subset $\{u, u+1, \dots, z\}$, $u = 0, 1, \dots, z$, at the moment t, $t \in (-\infty, \infty)$, while it was in the reliability state z at the moment $t = 0$.

Under this definition (Kołowrocki, 2014; Kołowrocki and Soszyńska-Budny, 2011)

$$\boldsymbol{R}(t, 0) \geq \boldsymbol{R}(t, 1) \geq \cdots \geq \boldsymbol{R}(t, z), \ \ t \in (-\infty, \infty),$$

and if

$$p(t,u) = P(s(t) = u | s(0) = z), \quad t \in (-\infty, \infty), u = 0, 1, \ldots, z, \tag{2.1.12}$$

is the probability that the system is in the reliability state u at the moment t, while it was in the state z at the moment $t=0$, then

$$\boldsymbol{R}(t,0) = 1, \boldsymbol{R}(t,z) = p(t,z), \quad t \in (-\infty, \infty), \tag{2.1.13}$$

and

$$p(t,u) = \boldsymbol{R}(t,u) - \boldsymbol{R}(t,u+1), \quad u = 0, 1, \ldots, z-1, t \in (-\infty, \infty). \tag{2.1.14}$$

Then, similarly as in the component reliability function, from (2.1.13), we replace $\boldsymbol{R}(t,0)$, existing in (2.1.10) by 1.

Moreover, if

$$\boldsymbol{R}(t,u) = 1 \text{ for } t \leq 0, u = 1, 2, \ldots, z, \tag{2.1.15}$$

then

$$\mu(u) = \int_0^\infty \boldsymbol{R}(t,u)dt, u = 1, 2, \ldots, z, \tag{2.1.16}$$

is the mean lifetime of the system in the reliability state subset $\{u, u+1, \ldots, z\}$,

$$\sigma(u) = \sqrt{n(u) - [\mu(u)]^2}, u = 1, 2, \ldots, z, \tag{2.1.17}$$

where

$$n(u) = 2\int_0^\infty t \cdot \boldsymbol{R}(t,u)dt, u = 1, 2, \ldots, z, \tag{2.1.18}$$

is the standard deviation of the system sojourn time in the reliability state subset $\{u, u+1, \ldots, z\}$ and moreover

$$\bar{\mu}(u) = \int_0^\infty p(t,u)dt, \quad u = 1, 2, \ldots, z, \tag{2.1.19}$$

is the mean lifetime of the system in the reliability state u while the integrals (2.1.16), (2.1.18) and (2.1.19) are convergent.

Additionally, according to (2.1.13), (2.1.14), (2.1.16), and (2.1.19), we get the following relationship

$$\bar{\mu}(u) = \mu(u) - \mu(u+1), u = 1, 2, \ldots, z-1, \bar{\mu}(z) = \mu(z). \tag{2.1.20}$$

Definition 2.1.3. The system risk function is a probability

$$\boldsymbol{r}(t) = P(s(t) < r | s(0) = z) = P(T(r) \leq t), t \in (-\infty, \infty),$$

that the system is in the subset of reliability states worse than the critical reliability state r, $r \in \{1, \ldots, z\}$ while it was in the reliability state z at the moment $t=0$.

Under this definition, from (2.1.11), we have

$$\boldsymbol{r}(t) = 1 - P(s(t) \geq r | s(0) = z) = 1 - \boldsymbol{R}(t, r), t \in (-\infty, \infty). \tag{2.1.21}$$

and if τ is the moment when the risk exceeds a permitted level δ, then

$$\tau = \boldsymbol{r}^{-1}(\delta), \tag{2.1.22}$$

where $\boldsymbol{r}^{-1}(t)$, if it exists, is the inverse function of the risk function $\boldsymbol{r}(t)$.

2.1.1 Reliability analysis of exemplary system

We consider a 4-state system ($z=3$) with the reliability function

$$\boldsymbol{R}(t, \cdot) = [1, \boldsymbol{R}(t, 1), \boldsymbol{R}(t, 2), \boldsymbol{R}(t, 3)], \quad t \geq 0, \tag{2.1.23}$$

where its coordinates are

$$\boldsymbol{R}(t, 1) = \exp\left[-0.01t^2\right], \boldsymbol{R}(t, 2) = \exp\left[-0.02t^2\right],$$
$$\boldsymbol{R}(t, 3) = \exp\left[-0.05t^2\right], \quad t \geq 0. \tag{2.1.24}$$

The reliability function coordinates of the exemplary system are illustrated in Fig. 2.3.

The mean lifetimes of the exemplary system in the reliability state subsets $\{1,2,3\}$, $\{2,3\}$, $\{3\}$, according to (2.1.16) and using (2.1.24), are determined from

$$\mu(1) = \int_0^\infty \exp\left[-0.01t^2\right] dt, \mu(2) = \int_0^\infty \exp\left[-0.02t^2\right] dt, \mu(3)$$
$$= \int_0^\infty \exp\left[-0.05t^2\right] dt. \tag{2.1.25}$$

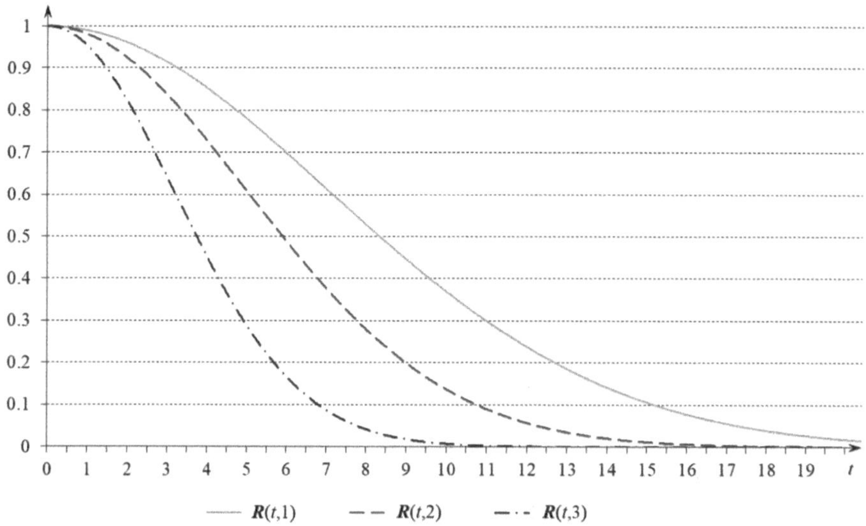

Fig. 2.3 The graphs of the exemplary system reliability function coordinates.

To calculate these integrals, we use the trapezoid rule for approximating integrals. The values of reliability functions and area under the graph of reliability functions, given in Tables 2.1 and 2.2, are calculated with a time step $h=2$ for reliability function coordinate $\boldsymbol{R}(t,1)$ and $h=1$ for reliability function coordinates $\boldsymbol{R}(t,2)$, $\boldsymbol{R}(t,3)$.

The number of steps depends on the accuracy of the calculation. We perform the calculations until the assumed accuracy of calculations is exceeded, for example for the exemplary system until the moment when area under the function graph on subintervals takes values less than 0.0001.

Then, using the values given in Table 2.1 we get the mean lifetime of the exemplary system in the reliability state subset $\{1,2,3\}$

$$\mu(1) \cong \frac{h}{2} \cdot [\boldsymbol{R}(0,1) + 2\boldsymbol{R}(h,1) + 2\boldsymbol{R}(2h,1) + \cdots + 2\boldsymbol{R}(15h,1) + \boldsymbol{R}(16h,1)] \cong 8.862. \tag{2.1.26}$$

Similarly, using the values given in Table 2.2 we obtain the mean lifetime of this system in the reliability state subsets $\{2,3\}$, $\{3\}$:

$$\mu(2) \cong \frac{h}{2} \cdot [\boldsymbol{R}(0,2) + 2\boldsymbol{R}(h,2) + \cdots + 2\boldsymbol{R}(20h,2) + \boldsymbol{R}(21h,2)] \cong 6.266, \tag{2.1.27}$$

Table 2.1 The values of reliability function $\boldsymbol{R}(t,1)$ and area under the graph of $\boldsymbol{R}(t,1)$ calculated with a time step $h=2$.

t	$R(t,1)$	$h \cdot \frac{R(t,1)+R(t+h,1)}{2}$
0	1.000000	1.960789
2	0.960789	1.812933
4	0.852144	1.549820
6	0.697676	1.224969
8	0.527292	0.895172
10	0.367879	0.604807
12	0.236928	0.377786
14	0.140858	0.218163
16	0.077305	0.116469
18	0.039164	0.057480
20	0.018316	0.026223
22	0.007907	0.011058
24	0.003151	0.004310
26	0.001159	0.001553
28	0.000394	0.000517
30	0.000123	0.000159
32	0.000036	
SUM		**8.862208**

Table 2.2 The values of reliability functions $R(t,2)$, $R(t,3)$ and area under the graph of these reliability functions calculated with a time step $h=1$.

t	$R(t,2)$	$h \cdot \frac{R(t,2)+R(t+h,2)}{2}$	$R(t,3)$	$h \cdot \frac{R(t,3)+R(t+h,3)}{2}$
0	1.000000	0.990099	1.000000	0.975615
1	0.980199	0.951658	0.951229	0.884980
2	0.923116	0.879193	0.818731	0.728179
3	0.835270	0.780710	0.637628	0.543479
4	0.726149	0.666340	0.449329	0.367917
5	0.606531	0.546641	0.286505	0.225902
6	0.486752	0.431032	0.165299	0.125796
7	0.375311	0.326674	0.086294	0.063528
8	0.278037	0.237968	0.040762	0.029092
9	0.197899	0.166617	0.017422	0.012080
10	0.135335	0.112128	0.006738	0.004548
11	0.088922	0.072528	0.002358	0.001552
12	0.056135	0.045091	0.000747	0.000480
13	0.034047	0.026944	0.000214	0.000135
14	0.019841	0.015475	0.000055	0.000034
15	0.011109	0.008543	0.000013	0.000008
16	0.005976	0.004532	0.000003	0.000002
17	0.003089	0.002311	0.000001	0.000000
18	0.001534	0.001133	0.000000	0.000000
19	0.000732	0.000534	0.000000	0.000000
20	0.000335	0.000242	0.000000	0.000000
21	0.000148		0.000000	
SUM		**6.266393**		**3.963327**

$$\mu(3) \cong \frac{h}{2} \cdot [R(0,3) + 2R(h,3) + \cdots + 2R(15h,3) + R(16h,3)]$$
$$\cong 3.963. \quad (2.1.28)$$

Applying (2.1.18) and using (2.1.24), and considering the formula

$$\int_0^\infty 2t \cdot \exp\left[-\lambda t^2\right] dt = \frac{1}{\lambda}, \quad (2.1.29)$$

we obtain:

$$n(1) = 2\int_0^\infty t \cdot \exp\left[-0.01t^2\right] dt = \frac{1}{0.01} = 100, \quad (2.1.30)$$

$$n(2) = 2\int_0^\infty t \cdot \exp\left[-0.02t^2\right] dt = \frac{1}{0.02} = 50, \quad (2.1.31)$$

$$n(3) = 2\int_0^\infty t \cdot \exp\left[-0.05t^2\right] dt = \frac{1}{0.05} = 20. \quad (2.1.32)$$

And then, substituting (2.1.26)–(2.1.28) and (2.1.30)–(2.1.32) for the formula (2.1.17), we determine the standard deviations of the exemplary system lifetimes:

$$\sigma(1) = \sqrt{n(1) - [\mu(1)]^2} \cong \sqrt{100 - 8.862^2} \cong 4.633, \quad (2.1.33)$$

$$\sigma(2) = \sqrt{n(2) - [\mu(2)]^2} \cong \sqrt{50 - 6.266^2} \cong 3.276, \quad (2.1.34)$$

$$\sigma(3) = \sqrt{n(3) - [\mu(3)]^2} \cong \sqrt{20 - 3.963^2} \cong 2.083. \quad (2.1.35)$$

The mean values of the exemplary system lifetimes in the particular reliability states 1,2,3, by (2.1.20) and using (2.1.26)–(2.1.28), are:

$$\bar{\mu}(1) = \mu(1) - \mu(2) \cong 8.862 - 6.266 = 2.596, \quad (2.1.36)$$

$$\bar{\mu}(2) = \mu(2) - \mu(3) \cong 6.266 - 3.963 = 2.303, \quad (2.1.37)$$

$$\bar{\mu}(3) = \mu(3) \cong 3.963. \quad (2.1.38)$$

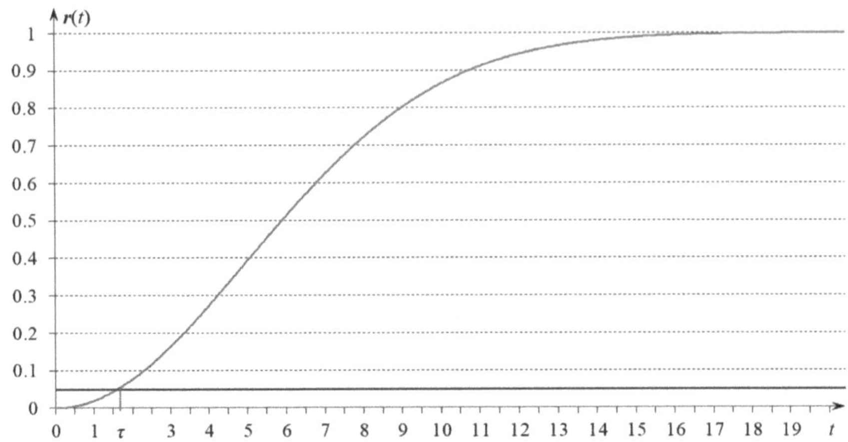

Fig. 2.4 The graph of the risk function ***r***(t) of the exemplary system.

Further, it is assumed that the critical reliability state is $r=2$, then under the Definition 2.1.3, applying (2.1.21), we obtain the risk function of the exemplary system

$$\boldsymbol{r}(t)=1-\boldsymbol{R}(t,2)=1-\exp\left[-0.02t^2\right], \quad t\geq 0, \tag{2.1.39}$$

and the moment of exceeding an acceptable level, for instance $\delta=0.05$, by (2.1.22), is

$$\tau=\boldsymbol{r}^{-1}(\delta)=\sqrt{\frac{\ln(1-\delta)}{-0.02}}\cong 1.601. \tag{2.1.40}$$

The exemplary system risk function is illustrated in Fig. 2.4.

2.2 Multistate approach in relation to two-state reliability analysis

Particularly, for $z=1$ proposed in Section 2.1 multistate approach becomes the two-state system reliability analysis. Namely, the assumptions take the form (Blokus-Roszkowska, 2007; Kołowrocki, 2014):
– n is the number of components in the system,
– E_i, $i=1,2,\ldots,n$, are components of a system,
– all components and a system under consideration have the reliability state set $\{0;1\}$,

- the state 1 is a state of component or system reliability and the state 0 is a reliability state of failure,
- T_i, $i=1,2, \ldots, n$, are independent random variables representing the lifetimes of components E_i, while they were in reliability state 1 at the moment $t=0$,
- $\boldsymbol{T}$ is a random variable representing the system lifetime, while it was in reliability state 1 at the moment $t=0$,
- the system reliability state degrades with time t without repair,
- $e_i(t)$ is a component reliability E_i state at the moment t, $t \in (-\infty, \infty)$, given that it was in reliability state 1 at the moment $t=0$,
- $s(t)$ is a system reliability state at the moment t, $t \in (-\infty, \infty)$, given that it was in reliability state 1 at the moment $t=0$.

The interpretations given in Figs. 2.1 and 2.2, for two-state systems are illustrated in Fig. 2.5, where $\boldsymbol{T}$ denotes the system lifetime.

Definition 2.2.1. The reliability function of a component E_i is

$$R_i(t) = P(e_i(t) = 1 | e_i(0) = 1) = P(T_i > t), t \in (-\infty, \infty),$$
$$i = 1, 2, \ldots, n, \tag{2.2.1}$$

is the probability that a component E_i is in the reliability state 1 at moment t, $t \in (-\infty, \infty)$, while it was in the reliability state 1 at the moment $t=0$.

The distribution function of component lifetime T_i is given by

$$F_i(t) = P(T_i \leq t) = 1 - R_i(t), t \in (-\infty, \infty), i = 1, 2, \ldots, n. \tag{2.2.2}$$

Definition 2.2.2. The reliability function of a two-state system is

$$\boldsymbol{R}(t) = P(s(t) = 1 | s(0) = 1) = P(\boldsymbol{T} > t), t \in (-\infty, \infty), \tag{2.2.3}$$

is the probability that a system is in the reliability state 1 at moment t, $t \in (-\infty, \infty)$, while it was in the reliability state 1 at the moment $t=0$.

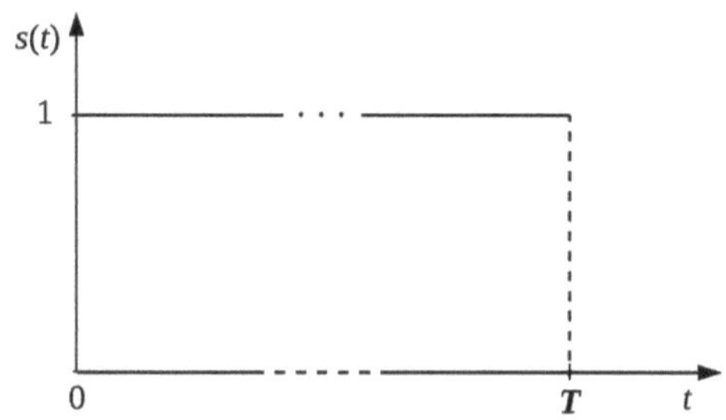

Fig. 2.5 The exemplary realization of reliability state changing in a two-state system.

The distribution function of system lifetime $\boldsymbol{T}$ is given by

$$\boldsymbol{F}(t) = P(\boldsymbol{T} \leq t) = 1 - \boldsymbol{R}(t), \quad t \in (-\infty, \infty), \tag{2.2.4}$$

If

$$\boldsymbol{R}(t) = 1 \text{ for } t \leq 0, \tag{2.2.5}$$

then

$$\mu = \int_0^\infty \boldsymbol{R}(t)dt, \tag{2.2.6}$$

is the system mean lifetime, and

$$\sigma = \sqrt{m_2 - \mu^2}, \tag{2.2.7}$$

where

$$m_2 = 2\int_0^\infty t \cdot \boldsymbol{R}(t)dt, \tag{2.2.8}$$

is the standard deviation of system lifetime.

To define the system risk function for a multistate system in Section 2.1, the critical reliability state r was given. In case of a two-state system, a reliability state when a system is not failed can be assumed as a critical state. Under this assumption, the system risk function, determined by the formula (2.1.21), is equivalent to the distribution function given by (2.2.4).

In that case, if τ is the moment when the risk exceeds a permitted level δ, then

$$\tau = \boldsymbol{r}^{-1}(\delta) = \boldsymbol{F}^{-1}(\delta), \tag{2.2.9}$$

where $\boldsymbol{F}^{-1}(t)$, if it exists, is the inverse function of the distribution function $\boldsymbol{F}(t)$.

2.2.1 Reliability analysis of exemplary system

In two-state reliability analysis, the example from Section 2.1.1 is analyzed as follows.

We consider a two-state system with the reliability function

$$\boldsymbol{R}(t) = \exp\left[-0.01t^2\right], t \geq 0. \tag{2.2.10}$$

In that case the system mean lifetime is

$$\mu = \int_0^\infty \exp\left[-0.01t^2\right] dt. \tag{2.2.11}$$

And this integral is calculated, similarly as in Section 2.1.1, using the trapezoid rule for approximating integrals from values given in Table 2.1. Subsequently, the mean value of exemplary system lifetime is

$$\mu \cong \frac{h}{2} \cdot [\boldsymbol{R}(0) + 2\boldsymbol{R}(h) + 2\boldsymbol{R}(2h) + \cdots + 2\boldsymbol{R}(15h) + \boldsymbol{R}(16h)] \cong 8.862. \tag{2.2.12}$$

The standard deviation of exemplary system lifetime, determined the same way as in Section 2.1.1, is

$$\sigma \cong \sqrt{100 - 8.862^2} \cong 4.633. \tag{2.2.13}$$

Further, we obtain the risk function of exemplary system

$$\boldsymbol{r}(t) = 1 - \boldsymbol{R}(t) = 1 - \exp\left[-0.01t^2\right], t \geq 0, \tag{2.2.14}$$

and the moment of exceeding an acceptable level, for instance $\delta = 0.05$, by (2.2.9), is

$$\tau = \boldsymbol{r}^{-1}(\delta) = \sqrt{\frac{\ln(1-\delta)}{-0.01}} \cong 2.265. \tag{2.2.15}$$

The exemplary system risk function is illustrated in Fig. 2.6.

Performing the above reliability analysis of exemplary two-state system, the reliability function coordinate $\boldsymbol{R}(t, 1)$ of exemplary 4-state system analyzed in Section 2.1.1 was assumed as the reliability function of two-state system in Section 2.2.1. However, it is more naturally practical to replace the state $z = 1$ for a two-state system with the critical state, assumed as $r = 2$ in Section 2.1.1. In that case, we assume that the reliability function coordinate $\boldsymbol{R}(t, 2)$ is the reliability function of two-state system, and from (2.1.24) we get:

$$\boldsymbol{R}(t) = \exp\left[-0.02t^2\right], t \geq 0. \tag{2.2.16}$$

In that case the system mean lifetime is

$$\mu = \int_0^\infty \exp\left[-0.02t^2\right] dt \cong 6.266, \tag{2.2.17}$$

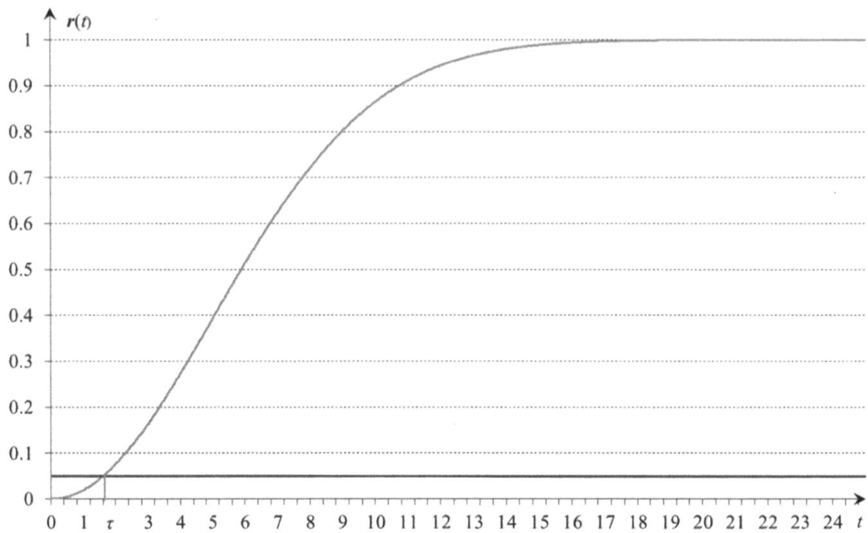

Fig. 2.6 The graph of the risk function ***r***(*t*) of exemplary two-state system.

and the standard deviation of exemplary system lifetime is

$$\sigma \cong \sqrt{50 - 6.266^2} \cong 3.276. \tag{2.2.18}$$

Further, we obtain the risk function of exemplary system

$$\boldsymbol{r}(t) = 1 - \boldsymbol{R}(t) = 1 - \exp\left[-0.02t^2\right], t \geq 0, \tag{2.2.19}$$

and the moment of exceeding an acceptable level, for instance $\delta = 0.05$, by (2.2.9), is

$$\tau = \boldsymbol{r}^{-1}(\delta) = \sqrt{\frac{\ln(1-\delta)}{-0.02}} \cong 1.601. \tag{2.2.20}$$

Evidently, these results are identical to those for the critical state in the multistate approach to reliability analysis of exemplary system, carried out in Section 2.1.1.

The simplification of the system model to the two-state system and selection of the system's state of functioning can significantly affect the value of basic system reliability characteristics. Assuming a critical state as the state of system's functioning (system not failed), the mean lifetime system is shorter by about 29% in relation to the lifetime received for state $z = 1$. Similarly, the moment of exceeding an acceptable level occurs 29% faster than for the state of system's functioning $z = 1$.

Comparing results of two-state approach to the reliability analysis of exemplary system with the multistate approach, we conclude that a multistate approach allows a more precise and closer to reality analysis of system reliability. In practice, we often find systems that adopt several reliability states, from a state of full system reliability, through a critical reliability state, the exceeding of which can be dangerous to the system environment, to a state of system failure.

An additional advantage of the presented multistate approach (Kołowrocki, 2001, 2008, 2014) is the possibility to receive the exact values of basic system reliability characteristics, independent of the number components and the number of their reliability states.

2.3 Basic reliability structures of multistate systems and their reliability functions

Definition 2.3.1. A multistate system is called series if its lifetime $\boldsymbol{T}(u)$ in the reliability state subset $\{u, u+1, \ldots, z\}$, $u=1,2, \ldots, z$, is given by (Kołowrocki, 2014)

$$\boldsymbol{T}(u) = \min_{1 \le i \le n} \{T_i(u)\}, u = 1, 2, \ldots, z. \tag{2.3.1}$$

This definition means that a multistate series system is in the reliability state subset $\{u, u+1, \ldots, z\}$, $u=1,2, \ldots, z$, if and only if all its components are in this subset of states.

It is easy to work out that the reliability function of a multistate series system is given by (Kołowrocki, 2014)

$$\boldsymbol{R}(t, \cdot) = [1, \boldsymbol{R}(t, 1), \ldots, \boldsymbol{R}(t, z)], t \in (-\infty, \infty), \tag{2.3.2}$$

where

$$\boldsymbol{R}(t, u) = \prod_{i=1}^{n} R_i(t, u), t \in (-\infty, \infty), u = 1, 2, \ldots, z. \tag{2.3.3}$$

Definition 2.3.2. A multistate system is called homogeneous if its component lifetimes $T_i(u)$ in the reliability state subsets have an identical distribution function (Kołowrocki, 2014)

$$F_i(t, u) = F(t, u), t \in (-\infty, \infty), u = 1, 2, \ldots, z, i = 1, 2, \ldots, n,$$

i.e., if its components E_i have the same reliability function

$$R_i(t,u) = R(t,u) = 1 - F(t,u), t \in (-\infty, \infty), u = 1,2,\ldots,z, i = 1,2,\ldots,n.$$

The reliability function of a homogeneous multistate series system, by (2.3.2)–(2.3.3), is

$$\boldsymbol{R}(t,\cdot) = [1, \boldsymbol{R}(t,1), \ldots, \boldsymbol{R}(t,z)], t \in (-\infty, \infty), \tag{2.3.4}$$

where

$$\boldsymbol{R}(t,u) = [R(t,u)]^n, t \in (-\infty, \infty), u = 1,2,\ldots,z. \tag{2.3.5}$$

Definition 2.3.3. A multistate system is called parallel if its lifetime $\boldsymbol{T}(u)$ in the reliability state subset $\{u, u+1, \ldots, z\}$, $u=1,2, \ldots, z$, is given by (Kołowrocki, 2014)

$$\boldsymbol{T}(u) = \max_{1 \le i \le n} \{T_i(u)\}, u = 1,2,\ldots,z. \tag{2.3.6}$$

The definition means that a multistate parallel system is in the reliability state subset $\{u, u+1, \ldots, z\}$, $u=1,2, \ldots, z$, if and only if at least one of its components is in this subset of states.

The reliability function of a multistate parallel system is given by (Kołowrocki, 2014)

$$\boldsymbol{R}(t,\cdot) = [1, \boldsymbol{R}(t,1), \ldots, \boldsymbol{R}(t,z)], t \in (-\infty, \infty), \tag{2.3.7}$$

with the coordinates

$$\boldsymbol{R}(t,u) = 1 - \prod_{i=1}^{n} F_i(t,u), t \in (-\infty, \infty), u = 1,2,\ldots,z, \tag{2.3.8}$$

where $F_i(t,u)$, $i=1,2, \ldots, n$, $u=1,2, \ldots, z$, is the distribution function of a component lifetime $T_i(u)$, given by

$$F_i(t,u) = 1 - R_i(t,u), t \in (-\infty, \infty), u = 1,2,\ldots,z, i = 1,2,\ldots,n. \tag{2.3.9}$$

The reliability function of a homogeneous multistate parallel system, from (2.3.7)–(2.3.8), is

$$\boldsymbol{R}(t,\cdot) = [1, \boldsymbol{R}(t,1), \ldots, \boldsymbol{R}(t,z)], t \in (-\infty, \infty), \tag{2.3.10}$$

with the coordinates

$$\boldsymbol{R}(t,u) = 1 - [F(t,u)]^n, t \in (-\infty, \infty), u = 1,2,\ldots,z, \tag{2.3.11}$$

where $F(t,u)$, $u=1,2, \ldots, z$, is the distribution function of a component lifetime $T(u)$, given by

$$F(t, u) = 1 - R(t, u), t \in (-\infty, \infty), u = 1, 2, \ldots, z. \tag{2.3.12}$$

Definition 2.3.4. A multistate system is called an "m out of n" system if its lifetime $\boldsymbol{T}(u)$ in the reliability state subset $\{u, u+1, \ldots, z\}$, $u=1,2, \ldots, z$, is given by (Kołowrocki, 2014)

$$\boldsymbol{T}(u) = T_{(n-m+1)}(u), m = 1, 2, \ldots, n, u = 1, 2, \ldots, z, \tag{2.3.13}$$

where $T_{(n-m+1)}(u)$ is the mth maximal order statistic in the sequence of the component lifetimes $T_1(u), T_2(u), \ldots, T_n(u)$.

This definition means that a multistate "m out of n" system is in the reliability state subset $\{u, u+1, \ldots, z\}$, $u=1,2, \ldots, z$, if and only if at least m out of its n components are in this state subset; and it is a multistate parallel system if $m=1$ and it is a multistate series system if $m=n$. The numbers m and n are called the system structure shape parameters.

It can be simply shown that the reliability function of a multistate "m out of n" system is given by the vector (Kołowrocki, 2014)

$$\boldsymbol{R}(t, \cdot) = [1, \boldsymbol{R}(t, 1), \ldots, \boldsymbol{R}(t, z)], t \in (-\infty, \infty), \tag{2.3.14}$$

with the coordinates either given by

$$\boldsymbol{R}(t, u) = \sum_{\substack{r_1, r_2, \ldots, r_n = 0 \\ m \le r_1 + r_2 + \ldots + r_n \le n}}^{1} \prod_{i=1}^{n} [R_i(t, u)]^{r_i} [F_i(t, u)]^{1-r_i},$$

$$t \in (-\infty, \infty), u = 1, 2, \ldots, z, \tag{2.3.15}$$

or with the coordinates given by

$$\boldsymbol{R}(t, u) = 1 - \sum_{\substack{r_1, r_2, \ldots, r_n = 0 \\ 0 \le r_1 + r_2 + \ldots + r_n \le m-1}}^{1} \prod_{i=1}^{n} [R_i(t, u)]^{r_i} [F_i(t, u)]^{1-r_i},$$

$$t \in (-\infty, \infty), u = 1, 2, \ldots, z. \tag{2.3.16}$$

The reliability function of a homogeneous multistate "m out of n" system, from (2.3.14)–(2.3.16), is

$$\boldsymbol{R}(t,\ \cdot\ ) = [1, \boldsymbol{R}(t, 1), \ldots, \boldsymbol{R}(t, z)], t \in (-\infty, \infty), \tag{2.3.17}$$

with the coordinates either given by

$$\boldsymbol{R}(t, u) = \sum_{i=m}^{n} \binom{n}{i} [R(t, u)]^i [F(t, u)]^{n-i}, t \in (-\infty, \infty), u = 1, 2, \ldots, z, \tag{2.3.18}$$

or by

$$\boldsymbol{R}(t, u) = 1 - \sum_{i=0}^{m-1} \binom{n}{i} [R(t, u)]^i [F(t, u)]^{n-i}, t \in (-\infty, \infty), u = 1, 2, \ldots, z. \tag{2.3.19}$$

Other basic multistate reliability structures with components degrading in time are series-parallel, parallel-series, series-"m out of k," and "m out of l"-series systems. To define them, we assume that (Kołowrocki, 2014; Kołowrocki and Soszyńska-Budny, 2011):

- k is the number of subsystems,
- l_i, $i = 1,2, \ldots, k$ is the number of components of the ith subsystem,
- E_{ij}, $i = 1,2, \ldots, k$, $j = 1,2, \ldots, l_i$, $k, l_1, l_2, \ldots, l_k \in N$, are components of a system,
- all components E_{ij} have the reliability state set $\{0,1, \ldots, z\}$,
- $T_{ij}(u)$, $i = 1,2, \ldots, k$, $j = 1,2, \ldots, l_i$, $k, l_1, l_2, \ldots, l_k \in N$, are random variables representing the lifetimes of components E_{ij} in the reliability state subset $\{u, u+1, \ldots, z\}$, $u = 1,2, \ldots, z$, while they were in the reliability state z at the moment $t = 0$,
- $e_{ij}(t)$ is a component E_{ij} reliability state at the moment t, $t \in <0, \infty)$, while it was in the reliability state z at the moment $t = 0$.

Definition 2.3.5. The reliability function of a multistate component $E_{i,j}$ is a vector (Kołowrocki, 2014; Kołowrocki and Soszyńska-Budny, 2011)

$$R_{ij}(t,\ \cdot\ ) = [R_{ij}(t, 0), R_{ij}(t, 1), \ldots, R_{ij}(t, z)], t \in (-\infty, \infty),$$
$$i = 1, 2, \ldots, k, j = 1, 2, \ldots, l_i, \tag{2.3.20}$$

where

$$R_{ij}(t, u) = P(e_{ij}(t) \geq u | e_{ij}(0) = z)$$
$$= P(T_{ij}(u) > t), t \in (-\infty, \infty), u = 0, 1, \ldots, z, \tag{2.3.21}$$

is the probability that the component $E_{i,j}$ is in the reliability state subset $\{u, u+1, \ldots, z\}$, $u=1,2, \ldots, z$, at the moment t, $t \in (-\infty, \infty)$, while it was in the reliability state z at the moment $t=0$.

Further, similarly as in the case of component reliability function (2.1.1), we replace $R_{i,j}(t,0)$, existing in (2.3.18), by 1.

Definition 2.3.6. A multistate system is called series-parallel if its lifetime $\boldsymbol{T}(u)$ in the reliability state subset $\{u, u+1, \ldots, z\}$, $u=1,2, \ldots, z$, is given by (Kołowrocki, 2014)

$$\boldsymbol{T}(u) = \max_{1 \le i \le k} \left\{ \min_{1 \le j \le l_i} \{ T_{i,j}(u) \} \right\}, u=0,1,\ldots,z, \tag{2.3.22}$$

where k is the number of series subsystems linked in parallel and l_i, $i=1, 2, \ldots, k$, are the numbers of components in these series subsystems.

The reliability function of a multistate series-parallel system is given by (Kołowrocki, 2014)

$$\boldsymbol{R}(t, \cdot) = [1, \boldsymbol{R}(t,1), \ldots, \boldsymbol{R}(t,z)], t \in (-\infty, \infty), \tag{2.3.23}$$

and

$$\boldsymbol{R}(t,u) = 1 - \prod_{i=1}^{k} \left[1 - \prod_{j=1}^{l_i} R_{ij}(t,u) \right], \quad t \in (-\infty, \infty), u=1,2,\ldots,z. \tag{2.3.24}$$

Definition 2.3.7. Multistate series-parallel, parallel-series, series-"m out of k," and "m_i out of l_i"-series systems, we call homogeneous if its lifetimes $T_{ij}(u)$, $i=1, 2, \ldots, k$, $j=1, 2, \ldots, l_i$, and $u=1,2, \ldots, z$, in the reliability state subset $\{u, u+1, \ldots, z\}$, $u=1,2, \ldots, z$, have an identical distribution function

$$F_{ij}(t,u) = F(t,u), t \in (-\infty, \infty), u=1,2,\ldots,z, i=1,2,\ldots,k, j=1,2,\ldots,l_i,$$

i.e., if its components E_{ij} have the same reliability function

$$R_{ij}(t,u) = R(t,u), t \in (-\infty, \infty), u=1,2,\ldots,z, i=1,2,\ldots,k, j=1,2,\ldots,l_i.$$

Definition 2.3.8. Multistate series-parallel, parallel-series, and series-"m out of k" systems are called regular if

$$l_1 = l_2 = \cdots l_k = l, l \in \mathbb{N},$$

where k is the number of subsystems and l is the number of components in each subsystem.

The reliability function of a multistate homogeneous regular series-parallel system, from (2.3.23)–(2.3.24), is given by

$$\boldsymbol{R}(t,\ \cdot\ ) = [1, \boldsymbol{R}(t, 1), \ldots, \boldsymbol{R}(t, z)], t \in (-\infty, \infty), \tag{2.3.25}$$

and

$$\boldsymbol{R}(t, u) = 1 - \left[1 - [R(t, u)]^l\right]^k, t \in (-\infty, \infty), u = 1, 2, \ldots, z, \tag{2.3.26}$$

where k is the number of series subsystems linked in parallel and l is the number of components in these series subsystems.

Definition 2.3.9. A multistate system is called parallel-series if its lifetime $\boldsymbol{T}(u)$ in the reliability state subset $\{u, u+1, \ldots, z\}$, $u = 1,2, \ldots, z$, is given by (Kołowrocki, 2014)

$$\boldsymbol{T}(u) = \min_{1 \le i \le k} \left\{ \max_{1 \le j \le l_i} \{T_{ij}(u)\} \right\}, u = 1, 2, \ldots, z, \tag{2.3.27}$$

where k is the number of its parallel subsystems linked in series and l_i, $i = 1,2, \ldots, k$, are the numbers of components in these parallel subsystems.

The reliability function of a multistate parallel-series system is given by (Kołowrocki, 2014)

$$\boldsymbol{R}(t,\ \cdot\ ) = [1, \boldsymbol{R}(t, 1), \ldots, \boldsymbol{R}(t, z)], t \in (-\infty, \infty), \tag{2.3.28}$$

and

$$\boldsymbol{R}(t, u) = \prod_{i=1}^{k} \left[1 - \prod_{j=1}^{l_i} F_{ij}(t, u)\right], \quad t \in (-\infty, \infty), u = 1, 2, \ldots, z, \tag{2.3.29}$$

where $F_{ij}(t,u)$, $i = 1,2, \ldots, k$, $j = 1,2, \ldots, l_i$, $u = 1,2, \ldots, z$, is the distribution function of a component lifetime $T_{ij}(u)$, given by

$$F_{ij}(t, u) = 1 - R_{ij}(t, u), t \in (-\infty, \infty), u = 1, 2, \ldots, z, i = 1, 2, \ldots, k, j = 1, 2, \ldots, l_i. \tag{2.3.30}$$

The reliability function of a multistate homogeneous regular parallel-series system, using (2.3.38)–(2.3.39), is given by

$$\boldsymbol{R}(t,\ \cdot\ ) = [1, \boldsymbol{R}(t, 1), \ldots, \boldsymbol{R}(t, z)], t \in (-\infty, \infty), \tag{2.3.31}$$

and

$$\boldsymbol{R}(t, u) = \left[1 - [F(t, u)]^l\right]^k, t \in (-\infty, \infty), u = 1, 2, \ldots, z, \qquad (2.3.32)$$

where $F(t,u)$, $u=1,2, \ldots, z$, is the distribution function of a component lifetime $T(u)$, given by (2.3.12), k is the number of parallel subsystems linked in series and l is the number of components in these parallel subsystems.

For each fixed i, $i=1,2, \ldots, k$, we denote by $T_{i(j)}(u)$, $j=1, 2, \ldots, l_i$, $u=1,2, \ldots, z$, the order statistics of lifetimes $T_{ij}(u)$ in the reliability state subset $\{u, u+1, \ldots, z\}$, $u=1,2, \ldots, z$, of ith parallel subsystem components E_{ij} satisfying inequalities

$$T_{i(1)}(u) \leq T_{i(2)}(u) \leq \cdots \leq T_{i(l_i)}(u), i = 1, 2, \ldots, k. \qquad (2.3.33)$$

Definition 2.3.10. A multistate system is called a series-"m out of k" system, if its lifetime $\boldsymbol{T}(u)$ in the reliability state subset $\{u, u+1, \ldots, z\}$, $u=1,2, \ldots, z$, is given by (Kołowrocki, 2014)

$$\boldsymbol{T}(u) = T_{(k-m+1)}(u), m = 1, 2, \ldots, k, u = 1, \ldots, z, \qquad (2.3.34)$$

where $T_{(k-m+1)}(u)$ is the mth maximal order statistic in the set of random variables

$$T_i(u) = \min_{1 \leq j \leq l_i} \{T_{ij}(u)\}, i = 1, 2, \ldots, k, u = 1, \ldots, z. \qquad (2.3.35)$$

In a multistate series-"m out of k" system, k is the number of its multistate series subsystems and l_i, $i=1,2, \ldots, k$, are the numbers of components in series subsystems. The numbers m, k and $l_1, l_2, \ldots, l_k$ are called the system structure shape parameters.

The definition means that a multistate series-"m out of k" system it is in the reliability state subset $\{u, u+1, \ldots, z\}$, $u=1,2, \ldots, z$, if and only if at least m out of its k series subsystems are in this state subset.

The reliability function of a multistate series-"m out of k" system is given by the vector (Kołowrocki, 2014)

$$\boldsymbol{R}(t, \cdot) = [1, \boldsymbol{R}(t, 1), \ldots, \boldsymbol{R}(t, z)], t \in (-\infty, \infty), \qquad (2.3.36)$$

with the coordinates either given by

$$\boldsymbol{R}(t,u)=\sum_{\substack{r_1,r_2,\ldots,r_k=0\\ m\le r_1+r_2+\ldots+r_k\le k}}^{1}\prod_{i=1}^{k}\left[\prod_{j=1}^{l_i}R_{ij}(t,u)\right]^{r_i}\left[1-\prod_{j=1}^{l_i}R_{ij}(t,u)\right]^{1-r_i},$$

$$t\in(-\infty,\infty),u=1,\ldots,z, \tag{2.3.37}$$

or with the coordinates given by

$$\boldsymbol{R}(t,u)=1-\sum_{\substack{r_1,r_2,\ldots,r_k=0\\ 0\le r_1+r_2+\ldots+r_k\le m-1}}^{1}\prod_{i=1}^{k}\left[\prod_{j=1}^{l_i}R_{ij}(t,u)\right]^{r_i}\left[1-\prod_{j=1}^{l_i}R_{ij}(t,u)\right]^{1-r_i},$$

$$t\in(-\infty,\infty),u=1,\ldots,z. \tag{2.3.38}$$

The reliability function of a multistate homogeneous regular series-"m out of k" system, by (2.3.36)–(2.3.38), is

$$\boldsymbol{R}(t,\cdot)=[1,\boldsymbol{R}(t,1),\ldots,\boldsymbol{R}(t,z)],t\in(-\infty,\infty), \tag{2.3.39}$$

with the coordinates either given by

$$\boldsymbol{R}(t,u)=\sum_{i=m}^{k}\binom{k}{i}[R(t,u)]^{l\cdot i}\left[1-[R(t,u)]^{l}\right]^{k-i},t\in(-\infty,\infty),$$

$$u=1,\ldots,z, \tag{2.3.40}$$

or by

$$\boldsymbol{R}(t,u)=1-\sum_{i=0}^{m-1}\binom{k}{i}[R(t,u)]^{l\cdot i}\left[1-[R(t,u)]^{l}\right]^{k-i},t\in(-\infty,\infty),$$

$$u=1,\ldots,z. \tag{2.3.41}$$

Definition 2.3.11. A multistate system is called an "m_i out of l_i"-series system, if its lifetime $\boldsymbol{T}(u)$ in the reliability state subset $\{u,u+1,\ldots,z\}$, $u=1,2,\ldots,z$, is given by (Kołowrocki, 2014).

$$\boldsymbol{T}(u) = \min_{1 \le i \le k} \{T_{i(l_i - m_i + 1)}(u)\}, m_i = 1, 2, \ldots, l_i, i = 1, 2, \ldots, k,$$
$$u = 1, \ldots, z, \qquad (2.3.42)$$

where $T_{i(l_i - mi+1)}(u)$ is the m_ith maximal order statistic in the set of random variables $T_{i1}(u)$, $T_{i2}(u)$, ..., $T_{il_i}(u)$, $i = 1,2, \ldots, k$, fulfilling the condition (2.3.33).

In a multistate "m_i out of l_i"-series system, k is the number of its "m_i out of l_i"subsystems linked in series and l_i, $i = 1,2, \ldots, k$, are the numbers of components in these "m_i out of l_i"subsystems. The numbers k, m_1, m_2,..., m_k and l_1, l_2, ..., l_k, are called the system structure shape parameters.

The above definition means that an "m_i out of l_i"-series system is in the reliability state subset $\{u, u+1, \ldots, z\}$, $u = 1,2, \ldots, z$, if and only if all its "m_i out of l_i"subsystems are in this subset of states.

The reliability function of a multistate "m_i out of l_i"-series system is given by the vector (Kołowrocki, 2014)

$$\boldsymbol{R}(t, \cdot) = [1, \boldsymbol{R}(t, 1), \ldots, \boldsymbol{R}(t, z)], t \in (-\infty, \infty), \qquad (2.3.43)$$

with the coordinates either given by

$$\boldsymbol{R}(t, u) = \prod_{i=1}^{k} \left[\sum_{\substack{r_1, r_2, \ldots, r_{l_i} = 0 \\ m_i \le r_1 + r_2 + \ldots + r_{l_i} \le l_i}}^{1} \left[\prod_{j=1}^{l_i} R_{ij}(t, u) \right]^{r_i} \left[1 - \prod_{j=1}^{l_i} R_{ij}(t, u) \right]^{1 - r_i} \right],$$
$$t \in (-\infty, \infty), u = 1, \ldots, z, \qquad (2.3.44)$$

or by

$$\boldsymbol{R}(t, u) = \prod_{i=1}^{k} \left[1 - \sum_{\substack{r_1, r_2, \ldots, r_{l_i} = 0 \\ 0 \le r_1 + r_2 + \ldots + r_{l_i} \le m_i - 1}}^{1} \left[\prod_{j=1}^{l_i} R_{ij}(t, u) \right]^{r_i} \left[1 - \prod_{j=1}^{l_i} R_{ij}(t, u) \right]^{1 - r_i} \right],$$
$$t \in (-\infty, \infty), u = 1, \ldots, z. \qquad (2.3.45)$$

Definition 2.3.12. A multistate "m_i out of l_i"-series system is called regular, if

$$l_1 = l_2 = \ldots = l_k = l \text{ and } m_1 = m_2 = \ldots = m_k = m, \text{where } m \le l, l, m \in N,$$

where k is the number of "m out of l" subsystems linked in series and l is the number of components in each "m out of l" subsystem.

From this definition and from definition of homogeneous system, using (2.3.43)–(2.3.45), it follows that the reliability function of a multistate homogeneous and regular "m out of l"-series system is given by the vector

$$\boldsymbol{R}(t, \cdot) = [1, \boldsymbol{R}(t, 1), \ldots, \boldsymbol{R}(t, z)], t \in (-\infty, \infty), \tag{2.3.46}$$

with the coordinates either given by

$$\boldsymbol{R}(t, u) = \left[\sum_{i=m}^{l} \binom{l}{i} [R(t, u)]^i [1 - R(t, u)]^{l-i}\right]^k, t \in (-\infty, \infty),$$
$$u = 1, \ldots, z, \tag{2.3.47}$$

or by

$$\boldsymbol{R}(t, u) = \left[1 - \sum_{i=0}^{m-1} \binom{l}{i} [R(t, u)]^i [1 - R(t, u)]^{l-i}\right]^k, t \in (-\infty, \infty),$$
$$u = 1, \ldots, z. \tag{2.3.48}$$

2.4 Applications

2.4.1 Reliability analysis of exemplary system

We consider a 3-state ($z = 2$) series system composed of three subsystems S_1, S_2, S_3. Its reliability structure is given in Fig. 2.7.

The subsystem S_1 is a homogeneous parallel system consisting of four components (Fig. 2.8).

The components $E_i^{(1)}$, $i = 1,2,3,4$, have the identical piecewise exponential reliability functions

$$R_i^{(1)}(t, \cdot) = [1, \exp(-0.02t), \exp(-0.04t)], t \ge 0, i = 1, 2, 3, 4. \tag{2.4.1}$$

Fig. 2.7 The scheme of the exemplary series system reliability structure.

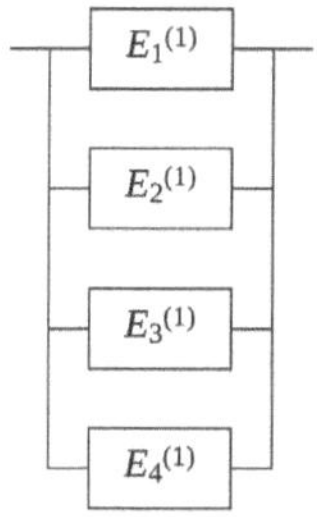

Fig. 2.8 The subsystem S_1 reliability structure.

Then, the reliability function of subsystem S_1, according to (2.3.10)–(2.3.12), is

$$\boldsymbol{R}^{(1)}(t,\ \cdot\ ) = \left[1, \boldsymbol{R}^{(1)}(t,1), \boldsymbol{R}^{(1)}(t,2)\right], t \geq 0, \tag{2.4.2}$$

where its coordinates are

$$\begin{aligned}\boldsymbol{R}^{(1)}(t,1) &= 1-[1-\exp(-0.02t)]^4\\ &= 4\exp(-0.02t) - 6\exp(-0.04t) + 4\exp(-0.06t)\\ &\quad -\exp(-0.08t),\end{aligned} \tag{2.4.3}$$

$$\begin{aligned}\boldsymbol{R}^{(1)}(t,2) &= 1-[1-\exp(-0.04t)]^4\\ &= 4\exp(-0.04t) - 6\exp(-0.08t) + 4\exp(-0.12t)\\ &\quad -\exp(-0.16t).\end{aligned} \tag{2.4.4}$$

The subsystem S_2 is a series of nonhomogeneous systems composed of five components (Fig. 2.9).

The components $E_i^{(2)}$, $i=1,2,3,4,5$, by the assumption, have piecewise exponential reliability functions

$$R_1^{(2)}(t,\ \cdot\ ) = [1,\ \exp(-0.005t),\ \exp(-0.015t)], t \geq 0, \tag{2.4.5}$$

$$R_i^{(2)}(t,\ \cdot\ ) = [1,\ \exp(-0.006t),\ \exp(-0.008t)], t \geq 0, i=2,3, \tag{2.4.6}$$

$$R_i^{(2)}(t,\ \cdot\ ) = [1,\ \exp(-0.008t),\ \exp(-0.011t)], t \geq 0, i=4,5. \tag{2.4.7}$$

Then, applying (2.3.2)–(2.3.3), the reliability function of subsystem S_2 is

$$\boldsymbol{R}^{(2)}(t,\ \cdot\ ) = \left[1, \boldsymbol{R}^{(2)}(t,1), \boldsymbol{R}^{(2)}(t,2)\right], t \geq 0, \tag{2.4.8}$$

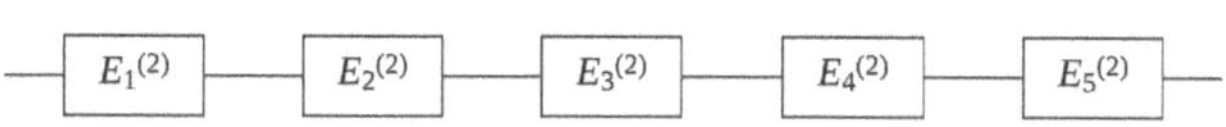

Fig. 2.9 The subsystem S_2 reliability structure.

with the coordinates

$$\boldsymbol{R}^{(2)}(t, 1) = \exp(-0.005t) \cdot [\exp(-0.006t)]^2 \cdot [\exp(-0.008t)]^2 \\ = \exp(-0.033t), \tag{2.4.9}$$

$$\boldsymbol{R}^{(2)}(t, 2) = \exp(-0.015t) \cdot [\exp(-0.008t)]^2 \cdot [\exp(-0.011t)]^2 \\ = \exp(-0.053t). \tag{2.4.10}$$

The subsystem S_3 is a homogeneous "3 out of 5" system consisting of five components (Fig. 2.10).

The components of subsystem S_3 have the identical, by the assumption, piecewise exponential reliability functions

$$R_i^{(3)}(t, \cdot) = [1, \exp(-0.035t), \exp(-0.045t)], t \geq 0, i \\ = 1, 2, 3, 4, 5. \tag{2.4.11}$$

Then, the reliability function of subsystem S_3, according to (2.3.17)–(2.3.18), is given by

$$\boldsymbol{R}^{(3)}(t, \cdot) = \left[1, \boldsymbol{R}^{(3)}(t, 1), \boldsymbol{R}^{(3)}(t, 2)\right], t \geq 0, \tag{2.4.12}$$

with the coordinates

$$\boldsymbol{R}^{(3)}(t, 1) = \sum_{i=3}^{5} \binom{5}{i} [\exp(-0.035t)]^i [1 - \exp(-0.035t)]^{5-i} \\ = 10\exp(-0.105t) - 15\exp(-0.14t) + 6\exp(-0.175t), \tag{2.4.13}$$

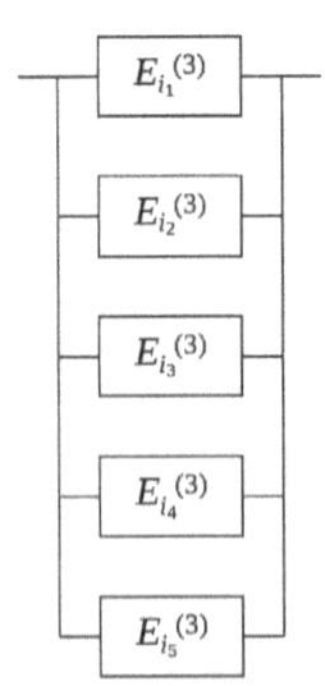

Fig. 2.10 The subsystem S_3 reliability structure.

$$\boldsymbol{R}^{(3)}(t,2)=\sum_{i=3}^{5}\binom{5}{i}\,[\exp(-0.045t)]^{i}[1-\exp(-0.045t)]^{5-i}$$
$$=10\exp(-0.135t)-15\exp(-0.18t)+6\exp(-0.225t). \tag{2.4.14}$$

Taking into account the reliability structure of the system S, presented in Fig. 2.7, and applying (2.3.2)–(2.3.3), the reliability function of system S is given by

$$\boldsymbol{R}(t,\ \cdot\ )=[1,\boldsymbol{R}(t,1),\boldsymbol{R}(t,2)],t\geq 0, \tag{2.4.15}$$

where

$$\boldsymbol{R}(t,1)=\boldsymbol{R}^{(1)}(t,1)\cdot\boldsymbol{R}^{(2)}(t,1)\cdot\boldsymbol{R}^{(3)}(t,1), \tag{2.4.16}$$
$$\boldsymbol{R}(t,2)=\boldsymbol{R}^{(1)}(t,2)\cdot\boldsymbol{R}^{(2)}(t,2)\cdot\boldsymbol{R}^{(3)}(t,2). \tag{2.4.17}$$

And consequently, using the results (2.4.3)–(2.4.4), (2.4.9)–(2.4.10), and (2.4.13)–(2.4.14), it takes following form

$$\boldsymbol{R}(t,\ \cdot\ )=[1,\boldsymbol{R}(t,1),\boldsymbol{R}(t,2)],t\geq 0, \tag{2.4.18}$$

where

$$\begin{aligned}\boldsymbol{R}(t,1)&=[4\exp(-0.02t)-6\exp(-0.04t)+4\exp(-0.06t)-\exp(-0.08t)]\\&\quad\cdot\exp(-0.033t)\cdot[10\exp(-0.105t)-15\exp(-0.14t)+6\exp(-0.175t)]\\&=40\exp(-0.158t)-60\exp(-0.178t)+40\exp(-0.198t)-10\exp(-0.218t)\\&\quad-60\exp(-0.193t)+90\exp(-0.213t)-60\exp(-0.233t)+15\exp(-0.253t)\\&\quad+24\exp(-0.228t)-36\exp(-0.248t)+24\exp(-0.268t)-6\exp(-0.288t),\end{aligned} \tag{2.4.19}$$

$$\begin{aligned}\boldsymbol{R}(t,2)&=[4\exp(-0.04t)-6\exp(-0.08t)+4\exp(-0.12t)-\exp(-0.16t)]\\&\quad\cdot\exp(-0.053t)\cdot[10\exp(-0.135t)-15\exp(-0.18t)+6\exp(-0.225t)]\\&=40\exp(-0.228t)-60\exp(-0.268t)+40\exp(-0.308t)-10\exp(-0.348t)\\&\quad-60\exp(-0.273t)+90\exp(-0.313t)-60\exp(-0.353t)+15\exp(-0.393t)\\&\quad+24\exp(-0.318t)-36\exp(-0.358t)+24\exp(-0.398t)-6\exp(-0.438t).\end{aligned} \tag{2.4.20}$$

The reliability function coordinates of the exemplary system S are illustrated in Fig. 2.11.

The mean lifetimes of the exemplary system S in the reliability state subsets $\{1,2\}$, $\{2\}$, according to (2.1.16), are determined from

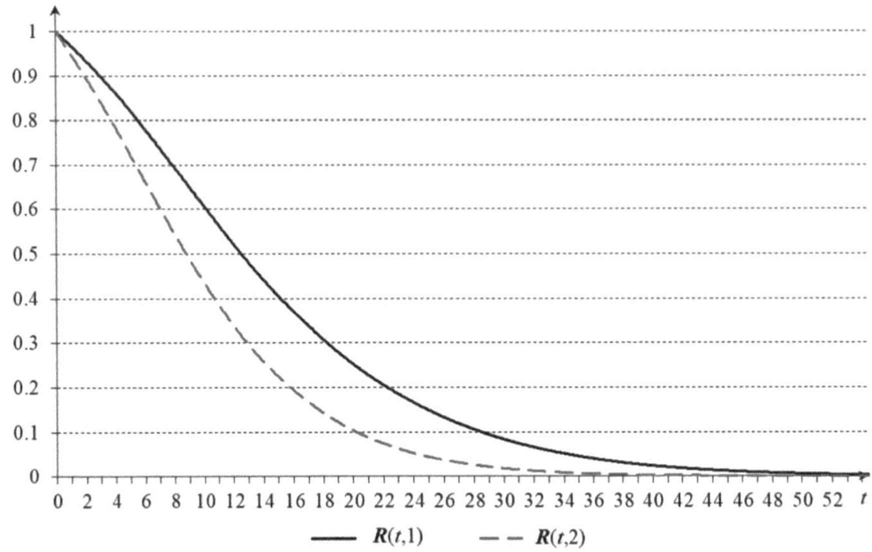

Fig. 2.11 The graphs of the exemplary system *S* reliability function coordinates.

$$\mu(1) = \int_0^\infty \mathbf{R}(t,1)dt, \mu(2) = \int_0^\infty \mathbf{R}(t,2)dt, \tag{2.4.21}$$

where the reliability function coordinates are given by (2.4.19)–(2.4.20).

Then, considering the formula

$$\int_0^\infty \exp[-\lambda t]dt = \frac{1}{\lambda}, \tag{2.4.22}$$

we get following values of the mean lifetimes of this system in the reliability state subsets $\{1,2\}$, $\{2\}$:

$$\mu(1) = \frac{40}{0.158} - \frac{60}{0.178} + \frac{40}{0.198} - \frac{10}{0.218} - \frac{60}{0.193} + \frac{90}{0.213} - \frac{60}{0.233} + \frac{15}{0.253}$$
$$+ \frac{24}{0.228} - \frac{36}{0.248} + \frac{24}{0.268} - \frac{6}{0.288} \cong 14.488, \tag{2.4.23}$$

$$\mu(2) = \frac{40}{0.228} - \frac{60}{0.268} + \frac{40}{0.308} - \frac{10}{0.348} - \frac{60}{0.273} + \frac{90}{0.313} - \frac{60}{0.353} + \frac{15}{0.393}$$
$$+ \frac{24}{0.318} - \frac{36}{0.358} + \frac{24}{0.398} - \frac{6}{0.438} \cong 10.164. \quad (2.4.24)$$

Similarly, applying (2.1.18) and using (2.4.19)–(2.4.20), and considering the formula

$$\int_0^\infty t \exp[-\lambda t] dt = \frac{1}{\lambda^2}, \quad (2.4.25)$$

we obtain:

$$n(1) = 2 \cdot \left[\frac{40}{0.158^2} - \frac{60}{0.178^2} + \frac{40}{0.198^2} - \frac{10}{0.218^2} - \frac{60}{0.193^2} + \frac{90}{0.213^2} - \frac{60}{0.233^2} \right.$$
$$\left. + \frac{15}{0.253^2} + \frac{24}{0.228^2} - \frac{36}{0.248^2} + \frac{24}{0.268^2} - \frac{6}{0.288^2}\right] \cong 317.507,$$
$$(2.4.26)$$

$$n(2) = 2 \cdot \left[\frac{40}{0.228^2} - \frac{60}{0.268^2} + \frac{40}{0.308^2} - \frac{10}{0.348^2} - \frac{60}{0.273^2} + \frac{90}{0.313^2} - \frac{60}{0.353^2} \right.$$
$$\left. + \frac{15}{0.393^2} + \frac{24}{0.318^2} - \frac{36}{0.358^2} + \frac{24}{0.398^2} - \frac{6}{0.438^2}\right] \cong 158.137.$$
$$(2.4.27)$$

And then, substituting (2.4.23)–(2.4.24) and (2.4.26)–(2.4.27) into (2.1.17), we determine the standard deviations of the exemplary system S lifetimes:

$$\sigma(1) = \sqrt{n(1) - [\mu(1)]^2} \cong \sqrt{317.507 - 14.488^2} \cong 10.373, \quad (2.4.28)$$

$$\sigma(2) = \sqrt{n(2) - [\mu(2)]^2} \cong \sqrt{158.137 - 10.164^2} \cong 7.405. \quad (2.4.29)$$

The mean values of the exemplary system lifetimes in the particular reliability states 1,2, by (2.1.20) and using (2.4.23)–(2.4.24), are:

$$\bar{\mu}(1) = \mu(1) - \mu(2) \cong 14.488 - 10.164 = 4.324, \quad (2.4.30)$$

$$\bar{\mu}(2) = \mu(2) \cong 10.164. \quad (2.4.31)$$

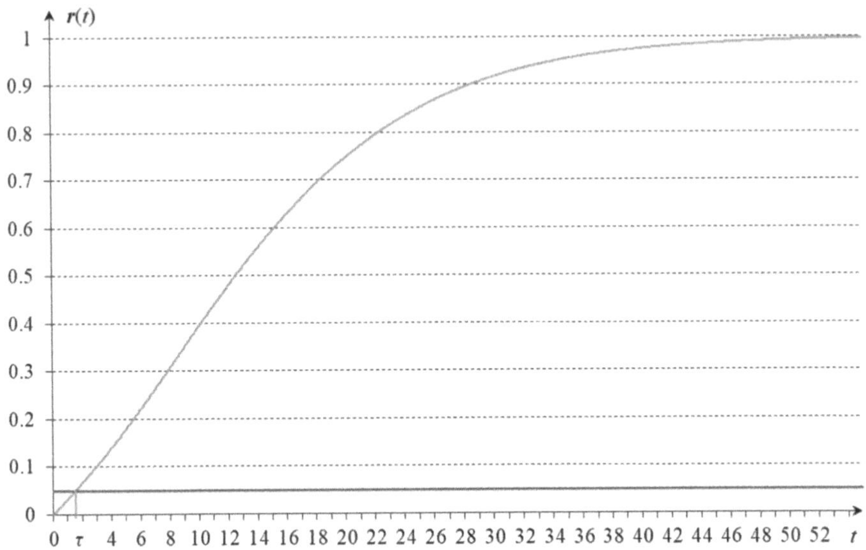

Fig. 2.12 The graph of the risk function ***r***(*t*) of exemplary system *S*.

Assuming that the critical reliability state is $r=1$ and applying (2.1.21), we obtain the risk function of the system *S*.

$$\boldsymbol{r}(t)=1-\boldsymbol{R}(t,1), t\geq 0, \tag{2.4.32}$$

where the reliability function coordinate $\boldsymbol{R}(t,1)$ is given by (2.4.19).

The risk function of the system *S* is illustrated in Fig. 2.12.

To find the moment of exceeding an acceptable level, for instance $\delta=0.05$, we determine the values of the system risk function, given in Table 2.3. Then, we find that the risk exceeds a permitted level $\delta=0.05$ for $t=1.516$, and by (2.1.22) we conclude that

$$\tau=\boldsymbol{r}^{-1}(\delta)\cong 1.516. \tag{2.4.33}$$

2.4.2 Reliability analysis of shipyard rope elevator

As a second application, the reliability of a shipyard rope elevator utilized in shipyard is analyzed (Blokus and Kołowrocki, 2019; Blokus-Roszkowska, 2006; Blokus-Roszkowska and Kołowrocki, 2014; Kołowrocki, 2014). The shipyard rope elevator is composed of a steel platform-carriage and 10 rope-hoisting winches. Fig. 2.13 shows the movement of the ship toward the lift platform.

Table 2.3 The values of the risk function $\boldsymbol{r}(t)$ of exemplary system.

t	$r(t)$
1.500	0.049473
1.501	0.049507
1.502	0.049540
1.503	0.049574
1.504	0.049608
1.505	0.049641
1.506	0.049675
1.507	0.049708
1.508	0.049742
1.509	0.049776
1.510	0.049809
1.511	0.049843
1.512	0.049877
1.513	0.049910
1.514	0.049944
1.515	0.049977
1.516	**0.050011**
1.517	0.050045
1.518	0.050078
1.519	0.050112
1.520	0.050146

Fig. 2.13 Movement of the ship to the lifting platform.

The platform is moved vertically using winches fed by electric motors and equipped in ropes with diameter 47 mm. Since the platform and electric motors are highly reliable in comparison to ropes, which are mainly exposed to external conditions, in further analysis the reliability of a rope system is only discussed.

The shipyard rope elevator is composed of $k=10$ subsystems; the ropes are linked in series, and each rope is composed of $l=22$ parallel-linked strands. In practice, a rope changes its reliability state subset after a fixed number of strands change their reliability state subsets. Therefore, a more realistic approach to the system reliability evaluation is the assumption that a rope is an "m out of l" system. Based on the rope's parameters (obtained from its technical certificate and expert opinion), we suppose that $m=5$ and that a rope is a "5 out of 22" system. Considering strands as the basic components given by E_{ij}, $i=1,2, \ldots, 10$, $j=1,2, \ldots, 22$, the shipyard rope elevator forms a regular "5 out of 22"-series system with the reliability structure presented in Fig. 2.14.

We consider the following four reliability states of strands (Blokus-Roszkowska, 2006):

- reliability state 3—the strand is new,
- reliability state 2—the extent of broken wires or corrosion is between 0% and 25%,
- reliability state 1—the extent of broken wires or corrosion is between 25% and 50%,
- reliability state 0—otherwise.

The critical reliability state is $r=2$.

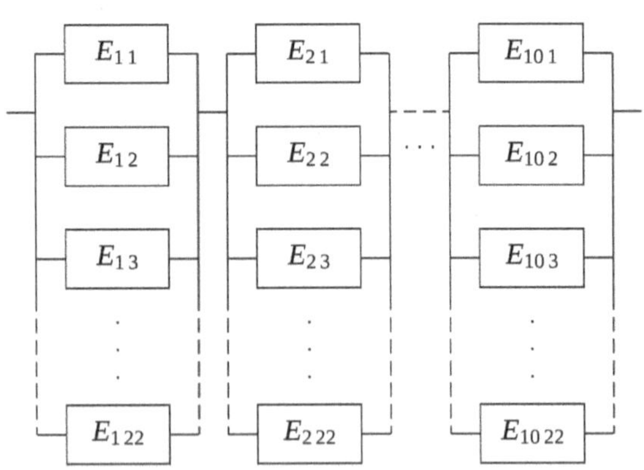

Fig. 2.14 The shipyard rope elevator reliability structure.

It is assumed that the components E_{ij}, $i=1,2, \ldots, 10$ and $j=1,2, \ldots, 22$, have identical piecewise exponential reliability functions

$$R(t, \cdot) = [1, R(t, 1), R(t, 2), R(t, 3)], t \geq 0, \tag{2.4.34}$$

with the coordinates (Blokus-Roszkowska, 2006).

$$R(t, 1) = \exp[-0.1613t], R(t, 2) = \exp[-0.2041t], R(t, 3) = \exp[-0.2326t], t \geq 0. \tag{2.4.35}$$

In (2.4.35), the intensities of leaving the reliability state subsets {1,2,3}, {2,3}, {3}, have been estimated based on the mean values of the component lifetimes in these subsets, expressed in years, originating from the expert, namely, the operator of the system.

Taking into account the reliability structure of the shipyard rope elevator, presented in Fig. 2.14, applying (2.3.46) and (2.3.48), and substituting the coordinates of the components' reliability functions (2.4.35), we obtain the reliability function of the shipyard rope elevator.

$$\boldsymbol{R}(t, \cdot) = [1, \boldsymbol{R}(t, 1), \boldsymbol{R}(t, 2), \boldsymbol{R}(t, 3)], t \geq 0, \tag{2.4.36}$$

where

$$\boldsymbol{R}(t, 1) = \left[1 - \sum_{i=1}^{4} \binom{22}{i} [\exp[-0.1613t]]^i [1 - \exp[-0.1613t]]^{22-i}\right]^{10}, \tag{2.4.37}$$

$$\boldsymbol{R}(t, 2) = \left[1 - \sum_{i=1}^{4} \binom{22}{i} [\exp[-0.2041t]]^i [1 - \exp[-0.2041t]]^{22-i}\right]^{10}, \tag{2.4.38}$$

$$\boldsymbol{R}(t, 3) = \left[1 - \sum_{i=1}^{4} \binom{22}{i} [\exp[-0.2326t]]^i [1 - \exp[-0.2326t]]^{22-i}\right]^{10}, \quad t \geq 0. \tag{2.4.39}$$

The reliability function coordinates of the shipyard rope elevator are illustrated in Fig. 2.15.

The mean lifetimes of the shipyard rope elevator in the reliability state subsets {1,2,3}, {2,3}, {3}, according to (2.1.16) and using (2.4.37)–(2.4.39), are determined from

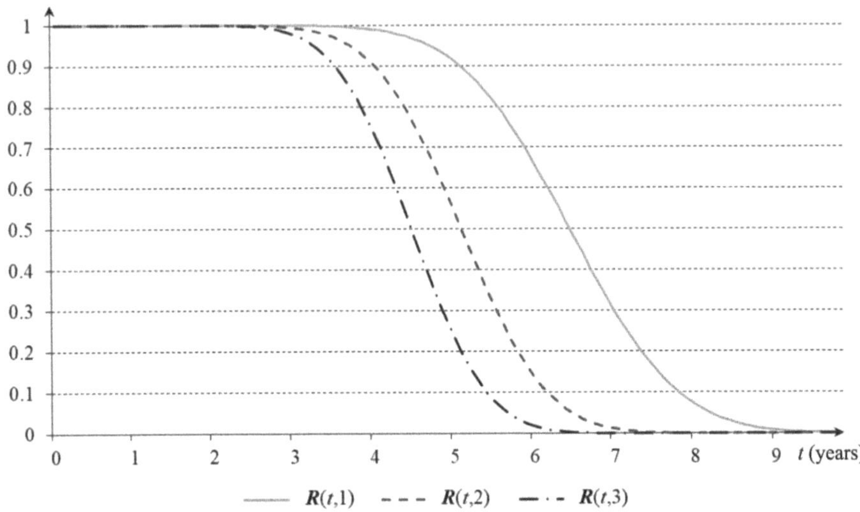

Fig. 2.15 The graphs of the shipyard rope elevator reliability function coordinates.

$$\mu(1) = \int_0^\infty \left[1 - \sum_{i=1}^{4} \binom{22}{i} [\exp[-0.1613t]]^i [1 - \exp[-0.1613t]]^{22-i}\right]^{10} dt, \tag{2.4.40}$$

$$\mu(2) = \int_0^\infty \left[1 - \sum_{i=1}^{4} \binom{22}{i} [\exp[-0.2041t]]^i [1 - \exp[-0.2041t]]^{22-i}\right]^{10} dt, \tag{2.4.41}$$

$$\mu(3) = \int_0^\infty \left[1 - \sum_{i=1}^{4} \binom{22}{i} [\exp[-0.2326t]]^i [1 - \exp[-0.2326t]]^{22-i}\right]^{10} dt. \tag{2.4.42}$$

To calculate these integrals, similarly as Section 2.1.1, we use the trapezoid rule for approximating integrals. The values of reliability function coordinates of the shipyard rope elevator and area under the graph of these functions, given in Table 2.4, are calculated with a time step $h=0.5$.

Table 2.4 The values of reliability functions $\boldsymbol{R}(t,1)$, $\boldsymbol{R}(t,2)$, $\boldsymbol{R}(t,3)$ and area under the graph of these reliability functions calculated with a time step $h = 0.5$.

t	$R(t,1)$	$R(t,1)$ area	$R(t,2)$	$R(t,2)$ area	$R(t,3)$	$R(t,3)$ area
0.0	1.000000	0.500000	1.000000	0.500000	1.000000	0.500000
0.5	1.000000	0.500000	1.000000	0.500000	1.000000	0.500000
1.0	1.000000	0.500000	1.000000	0.500000	1.000000	0.499999
1.5	1.000000	0.500000	0.999999	0.499988	0.999994	0.499938
2.0	0.999998	0.499990	0.999955	0.499803	0.999759	0.499099
2.5	0.999961	0.499893	0.999258	0.498332	0.996637	0.493415
3.0	0.999609	0.499300	0.994068	0.491294	0.977024	0.470582
3.5	0.997589	0.496809	0.971110	0.468535	0.905305	0.410812
4.0	0.989646	0.488991	0.903029	0.415769	0.737942	0.305320
4.5	0.966318	0.469717	0.760047	0.326127	0.483338	0.179287
5.0	0.912548	0.431332	0.544461	0.214333	0.233812	0.078155
5.5	0.812779	0.368827	0.312870	0.112670	0.078810	0.024219
6.0	0.662527	0.285444	0.137810	0.045786	0.018066	0.005228
6.5	0.479249	0.194644	0.045333	0.014103	0.002846	0.000792
7.0	0.299325	0.114345	0.011078	0.003280	0.000323	0.000088
7.5	0.158055	0.056921	0.002042	0.000584	0.000029	0.000008
8.0	0.069629	0.023772	0.000295	0.000083	0.000002	0.000001
8.5	0.025459	0.008302	0.000036	0.000010	0.000000	0.000000
9.0	0.007748	0.002433	0.000004	0.000001	0.000000	0.000000
9.5	0.001984	0.000605	0.000000	0.000000	0.000000	0.000000
10.0	0.000436		0.000000		0.000000	
SUM		**6.441322**		**5.090697**		**4.466944**

Then, the mean lifetimes of the shipyard rope elevator in the reliability state subsets {1,2,3}, {2,3}, {3}, calculated using the values given in Table 2.4, in years, are, respectively, given by:

$$\mu(1) \cong \frac{h}{2} \cdot [\boldsymbol{R}(0,1) + 2\boldsymbol{R}(h,1) + \cdots + 2\boldsymbol{R}(19h,1) + \boldsymbol{R}(20h,1)] \cong 6.441, \tag{2.4.43}$$

$$\mu(2) \cong \frac{h}{2} \cdot [\boldsymbol{R}(0,2) + 2\boldsymbol{R}(h,2) + \cdots + 2\boldsymbol{R}(19h,2) + \boldsymbol{R}(20h,2)] \cong 5.091, \tag{2.4.44}$$

$$\mu(3) \cong \frac{h}{2} \cdot [\boldsymbol{R}(0,3) + 2\boldsymbol{R}(h,3) + \cdots + 2\boldsymbol{R}(19h,3) + \boldsymbol{R}(20h,3)] \cong 4.467. \tag{2.4.45}$$

Similarly, applying (2.1.18) and using (2.4.37)–(2.4.39), we obtain:

$$n(1) = 2\int_0^{\infty} t \cdot \left[1 - \sum_{i=1}^{4} \binom{22}{i} [\exp[-0.1613t]]^i [1 - \exp[-0.1613t]]^{22-i}\right]^{10} dt, \tag{2.4.46}$$

$$n(2) = 2\int_0^{\infty} t \cdot \left[1 - \sum_{i=1}^{4} \binom{22}{i} [\exp[-0.2041t]]^i [1 - \exp[-0.2041t]]^{22-i}\right]^{10} dt, \tag{2.4.47}$$

$$n(3) = 2\int_0^{\infty} t \cdot \left[1 - \sum_{i=1}^{4} \binom{22}{i} [\exp[-0.2326t]]^i [1 - \exp[-0.2326t]]^{22-i}\right]^{10} dt. \tag{2.4.48}$$

To calculate these integrals and determine the standard deviation of system lifetimes, we use the trapezoid rule for approximating integrals and the values given in Table 2.5.

Then, using the values given in Table 2.5, we get:

$$n(1) = 2\int_0^{\infty} t \cdot \boldsymbol{R}(t, 1)dt \cong 2 \cdot 21.2844 = 42.5688, \tag{2.4.49}$$

$$n(2) = 2\int_0^{\infty} t \cdot \boldsymbol{R}(t, 2)dt \cong 2 \cdot 13.2858 = 26.5716, \tag{2.4.50}$$

$$n(3) = 2\int_0^{\infty} t \cdot \boldsymbol{R}(t, 3)dt \cong 2 \cdot 10.2247 = 20.4494, \tag{2.4.51}$$

and, substituting (2.4.43)–(2.4.45) and (2.4.49)–(2.4.51) for the formula (2.1.17), we determine the standard deviations of the shipyard rope elevator lifetimes in the reliability state subsets $\{1,2,3\}$, $\{2,3\}$, $\{3\}$, in years:

$$\sigma(1) = \sqrt{n(1) - [\mu(1)]^2} \cong \sqrt{42.5688 - 6.441^2} \cong 1.040, \tag{2.4.52}$$

$$\sigma(2) = \sqrt{n(2) - [\mu(2)]^2} \cong \sqrt{26.5716 - 5.091^2} \cong 0.808, \tag{2.4.53}$$

$$\sigma(3) = \sqrt{n(3) - [\mu(3)]^2} \cong \sqrt{20.4494 - 4.467^2} \cong 0.704. \tag{2.4.54}$$

Table 2.5 The values of functions $t\,\boldsymbol{R}(t,1)$, $t\,\boldsymbol{R}(t,2)$, $t\,\boldsymbol{R}(t,3)$ and area under the graph of these functions calculated with a time step $h=0.5$.

t	$t\cdot R(t,1)$	$t\cdot R(t,1)$ area	$t\cdot R(t,2)$	$t\cdot R(t,2)$ area	$t\cdot R(t,3)$	$t\cdot R(t,3)$ area
0.0	0.000000	0.125000	0.000000	0.125000	0.000000	0.125000
0.5	0.500000	0.375000	0.500000	0.375000	0.500000	0.375000
1.0	1.000000	0.625000	1.000000	0.625000	1.000000	0.624998
1.5	1.500000	0.874999	1.499999	0.874977	1.499992	0.874878
2.0	1.999996	1.124975	1.999909	1.124513	1.999519	1.122778
2.5	2.499903	1.374683	2.498144	1.370087	2.491592	1.355666
3.0	2.998827	1.622597	2.982205	1.595272	2.931072	1.524910
3.5	3.491562	1.862536	3.398883	1.752750	3.168567	1.530084
4.0	3.958583	2.076753	3.612116	1.758082	2.951768	1.281697
4.5	4.348431	2.227793	3.420213	1.535629	2.175022	0.836020
5.0	4.562742	2.258257	2.722305	1.110773	1.169058	0.400628
5.5	4.470286	2.111362	1.720785	0.636911	0.433455	0.135463
6.0	3.975163	1.772570	0.826858	0.280381	0.108398	0.031725
6.5	3.115119	1.302599	0.294667	0.093053	0.018500	0.005190
7.0	2.095278	0.820173	0.077544	0.023214	0.002260	0.000619
7.5	1.185414	0.435613	0.015312	0.004417	0.000217	0.000059
8.0	0.557036	0.193360	0.002357	0.000665	0.000019	0.000005
8.5	0.216405	0.071534	0.000302	0.000085	0.000002	0.000001
9.0	0.069732	0.022145	0.000036	0.000010	0.000000	0.000000
9.5	0.018848	0.005802	0.000005	0.000001	0.000000	0.000000
10.0	0.004360	0.001313	0.000001		0.000000	
10.5	0.000891	0.000265				
11.0	0.000168	0.000050				
11.5	0.000031	0.000009				
12.0	0.000006					
SUM		**21.284390**		**13.285820**		**10.224721**

Considering (2.1.20) and using (2.4.43)–(2.4.45), the mean values of the shipyard rope elevator lifetimes in the particular reliability states 1, 2, 3, in years, respectively, are:

$$\bar{\mu}(1)=\mu(1)-\mu(2)\cong 6.441-5.091=1.350, \tag{2.4.55}$$

$$\bar{\mu}(2)=\mu(2)-\mu(3)\cong 5.091-4.467=0.624, \tag{2.4.56}$$

$$\bar{\mu}(3)=\mu(3)\cong 4.467. \tag{2.4.57}$$

Assuming that $r=2$ is the critical reliability state, then under the Definition 2.1.3, applying (2.1.21), we obtain the risk function of the shipyard rope elevator

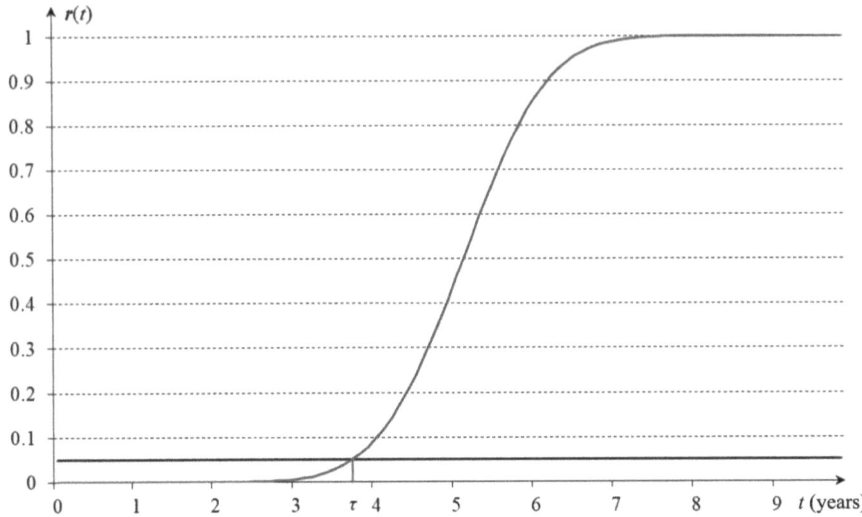

Fig. 2.16 The graph of the risk function $\boldsymbol{r}(t)$ of shipyard rope elevator.

$$\begin{aligned} \boldsymbol{r}(t) &= 1 - \boldsymbol{R}(t, 2) \\ &= 1 - \left[1 - \sum_{i=1}^{4} \binom{22}{i} [\exp[-0.2041t]]^{i} [1 - \exp[-0.2041t]]^{22-i}\right]^{10}, \qquad (2.4.58) \\ & t \geq 0, \end{aligned}$$

and the moment of exceeding an acceptable level, for instance $\delta = 0.05$, by (2.1.22), is

$$\tau = \boldsymbol{r}^{-1}(\delta) \cong 3.71 \text{ years} \cong 3 \text{ years } 259 \text{ days}. \qquad (2.4.59)$$

The shipyard rope elevator risk function is illustrated in Fig. 2.16.

References

Blokus, A., Kołowrocki, K., 2019. Reliability and maintenance strategy for systems with aging-dependent components. Qual. Reliab. Eng. Int. 35, 2709–2731. https://doi.org/10.1002/qre.2552.

Blokus-Roszkowska, A., 2006. Reliability analysis of multi-state rope transportation system with dependent components. In: Soares, C.G., Zio, E. (Eds.), Safety and Reliability for Managing Risk – Proceedings of the European Safety and Reliability Conference, ESREL 2006. CRC Press/Balkema, Estoril, pp. 1561–1568.

Blokus-Roszkowska, A., 2007. Reliability Analysis of Homogenous Large Systems With Component Dependent Failures. PhD Thesis, Gdynia Maritime University – Polish Academy of Sciences, System Research Institute, Warsaw (in Polish).

Blokus-Roszkowska, A., Kołowrocki, K., 2014. Reliability analysis of complex shipyard transportation system with dependent components. J. Pol. Saf. Reliab. Assoc. Summer Saf. Reliab. Semin. 5 (1), 21–31.
Blokus-Roszkowska, A., Kolowrocki, K., 2017. Modelling safety of multistate systems with dependent components and subsystems. J. Pol. Saf. Reliab. Assoc. Summer Saf. Reliab. Semin. 8 (3), 23–41.
Kołowrocki, K., 2001. On limit reliability functions of large multi-state systems with ageing components. Appl. Math. Comput. 121, 313–361.
Kołowrocki, K., 2008. Reliability and risk analysis of large systems with ageing components. Reliab. Eng. Syst. Saf. 93, 1821–1829.
Kołowrocki, K., 2014. Reliability of Large and Complex Systems, second ed. Elsevier, London.
Kołowrocki, K., Soszyńska, J., 2006. Reliability and availability of complex systems. Qual. Reliab. Eng. Int. 22 (1), 79–99. J. Wiley & Sons Ltd.
Kołowrocki, K., Soszyńska-Budny, J., 2011. Reliability and Safety of Complex Technical Systems and Processes: Modeling – Identification – Prediction – Optimization, first ed. Springer-Verlag, London.
Xue, J., 1985. On multi-state system analysis. IEEE Trans. Reliab. 34, 329–337.
Xue, J., Yang, K., 1995a. Dynamic reliability analysis of coherent multi-state systems. IEEE Trans. Reliab. 44 (4), 683–688.
Xue, J., Yang, K., 1995b. Symmetric relations in multi-state systems. IEEE Trans. Reliab. 44 (4), 689–693.
Yu, K., Koren, I., Guo, Y., 1994. Generalised multistate monotone coherent systems. IEEE Trans. Reliab. 43, 242–250.

CHAPTER THREE

Reliability of aging multistate dependent systems

Multistate aging systems with dependent components and dependent subsystems are analyzed and their basic reliability characteristics are determined. Based on dependencies among components and subsystems, a system structure, specificity of its operation and the specificity of its components, three dependency models in a multistate aging system are considered: the local load-sharing (LLS) dependency model, the equal load-sharing (ELS) dependency model, and the mixed load-sharing (MLS) dependency model. The reliability analysis of systems with those three dependency models is performed. Finally, to illustrate the way of theoretical results' practical applications, the reliability evaluations of an exemplary system and a shipyard rope elevator are presented.

3.1 Local load-sharing model of system components dependency

The LLS dependency model of components is introduced. An approach to reliability analysis of multistate series systems with components following the LLS dependency rule is described. The reliability functions of such systems are determined in a general case and in particular case under fixed their components' reliability functions. Namely, for a series system with components having piecewise exponential distribution, Rayleigh distribution and Erlang distribution its reliability functions are determined. Additionally, for a piecewise exponential, aging series system, following the LLS dependency rule, its mean lifetime in the reliability state subset, the standard deviation of system lifetime, and the intensities of system's departures from the reliability state subsets are given. Next, the LLS dependency model is extended for multistate aging series-parallel and series-"m out of k" systems, assuming dependence according to the LLS rule among components in their series subsystems. For those systems, the reliability functions in general case when their components have any reliability functions and in particular case when components have piecewise exponential reliability functions are determined. As a special case, a homogeneous and regular series-parallel system and a homogeneous and regular series-"m out of k" system are considered as well.

Multistate System Reliability with Dependencies
https://doi.org/10.1016/B978-0-12-821260-8.00003-8

3.1.1 Approach description

We consider a multistate aging (Blokus-Roszkowska and Kołowrocki, 2007, 2015a,b,c; Kołowrocki, 2014; Kołowrocki and Soszyńska-Budny, 2011; Lai and Xie, 2006; Xue, 1985; Xue and Yang, 1995a,b) series system, defined in Chapter 2, composed of n aging and independent components E_i, $i=1, 2, \ldots, n$. We apply the same assumptions as in Chapter 2, that all components and a system under consideration have the reliability state set $\{0, 1, \ldots, z\}$, $z \geq 1$, where state 0 is the worst and state z is the best. Therefore, the reliability function of a component E_i is defined by (2.1.1), (2.1.2) and the reliability function of a system is given by (2.1.10), (2.1.11).

Local load-sharing model of dependency was introduced by Harlow and Phoenix (1978, 1982), and further analyzed by Phoenix and Smith (1983); however, they apply to two-state systems. In a multistate system with components dependent according to the LLS rule, we assume that if the component E_j, $j=1, 2, \ldots, n$, leaves the reliability state subset $\{u, u+1, \ldots, z\}$, $u=1, 2, \ldots, z$, then the reliability parameters of other components E_i, $i=1, 2, \ldots, n$, $i \neq j$, worsen depending on their exposure to the component E_j reliability state change. This influence can be expressed by the coefficients $q(v, E_j, E_i)$, $v=u, u-1, \ldots, 1$, $u=1, 2, \ldots, z-1$, of component E_j, $j=1, 2, \ldots, n$, impact on other components E_i, $i=1, 2, \ldots, n$, $i \neq j$, lifetimes in the reliability state subsets. Further, it is assumed that $0 < q(v, E_j, E_i) \leq 1$, $i=1, 2, \ldots, n$, $i \neq j$ and $q(v, E_j, E_j)=1$, $j=1, 2, \ldots, n$. More exactly, in the LLS dependency rule, we assume that the system component lifetimes and their mean values in the reliability state subset $\{v, v+1, \ldots, z\}$, $v=u, u-1, \ldots, 1$ and $u=1, 2, \ldots, z-1$, decrease according to the formulas (Blokus and Kołowrocki, 2019)

$$
\begin{aligned}
T_{i/j}(v) &= q(v, E_j, E_i) \cdot T_i(v), \\
E\left[T_{i/j}(v)\right] &= q(v, E_j, E_i) \cdot E[T_i(v)], \quad i=1,2,\ldots,n, \quad j=1,2,\ldots,n,
\end{aligned}
\tag{3.1.1}
$$

where $T_{i/j}(v)$ denotes the lifetime in reliability state subset $\{v, v+1, \ldots, z\}$, $v=u, u-1, \ldots, 1$, $u=1, 2, \ldots, z-1$, of component E_i, $i=1, 2, \ldots, n$, after departure of E_j, $j=1, 2, \ldots, n$, from the subset $\{u, u+1, \ldots, z\}$, $u=1, 2, \ldots, z$.

Formula (3.1.1) means that, if the component E_j, $j=1, 2, \ldots, n$, leaves the reliability state subset $\{u, u+1, \ldots, z\}$, $u=1, 2, \ldots, z$, then system components E_i, $i=1, 2, \ldots, n$, lifetimes in a subset of states not worse than $v=u$, i.e., in $\{u, u+1, \ldots, z\}$, $u=1, 2, \ldots, z-1$, will decrease, but also their lifetimes in a subset of states not worse than $v=u-1$, i.e., in $\{u-1, u, \ldots, z\}$, and their lifetimes in a subset of states not worse than $v=u-2$, i.e., in

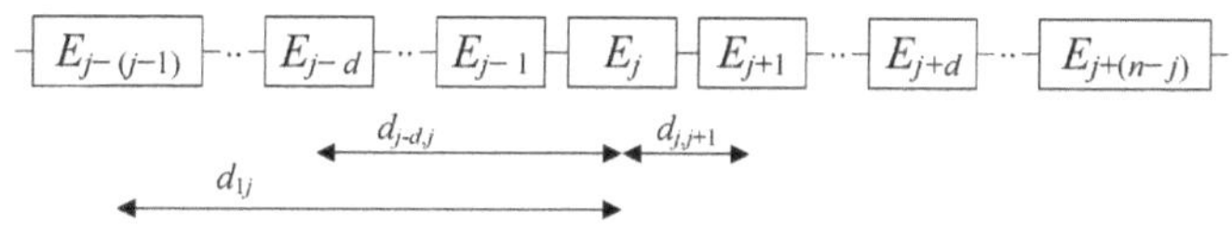

Fig. 3.1 Illustration of LLS rule (Blokus and Kołowrocki, 2019).

$\{u-2, u-1, \ldots, z\}$, and so on, will be reduced. These subsets of reliability states, to which the changes apply, are noted in general by $\{v, v+1, \ldots, z\}$ for $v=u, u-1, \ldots, 1$ and $u=1, 2, \ldots, z-1$, what is expressed in (3.1.1).

The coefficients $q(v, E_j, E_i)$, in particular case, can be functions of a distance d_{ij} among components E_i and E_j, $i, j=1, 2, \ldots, n$, expressed by $q(v, d_{ij})$, where $d_{ij}=|i-j|$ and interpretation of the distance index is illustrated in Fig. 3.1.

In that case, the component lifetimes and their mean values in the subset $\{v, v+1, \ldots, z\}$, $v=u, u-1, \ldots, 1$ and $u=1, 2, \ldots, z-1$, decrease according to the formulas (Blokus-Roszkowska and Kołowrocki, 2015b)

$$T_{i/j}(v)=q(v, d_{ij}) \cdot T_i(v),$$
$$E[T_{i/j}(v)]=q(v, d_{ij}) \cdot E[T_i(v)], \quad i=1,2,\ldots,n, \quad j=1,2,\ldots,n, \quad (3.1.2)$$

where the coefficients $q(v, d_{ij})$ are functions of component distance d_{ij} and fulfill conditions $0<q(v, d_{ij})\le 1$ and $q(v, 0)=1$, $v=u, u-1, \ldots, 1$, $u=1, 2, \ldots, z-1$.

Consequently, we define the reliability function of E_i, $i=1, \ldots, n$, after departure of E_j ($j=1, 2, \ldots, n$) from the subset $\{u, u+1, \ldots, z\}$, $u=1, 2, \ldots, z$,

$$R_{i/j}(t, \cdot)=[1, R_{i/j}(t, 1), \ldots, R_{i/j}(t, z)], \quad t\ge 0, \quad i=1,2,\ldots,n, \quad j=1,2,\ldots,n, \quad (3.1.3)$$

with the coordinates given by

$$R_{i/j}(t, v)=P(T_{i/j}(v)>t), \quad t\ge 0, \quad v=u,u-1,\ldots,1, \quad u=1,2,\ldots,z-1, \quad (3.1.4)$$

$$R_{i/j}(t, v)=P(T_{i/j}(v)>t)=P(T_i(v)>t)=R_i(t, v), \quad v=u+1,\ldots,z, \quad u=1,2,\ldots,z-1. \quad (3.1.5)$$

3.1.2 Reliability of multistate series system with dependent components

We consider a multistate aging series system composed of components dependent according to the LLS rule. These components have reliability functions, defined by (2.1.1)–(2.1.3), and given by the vector

$$R_i(t,\ \cdot\ ) = [1, R_i(t, 1), \ldots, R_i(t, z)], \quad t \geq 0, i = 1,2,\ldots,n. \tag{3.1.6}$$

We can notice that at the moment t, $t \geq 0$, the system is in the reliability state subset $\{u, u+1, \ldots, z\}$, if at that moment either all its components are in the subset $\{u+1, \ldots, z\}$, $u = 1, 2, \ldots, z-1$ (what is expressed in the first part of the formula (3.1.8) given below), or any of its components has left the state subset $\{u+1, \ldots, z\}$, $u = 1, 2, \ldots, z-1$, at any moment a, $0 < a < t$, and simultaneously, the remaining $n-1$ components have not left the subset $\{u+1, \ldots, z\}$, $u = 1, 2, \ldots, z-1$, during time a, while the component that has left the state subset $\{u+1, \ldots, z\}$ remains in the subset $\{u, u+1, \ldots, z\}$ during time a, and all system components remain in the subset $\{u, u+1, \ldots, z\}$ during the remaining time $t-a$, $0 < a < t$, with changed reliability functions according to LLS rule (what is expressed, respectively, by subsequent parts of the expression occurring under the integral in (3.1.8)).

Finally, as a multistate series system is in the best state z if and only if all its components are in this state, the zth coordinate of a system reliability function takes the same form as for a multistate series system with independent components (what is expressed in formula (3.1.9)).

This idea allows to formulate the following new proposition, core in investigating jointly aging and dependences in system reliability analysis. Its formulation first version with the proof is presented in Blokus-Roszkowska and Kołowrocki (2015b), while their modified final versions, given in Blokus and Kołowrocki (2019), are presented now.

Proposition 3.1.1. *If, in an aging series system following the LLS dependency rule, the components have reliability functions* (3.1.6), *then the system's reliability function is*

$$\boldsymbol{R}_{LLS}(t,\ \cdot\ ) = [1, \boldsymbol{R}_{LLS}(t, 1), \ldots, \boldsymbol{R}_{LLS}(t, z)], \quad t \geq 0, \tag{3.1.7}$$

with the coordinates

$$\boldsymbol{R}_{LLS}(t, u) = \prod_{i=1}^{n} R_i(t, u+1) + \int_0^t \sum_{j=1}^{n} \left[\widetilde{f}_j(a, u+1) \cdot \prod_{\substack{i=1 \\ i \neq j}}^{n} R_i(a, u+1) \cdot R_j(a, u) \cdot \prod_{i=1}^{n} R_{i/j}(t-a, u) \right] da,$$
$$u = 1, 2, \ldots, z-1, \tag{3.1.8}$$

$$\boldsymbol{R}_{LLS}(t, z) = \prod_{i=1}^{n} R_i(t, z), \tag{3.1.9}$$

where:

$R_i(t, u+1)$—*the reliability function coordinate of a component* E_i, $i=1, \ldots, n$,
$\widetilde{f}_j(t, u+1)$—*the density function coordinate of a component* E_j, $j=1, \ldots, n$, *corresponding to the distribution function* $\widetilde{F}_j(t, u+1)$ *given by*

$$\widetilde{F}_j(t, u+1) = 1 - \frac{R_j(t, u+1)}{R_j(t, u)}, \quad u = 1, 2, \ldots, z-1, \quad t \geq 0, \tag{3.1.10}$$

$R_j(t, u)$—*the reliability function coordinate of a component* E_j, $j=1, \ldots, n$,
$R_{i/j}(t, u)$—*the reliability function coordinate of a component* E_i, $i=1, \ldots, n$, *after departure of the component* E_j, $j=1, \ldots, n$, *from the reliability state subset* $\{u+1, \ldots, z\}$, $u=1, 2, \ldots, z-1$, *such that*

$$R_{i/j}(t-a, u) = \frac{R_{i/j}(t, u)}{R_i(a, u)}, \quad u = 1, 2, \ldots, z-1, \ 0 < a < t, \quad t \geq 0. \tag{3.1.11}$$

Proof. We write the coordinate $\boldsymbol{R}_{LLS}(t, u)$, $u=1, 2, \ldots, z-1$ and $t \geq 0$, of a system reliability function $\boldsymbol{R}_{LLS}(t, \cdot)$, in the form

$$\begin{aligned}
\boldsymbol{R}_{LLS}(t, u) &= P\{[(\boldsymbol{T}(u+1) > t) \cap (\boldsymbol{T}(u) > t)] \cup [(\boldsymbol{T}(u+1) \leq t) \cap (\boldsymbol{T}(u) > t)]\} \\
&= P[(\boldsymbol{T}(u+1) > t) \cap (\boldsymbol{T}(u) > t)] + P[(\boldsymbol{T}(u+1) \leq t) \cap (\boldsymbol{T}(u) > t)] \\
&= P(\boldsymbol{T}(u+1) > t) \cdot P[(\boldsymbol{T}(u) > t)/(\boldsymbol{T}(u+1) > t)] \\
&\quad + P[(\boldsymbol{T}(u+1) \leq t)/(\boldsymbol{T}(u) > t)] \cdot P(\boldsymbol{T}(u) > t), \quad t \geq 0.
\end{aligned} \tag{3.1.12}$$

Since the condition that the system's lifetime in the reliability state subset $\{u, u+1, \ldots, z\}$ is greater than t, under the condition that the system lifetime in the subset $\{u+1, \ldots, z\}$, $u=1, 2, \ldots, z-1$, is greater than t, is always fulfilled, then

$$P(\boldsymbol{T}(u) > t/\boldsymbol{T}(u+1) > t) = 1 \ \text{ for } u = 1, 2, \ldots, z-1, \quad t \geq 0 \tag{3.1.13}$$

and from (3.1.12), the coordinate $\boldsymbol{R}_{LLS}(t, u)$, $u=1, 2, \ldots, z-1$, of a system reliability function is given by

$$\begin{aligned}
\boldsymbol{R}_{LLS}(t, u) &= P(\boldsymbol{T}(u+1) > t) + P(\boldsymbol{T}(u+1) \leq t/\boldsymbol{T}(u) > t) \cdot P(\boldsymbol{T}(u) > t), \\
&\quad u = 1, 2, \ldots, z-1, \quad t \geq 0.
\end{aligned} \tag{3.1.14}$$

Thus, from the definition of the multistate reliability function and its coordinates, we have

$$\boldsymbol{R}_{LLS}(t,u)=\boldsymbol{R}_{LLS}(t,u+1)+P(\boldsymbol{T}(u+1)\le t/\boldsymbol{T}(u)>t)\cdot P(\boldsymbol{T}(u)>t),$$
$$u=1,2,\ldots,z-1,\quad t\ge 0. \tag{3.1.15}$$

Further, we will prove that equation (3.1.15) is equivalent to (3.1.8).

In the case where none of the system components have left the reliability state subset $\{u+1, \ldots, z\}$, $u=1, 2, \ldots, z-1$, then the intensities of the departure from this subset of all the system components remain unchanged. Thus, in this case, the reliability function coordinate $\boldsymbol{R}_{LLS}(t, u+1)$ in (3.1.15) of the multistate series system, by (2.3.3), is given by

$$\boldsymbol{R}_{LLS}(t,u+1)=\boldsymbol{R}(t,u+1)=\prod_{i=1}^{n}R_i(t,u+1),$$
$$u=1,2,\ldots,z-1,\quad t\ge 0. \tag{3.1.16}$$

The probability

$$P(\boldsymbol{T}(u+1)\le t/\boldsymbol{T}(u)>t),\quad t\ge 0. \tag{3.1.17}$$

in (3.1.15) implies that the system leaves the subset $\{u+1, \ldots, z\}$ under the condition that it has not left the subset $\{u, u+1, \ldots, z\}$, $u=1, 2, \ldots, z-1$.

Consequently, the conditional distribution function, corresponding to the component E_j, $j=1, \ldots, n$, of the lifetime $T_j(u+1)$ in the reliability state subset $\{u+1, \ldots, z\}$, $u=1, 2, \ldots, z-1$, is given by

$$\widetilde{F}_j(t,u+1)=P\big(T_j(u+1)\le t/T_j(u)>t\big)=1-P\big(T_j(u+1)>t/T_j(u)>t\big)$$
$$=1-\frac{P\big(\big(T_j(u+1)>t\big)\cap\big(T_j(u)>t\big)\big)}{P\big(T_j(u)>t\big)}$$

and from the fact that $P[(T_j(u+1)>t)\cap(T_j(u)>t)]=P(T_j(u+1)>t)$, this distribution function for a homogeneous series system, if its components have reliability functions (3.1.6), takes the form

$$\widetilde{F}_j(t,u+1)=1-\frac{P\big(T_j(u+1)>t\big)}{P\big(T_j(u)>t\big)}=1-\frac{R_j(t,u+1)}{R_j(t,u)},$$
$$u=1,2,\ldots,z-1,\quad t\ge 0. \tag{3.1.18}$$

Therefore, the density function corresponding to this distribution function is

$$\widetilde{f}_j(t, u+1) = \frac{d}{dt}\widetilde{F}_j(t, u+1) = -\frac{d}{dt}\left(\frac{R_j(t, u+1)}{R_j(t, u)}\right),$$
$$u = 1, 2, \ldots, z-1, \quad t \geq 0. \qquad (3.1.19)$$

Definition 2.2.1 of a multistate series system implies that a system leaves the reliability state subset $\{u+1, \ldots, z\}$, $u=1, 2, \ldots, z-1$, if one of the system components leaves this subset. Taking into account this definition, a multistate series system composed of n aging components can leave the subset $\{u+1, \ldots, z\}$ under the condition that it has not left the subset $\{u, u+1, \ldots, z\}$, $u=1, 2, \ldots, z-1$, if any of these components leaves; for instance, the component E_j, $j=1, \ldots, n$, leaves the subset $\{u+1, \ldots, z\}$ at any moment a $(0<a<t$, or, more exactly, in a small time interval $\langle a, a+da))$, with probability

$$\widetilde{f}_j(a, u+1)da, \qquad (3.1.20)$$

where $\widetilde{f}_j(a, u+1)$ is given by (3.1.19). Simultaneously, the remaining $n-1$ components E_i, $i=1, \ldots, n$, $i \neq j$, have not left the subset $\{u+1, \ldots, z\}$, $u=1, 2, \ldots, z-1$, during time a with probability

$$\prod_{\substack{i=1 \\ i \neq j}}^{n} P(T_i(u+1) > a) = \prod_{\substack{i=1 \\ i \neq j}}^{n} R_i(a, u+1), \quad u = 1, 2, \ldots, z-1,$$
$$0 < a < t, \quad t \geq 0. \qquad (3.1.21)$$

Moreover, since the inequality $\boldsymbol{T}(u) > t$ implies that the system has not left the reliability state subset $\{u, u+1, \ldots, z\}$, $u=1, 2, \ldots, z-1$, then from the performed analysis, the inequality is fulfilled when the component E_j, $j=1, \ldots, n$ (which has left the subset $\{u+1, \ldots, z\}$ at any moment a, $0<a<t$, with the probability given by (3.1.20) and has not left the subset $\{u, u+1, \ldots, z\}$, $u=1, 2, \ldots, z-1$, during time a) and all the components after the departure of E_j, $j=1, \ldots, n$, from the subset $\{u+1, \ldots, z\}$, $u=1, 2, \ldots, z-1$, are in the subset $\{u, u+1, \ldots, z\}$, $u=1, 2, \ldots, z-1$, for times greater than $t-a$, $0<a<t$.

By considering the theorem's assumption that the components are dependent according to the LLS rule (3.1.1) and after the departure of E_j, $j=1, \ldots, n$, from the subset $\{u+1, \ldots, z\}$, $u=1, 2, \ldots, z-1$, then the reliability function coordinates (3.1.8), (3.1.9) in terms of the probability of the first condition are given by

$$P\big(T_j(u) > a\big) = R_j(a, u), \quad u = 1, 2, \ldots, z-1, \quad 0 < a < t, \quad t \geq 0, \qquad (3.1.22)$$

and in terms of the probability of the second condition are given by

$$\prod_{i=1}^{n} P\left(T_{i/j}(u) > t - a\right) = \prod_{i=1}^{n} R_{i/j}(t - a, u), \quad u = 1, 2, \ldots, z - 1, \quad 0 < a < t, \quad t \geq 0, \tag{3.1.23}$$

and the probability

$$P(\boldsymbol{T}(u) > t), \quad t \geq 0, \tag{3.1.24}$$

in (3.1.15) is given by

$$P(\boldsymbol{T}(u) > t) = R_j(a, u) \cdot \prod_{i=1}^{n} R_{i/j}(t - a, u), \quad u = 1, 2, \ldots, z - 1, \quad 0 < a < t, \quad t \geq 0. \tag{3.1.25}$$

Thus, by considering (3.1.20), (3.1.21), and (3.1.25), the second component of the sum existing in (3.1.15) can be expressed by

$$P(\boldsymbol{T}(u+1) \leq t / \boldsymbol{T}(u) > t) \cdot P(\boldsymbol{T}(u) > t)$$
$$= \int_0^t \sum_{j=1}^{n} \left[\tilde{f}_j(a, u+1) \cdot \prod_{\substack{i=1 \\ i \neq j}}^{n} R_i(a, u+1) \cdot R_j(a, u) \cdot \prod_{i=1}^{n} R_{i/j}(t - a, u) \right] da,$$
$$\text{for } t \geq 0, \quad u = 1, 2, \ldots, z - 1. \tag{3.1.26}$$

Substituting (3.1.16), (3.1.26) into (3.1.15), we obtain (3.1.8).

To prove equation (3.1.8) completely, it remains to justify that

$$R_{i/j}(t - a, u), \quad u = 1, 2, \ldots, z - 1, \quad 0 < a < t, \quad t \geq 0,$$

in (3.1.26) is given by (3.1.11). For this justification, note that the probability that the lifetime of E_i, $i = 1, \ldots, n$, in the reliability state subset $\{u, u+1, \ldots, z\}$, $u = 1, 2, \ldots, z - 1$, is greater than t under the condition that its lifetime in this subset is greater than a, $0 < a < t$, can be written in the form

$$R_{i/j}(t - a, u) = P\left(T_{i/j}(u) > t \mid T_i(u) > a\right)$$
$$= \frac{P\left[\left(T_{i/j}(u) > t\right) \cap \left(T_i(u) > a\right)\right]}{P(T_i(u) > a)}, \quad u = 1, 2, \ldots, z - 1, \quad 0 < a < t, \quad t \geq 0. \tag{3.1.27}$$

Since the condition that the lifetime of E_i, $i = 1, \ldots, n$, in the subset $\{u, u+1, \ldots, z\}$, $u = 1, 2, \ldots, z - 1$, is greater than a, $0 < a < t$, is always fulfilled if its lifetime in this subset is greater than t, then the probability (3.1.27) takes the form

$$P\left(T_{i/j}(u) > t \mid T_i(u) > a\right) = \frac{P\left(T_{i/j}(u) > t\right)}{P(T_i(u) > a)} = \frac{R_{i/j}(t,u)}{R_i(a,u)}, \quad t \geq 0. \quad (3.1.28)$$

Thus, equation (3.1.11) is justified, and equation (3.1.8) is completely proved.

To complete the proof of Proposition 3.1.1, equation (3.1.9) must be validated. Because a multistate series system is in the best state z if and only if all its components are in this state, we conclude that the zth coordinate of a system reliability function takes the form of (3.1.9), which completes the proof. □

In particular case, if the coefficients $q(v, E_j, E_i)$, $v = u, u-1, \ldots, 1$, $u = 1, 2, \ldots, z-1$, of the component E_j, $j = 1, 2, \ldots, n$, impact on other components E_i, $i = 1, 2, \ldots, n$, $i \neq j$, lifetimes in the reliability state subsets, are equal to 1, then the considered system becomes a series system with independent components. If this condition holds, then

$$R_{i/j}(t,u) = R_i(t,u), \quad i = 1, \ldots, n, \quad j = 1, \ldots, n, \quad t \geq 0, \quad u = 1, 2, \ldots, z-1,$$

i.e., $\boldsymbol{R}_{LLS}(t,u) = \boldsymbol{R}(t,u)$, $t \geq 0$, $u = 1, 2, \ldots, z-1$ and the formula (3.1.8) takes form of the formula (2.3.3) for $u = 1, \ldots, z-1$. More exactly, we obtain

$$\begin{aligned}
\boldsymbol{R}_{LLS}(t,u) &= \prod_{i=1}^{n} R_i(t,u+1) \\
&+ \int_0^t \sum_{j=1}^{n} \left[\tilde{f}_j(a,u+1) \cdot \prod_{\substack{i=1 \\ i \neq j}}^{n} R_i(a,u+1) \cdot R_j(a,u) \cdot \prod_{i=1}^{n} R_i(t-a,u) \right] da \\
&= \prod_{i=1}^{n} R_i(t,u+1) \\
&+ \int_0^t \sum_{j=1}^{n} \left[\tilde{f}_j(a,u+1) \cdot \prod_{\substack{i=1 \\ i \neq j}}^{n} R_i(a,u+1) \cdot \prod_{\substack{i=1 \\ i \neq j}}^{n} R_i(t-a,u) \cdot R_j(a,u) \cdot R_j(t-a,u) \right] da \\
&= \prod_{i=1}^{n} R_i(t,u+1) \\
&+ \int_0^t \sum_{j=1}^{n} \left[\tilde{f}_j(a,u+1) \cdot \prod_{\substack{i=1 \\ i \neq j}}^{n} R_i(a,u+1) \cdot \prod_{\substack{i=1 \\ i \neq j}}^{n} R_i(t-a,u) \cdot R_j(t,u) \right] da, \quad (3.1.29)
\end{aligned}$$

$$u = 1, 2, \ldots, z-1, \quad t \geq 0.$$

Next, toward (3.1.11), we get

$$\boldsymbol{R}_{LLS}(t,u)=\prod_{i=1}^{n}R_i(t,u+1)$$
$$+\int_0^t\sum_{j=1}^{n}\left[\widetilde{f}_j(a,u+1)\cdot\prod_{\substack{i=1\\ i\neq j}}^{n}R_i(a,u+1)\cdot\prod_{\substack{i=1\\ i\neq j}}^{n}\frac{R_i(t,u)}{R_i(a,u)}\cdot R_j(t,u)\right]da$$
$$=\prod_{i=1}^{n}R_i(t,u+1)$$
$$+\int_0^t\sum_{j=1}^{n}\left[\widetilde{f}_j(a,u+1)\cdot\prod_{\substack{i=1\\ i\neq j}}^{n}\frac{R_i(a,u+1)}{R_i(a,u)}\cdot\prod_{i=1}^{n}R_i(t,u)\right]da$$
$$=\prod_{i=1}^{n}R_i(t,u+1)+\prod_{i=1}^{n}R_i(t,u)\int_0^t\sum_{j=1}^{n}\left[\widetilde{f}_j(a,u+1)\cdot\prod_{\substack{i=1\\ i\neq j}}^{n}\frac{R_i(a,u+1)}{R_i(a,u)}\right]da,$$
$$u=1,2,\ldots,z-1,\quad t\geq 0. \tag{3.1.30}$$

Further substituting for the density function $\widetilde{f}_j(t,u+1)$ coordinate, corresponding to the distribution function $\widetilde{F}_j(t,u+1)$ given by (3.1.10), the following expression

$$\widetilde{f}_j(t,u+1)=\frac{d}{dt}\left(-\frac{R_j(t,u+1)}{R_j(t,u)}\right),\quad u=1,2,\ldots,z-1,\quad t\geq 0, \tag{3.1.31}$$

formula (3.1.30) takes form

$$\boldsymbol{R}_{LLS}(t,u)=\prod_{i=1}^{n}R_i(t,u+1)+\prod_{i=1}^{n}R_i(t,u)\int_0^t\sum_{j=1}^{n}\left[\frac{d}{dt}\left(-\frac{R_j(a,u+1)}{R_j(a,u)}\right)\cdot\prod_{\substack{i=1\\ i\neq j}}^{n}\frac{R_i(a,u+1)}{R_i(a,u)}\right]da$$
$$=\prod_{i=1}^{n}R_i(t,u+1)+\prod_{i=1}^{n}R_i(t,u)\int_0^t\frac{d}{dt}\left(-\prod_{i=1}^{n}\frac{R_i(a,u+1)}{R_i(a,u)}\right)da,\quad u=1,2,\ldots,z-1,\quad t\geq 0.$$
$$\tag{3.1.32}$$

After counting the integral in (3.1.32), the result is

$$\begin{aligned}\boldsymbol{R}_{LLS}(t,u) &= \prod_{i=1}^{n} R_i(t,u+1) + \prod_{i=1}^{n} R_i(t,u)\left[-\prod_{i=1}^{n}\frac{R_i(a,u+1)}{R_i(a,u)}\right]\Bigg|_0^t \\ &= \prod_{i=1}^{n} R_i(t,u+1) + \prod_{i=1}^{n} R_i(t,u)\left[-\prod_{i=1}^{n}\frac{R_i(t,u+1)}{R_i(t,u)}+1\right] \\ &= \prod_{i=1}^{n} R_i(t,u+1) - \prod_{i=1}^{n} R_i(t,u+1) + \prod_{i=1}^{n} R_i(t,u) = \prod_{i=1}^{n} R_i(t,u), \\ & u = 1,2,\ldots,z-1, \quad t \geq 0. \end{aligned} \tag{3.1.33}$$

Taking into account the result (3.1.33) and the formula (3.1.9), we get $\boldsymbol{R}_{LLS}(t,u) = \boldsymbol{R}(t,u)$, $u = 1, 2, \ldots, z$, and we conclude that formulas (3.1.7)–(3.1.9) present the reliability function of a multistate series system with independent components.

As a particular case, we consider a homogeneous, aging series system composed of dependent components, which have identical reliability functions

$$R(t,\cdot) = [1, R(t,1), \ldots, R(t,z)], \quad t \geq 0. \tag{3.1.34}$$

In that case, using the result of Proposition 3.1.1, the system reliability function, assuming the LLS dependency model among its components, is determined in the following proposition.

Proposition 3.1.2. *If, in a homogeneous and aging series system following the LLS dependency rule, the components have reliability functions* (3.1.34), *then the system's reliability function is*

$$\boldsymbol{R}_{LLS}(t,\cdot) = [1, \boldsymbol{R}_{LLS}(t,1), \ldots, \boldsymbol{R}_{LLS}(t,z)], t \geq 0, \tag{3.1.35}$$

with the coordinates

$$\begin{aligned}\boldsymbol{R}_{LLS}(t,u) &= [R(t,u+1)]^n \\ &+ \int_0^t \sum_{j=1}^{n}\left[\widetilde{f}(a,u+1)\cdot[R(a,u+1)]^{n-1}\cdot R(a,u)\cdot\prod_{i=1}^{n} R_{i/j}(t-a,u)\right] da, \\ & u = 1,2,\ldots,z-1, \end{aligned} \tag{3.1.36}$$

$$\boldsymbol{R}_{LLS}(t, z) = [R(t, z)]^n, \tag{3.1.37}$$

where

$\widetilde{F}(t, u+1)$ *is the distribution function, defined as the probability of component's exit from the subset* $\{u+1, \ldots, z\}$, $u=1, 2, \ldots, z-1$, *before the time* t, *given that its lifetime in the subset* $\{u, u+1, \ldots, z\}$, $u=1, 2, \ldots, z$, *is greater than* t, *is given by*

$$\widetilde{F}(t, u+1) = 1 - \frac{R(t, u+1)}{R(t, u)}, \tag{3.1.38}$$

$\widetilde{f}(t, u+1)$ *is the density function coordinate of a component corresponding to* $\widetilde{F}(t, u+1)$,

$R_{i/j}(t, u)$ *is the reliability function coordinate of* E_i, $i=1, \ldots, n$, *after departure of* E_j, $j=1, \ldots, n$, *from the subset* $\{u+1, \ldots, z\}$, $u=1, 2, \ldots, z-1$, *such that*

$$R_{i/j}(t-a, u) = \frac{R_{i/j}(t, u)}{R(a, u)}, \quad 0 < a < t, \quad t \geq 0. \tag{3.1.39}$$

Proof. Considering the homogeneous multistate series system with components having identical reliability functions, we substitute into the formulas (3.1.8), (3.1.9) the reliability function coordinates $R_i(t, u) = R(t, u)$ for $i=1, \ldots, n$, $u=1, 2, \ldots, z$ and $t \geq 0$, of system components E_i, $i=1, \ldots, n$. Similarly, we replace the density function coordinate $\widetilde{f}_j(t, u+1)$ of the component E_j, $j=1, \ldots, n$, that has got out of the reliability state subset $\{u+1, \ldots, z\}$ by $\widetilde{f}(t, u+1)$, $u=1, 2, \ldots, z-1$.

The reliability function coordinates $R_{i/j}(t, u)$ of components E_i, $i=1, \ldots, n$, have been changed according to the LLS rule after departure of the component E_j, $j=1, \ldots, n$, from the reliability state subset $\{u+1, \ldots, z\}$, $u=1, 2, \ldots, z-1$. They are expressed in (3.1.4), (3.1.5), and depend on the coefficients $q(v, E_j, E_i)$, $v=u, u-1, \ldots, 1$, $u=1, 2, \ldots, z-1$, of the component E_j, $j=1, 2, \ldots, n$, impact on other components E_i, $i=1, 2, \ldots, n$, $i \neq j$, lifetimes in the reliability state subsets. As the way in which the reliability function coordinates $R_{i/j}(t, u)$ worsen, depends on indices i and j, these indices cannot be ignored in formula (3.1.36), although components have identical reliability functions (3.1.34). Thereby, the coordinates $\boldsymbol{R}_{LLS}(t, u)$, $u=1, 2, \ldots, z-1$ of the reliability function of a series system with components following the LLS rule are given by (3.1.36). The zth coordinate of a system reliability function given by formula (3.1.37) follows immediately from (3.1.9) in case system components have identical reliability functions (3.1.34). □

We assume that components E_i, $i=1, \ldots, n$, have piecewise exponential reliability functions (3.1.6) with the coordinates

$$R_i(t, u) = \exp\left[-\lambda_i(u)t\right], \quad t \geq 0, \quad u = 1, 2, \ldots, z, \tag{3.1.40}$$

where $\lambda_i(u)$, $i=1, \ldots, n$, $\lambda_i(u) \geq 0$, are components' intensities of departure from the reliability state subset $\{u, u+1, \ldots, z\}$, $u=1, 2, \ldots, z$. Under this assumption, the conditional intensities $\lambda_{i/j}(v)$, $i=1, \ldots, n$, $j=1, \ldots, n$, of the component E_i departure from the subset $\{v, v+1, \ldots, z\}$, $v=u, u-1, \ldots, 1$, $u=1, 2, \ldots, z-1$, after the departure of the component E_j, by (3.1.1), are

$$\lambda_{i/j}(v) = \frac{\lambda_i(v)}{q(v, E_j, E_i)}. \tag{3.1.41}$$

Thus, considering (3.1.3)–(3.1.5) and (3.1.40), (3.1.41), the components E_i, $i=1, \ldots, n$, after the departure of E_j, $j=1, \ldots, n$, from that subset $\{u, u+1, \ldots, z\}$, $u=1, 2, \ldots, z$, have the reliability functions (3.1.3) with the coordinates

$$R_{i/j}(t, v) = \exp\left[-\frac{\lambda_i(v)}{q(v, E_j, E_i)}t\right], \quad v = u, u-1, \ldots, 1,$$
$$u = 1, 2, \ldots, z-1, \tag{3.1.42}$$

$$R_{i/j}(t, v) = \exp\left[-\lambda_i(v)t\right], \quad v = u+1, \ldots, z, \quad u = 1, 2, \ldots, z-1. \tag{3.1.43}$$

Further for the piecewise exponential multistate series system with dependent components, the distribution function corresponding to the system component E_j, given by (3.1.10), takes form

$$\tilde{F}_j(t, u+1) = 1 - \exp\left[-\left(\lambda_j(u+1) - \lambda_j(u)\right)t\right],$$

and its corresponding density function is

$$\tilde{f}_j(t, u+1) = \left(\lambda_j(u+1) - \lambda_j(u)\right) \cdot \exp\left[-\left(\lambda_j(u+1) - \lambda_j(u)\right)t\right],$$
$$u = 1, 2, \ldots, z-1, \quad t \geq 0. \tag{3.1.44}$$

Considering (3.1.42)–(3.1.44), in case the system components have piecewise exponential reliability functions from Proposition 3.1.1, we obtain the following result (Blokus-Roszkowska and Kołowrocki, 2015b).

Proposition 3.1.3. *If, in an aging series system following the LLS dependency rule, the components have piecewise exponential reliability functions with the coordinates* (3.1.40), *then the system's reliability function is*

$$\boldsymbol{R}_{LLS}(t, \cdot) = [1, \boldsymbol{R}_{LLS}(t, 1), \ldots, \boldsymbol{R}_{LLS}(t, z)], \quad t \geq 0, \tag{3.1.45}$$

with the coordinates

$$\boldsymbol{R}_{LLS}(t,u)=\exp\left[-\sum_{i=1}^{n}\lambda_i(u+1)t\right]+\sum_{j=1}^{n}\frac{\lambda_j(u+1)-\lambda_j(u)}{\sum_{i=1}^{n}(\lambda_i(u+1)-\lambda_i(u))}$$

$$\left[\exp\left[-\sum_{i=1}^{n}\frac{\lambda_i(u)}{q\left(u,E_j,E_i\right)}t\right]-\exp\left[-\sum_{i=1}^{n}\left(\lambda_i(u+1)-\lambda_i(u)+\frac{\lambda_i(u)}{q\left(u,E_j,E_i\right)}\right)\cdot t\right]\right],$$

$$u=1,2,\ldots,z-1, \tag{3.1.46}$$

$$\boldsymbol{R}_{LLS}(t,z)=\exp\left[-\sum_{i=1}^{n}\lambda_i(z)t\right]. \tag{3.1.47}$$

Proof. Considering (3.1.40), we have

$$\prod_{i=1}^{n}R_i(t,u+1)=\exp\left[-\sum_{i=1}^{n}\lambda_i(u+1)t\right],\quad u=1,2,\ldots,z-1, \tag{3.1.48}$$

and additionally by (3.1.42)–(3.1.44), we get

$$\int_0^t\sum_{j=1}^{n}\left[\tilde{f}_j(a,u+1)\cdot\prod_{\substack{i=1\\i\neq j}}^{n}R_i(a,u+1)\cdot R_j(a,u)\cdot\prod_{i=1}^{n}R_{i/j}(t-a,u)\right]da$$

$$=\int_0^t\sum_{j=1}^{n}\left[\left(\lambda_j(u+1)-\lambda_j(u)\right)\cdot\exp\left[-\lambda_j(u+1)a+\lambda_j(u)a-\sum_{\substack{i=1\\i\neq j}}^{n}\lambda_i(u+1)a-\lambda_j(u)a\right.\right.$$

$$\left.\left.-\sum_{i=1}^{n}\frac{\lambda_i(u)}{q\left(u,E_j,E_i\right)}t+\sum_{i=1}^{n}\lambda_i(u)a\right]\right]da=\int_0^t\sum_{j=1}^{n}\left[\left(\lambda_j(u+1)-\lambda_j(u)\right)\cdot\exp\left[-\sum_{i=1}^{n}\lambda_i(u+1)a\right.\right.$$

$$\left.\left.-\sum_{i=1}^{n}\frac{\lambda_i(u)}{q\left(u,E_j,E_i\right)}t+\sum_{i=1}^{n}\lambda_i(u)a\right]\right]da=\sum_{j=1}^{n}\left[\left(\lambda_j(u+1)-\lambda_j(u)\right)\cdot\exp\left[-\sum_{i=1}^{n}\frac{\lambda_i(u)}{q\left(u,E_j,E_i\right)}t\right]\right.$$

$$\left.\cdot\frac{\exp\left[-(\sum_{i=1}^{n}\lambda_i(u+1)-\sum_{i=1}^{n}\lambda_i(u)a\right]}{-\left(\sum_{i=1}^{n}\lambda_i(u+1)-\sum_{i=1}^{n}\lambda_i(u)\right)}\right|_0^t\Bigg]=\sum_{j=1}^{n}\frac{\lambda_j(u+1)-\lambda_j(u)}{\sum_{i=1}^{n}\lambda_i(u+1)-\sum_{i=1}^{n}\lambda_i(u)}$$

$$\cdot\left[\exp\left[-\sum_{i=1}^{n}\frac{\lambda_i(u)}{q\left(u,E_j,E_i\right)}t\right]-\exp\left[-\left(\sum_{i=1}^{n}\lambda_i(u+1)-\sum_{i=1}^{n}\lambda_i(u)+\sum_{i=1}^{n}\frac{\lambda_i(u)}{q\left(u,E_j,E_i\right)}\right)t\right]\right],$$

$$u=1,2,\ldots,z-1. \tag{3.1.49}$$

Next, substituting these results (3.1.48), (3.1.49) into (3.1.8), we get (3.1.46).

Moreover, considering (3.1.40) for $u=z$, from (3.1.9) we get (3.1.47), which completes the proof. □

In particular case, if the coefficients $q(v, E_j, E_i)$, $v=u, u-1, \ldots, 1$, $u=1, 2, \ldots, z-1$ of the component E_j, $j=1, 2, \ldots, n$, impact on other components E_i, $i=1, 2, \ldots, n$, $i \neq j$, lifetimes in the reliability state subsets, are equal to 1, then we obtain the results identical with those for a series system with independent components and the formulas (3.1.45)–(3.1.47) represent the reliability function of an piecewise exponential multistate series system with independent components.

From Proposition 3.1.3, we immediately obtain a corollary concerned with the mean values and standard deviations of the lifetimes in the reliability state subsets of a multistate series system (Blokus-Roszkowska and Kołowrocki, 2015b).

Corollary 3.1.1. *If, in an aging series system following the LLS dependency rule, the components have piecewise exponential reliability functions with the coordinates* (3.1.40)*, then the system's mean lifetime in the reliability state subset* $\{u, u+1, \ldots, z\}$, $u=1, 2, \ldots, z$, *is given by*

$$\mu_{LLS}(u) = \frac{1}{\sum_{i=1}^{n} \lambda_i(u+1)} + \sum_{j=1}^{n} \frac{\lambda_j(u+1) - \lambda_j(u)}{\sum_{i=1}^{n} (\lambda_i(u+1) - \lambda_i(u))}$$

$$\cdot \left[\frac{1}{\sum_{i=1}^{n} \frac{\lambda_i(u)}{q(u, E_j, E_i)}} - \frac{1}{\sum_{i=1}^{n} \left(\lambda_i(u+1) - \lambda_i(u) + \frac{\lambda_i(u)}{q(u, E_j, E_i)} \right)} \right],$$

$$u=1,2,\ldots,z-1, \tag{3.1.50}$$

$$\mu_{LLS}(z) = \frac{1}{\sum_{i=1}^{n} \lambda_i(z)}, \tag{3.1.51}$$

and the standard deviation of the system lifetime is

$$\sigma_{LLS}(u) = \sqrt{n_{LLS}(u) - [\mu_{LLS}(u)]^2}, \quad u=1,2,\ldots,z-1, \tag{3.1.52}$$

where

$$n_{LLS}(u) = \frac{2}{\left[\sum_{i=1}^{n} \lambda_i(u+1)\right]^2} + 2\sum_{j=1}^{n} \frac{\lambda_j(u+1) - \lambda_j(u)}{\sum_{i=1}^{n}(\lambda_i(u+1) - \lambda_i(u))}$$

$$\left[\frac{1}{\left[\sum_{i=1}^{n} \frac{\lambda_i(u)}{q(u, E_j, E_i)}\right]^2} - \frac{1}{\left[\sum_{i=1}^{n}\left(\lambda_i(u+1) - \lambda_i(u) + \frac{\lambda_i(u)}{q(u, E_j, E_i)}\right)\right]^2}\right],$$

$$u = 1, 2, \ldots, z-1, \tag{3.1.53}$$

and

$$\sigma_{LLS}(z) = \frac{1}{\sum_{i=1}^{n} \lambda_i(z)}. \tag{3.1.54}$$

Proof. Substituting the reliability function coordinates $\boldsymbol{R}_{LLS}(t, u)$, $u = 1, 2, \ldots, z$, given by (3.1.46), (3.1.47), into the formula for the mean lifetime of the system in the state subset $\{u, u+1, \ldots, z\}$

$$\mu_{LLS}(u) = \int_0^{+\infty} \boldsymbol{R}_{LLS}(t, u)dt, \quad u = 1, 2, \ldots, z,$$

given by (2.1.16) and applying the formula

$$\int_0^{+\infty} \exp[-\lambda t]dt = \frac{1}{\lambda},$$

we obtain (3.1.50), (3.1.51). Similarly, substituting $\boldsymbol{R}_{LLS}(t, u)$, $u = 1, 2, \ldots, z$, given by (3.1.46), (3.1.47), into the formula for the second-order moment of the lifetime of the system in the state subset $\{u, u+1, \ldots, z\}$, by (2.1.18), we have

$$n_{LLS}(u) = 2\int_0^{+\infty} t \cdot \boldsymbol{R}_{LLS}(t, u)dt, \quad u = 1, 2, \ldots, z,$$

and applying the formula

$$2\int_{0}^{+\infty} t\exp[-\lambda t]dt = \frac{2}{\lambda^2},$$

we obtain (3.1.53). Next, considering a formula for the variance and the result (3.1.51), we get the formula for the standard deviation of the system sojourn time in the reliability state subset $\{u, u+1, ..., z\}$, $u=1, 2, ..., z$, given by (3.1.52), (3.1.54). □

Further, we consider a homogeneous, aging series system with components having identical piecewise exponential reliability functions (3.1.34) with the coordinates

$$R(t, u) = \exp[-\lambda(u)t], \quad t \geq 0, \quad \lambda(u) \geq 0, \quad u = 1, 2, ..., z. \tag{3.1.55}$$

In that case, the conditional intensities $\lambda_{i/j}(v)$, $i=1, ..., n, j=1, ..., n$, of the component E_i departure from the subset $\{v, v+1, ..., z\}$, $v=u, u-1, ..., 1$, $u=1, 2, ..., z-1$, after the departure of E_j, according to the formula (3.1.41), are of the form

$$\lambda_{i/j}(v) = \frac{\lambda(v)}{q(v, E_j, E_i)} \tag{3.1.56}$$

and Proposition 3.1.3 takes following form (Blokus-Roszkowska and Kołowrocki, 2015c).

Proposition 3.1.4. *If, in a homogeneous and aging series system following the LLS dependency rule, the components have piecewise exponential reliability functions* (3.1.34) *with the coordinates* (3.1.55), *then the system's reliability function is given by the vector*

$$\boldsymbol{R}_{LLS}(t, \cdot) = [1, \boldsymbol{R}_{LLS}(t, 1), ..., \boldsymbol{R}_{LLS}(t, z)], \quad t \geq 0, \tag{3.1.57}$$

with the coordinates

$$\boldsymbol{R}_{LLS}(t, u) = \exp[-n\lambda(u+1)t] + \sum_{j=1}^{n}\frac{1}{n}\cdot\left[\exp\left[-\lambda(u)\sum_{i=1}^{n}\frac{1}{q(u, E_j, E_i)}t\right]\right.$$
$$\left.-\exp\left[-\left(n(\lambda(u+1)-\lambda(u)) + \lambda(u)\sum_{i=1}^{n}\frac{1}{q(u, E_j, E_i)}\right)t\right]\right], \quad u = 1, 2, ..., z-1, \tag{3.1.58}$$

$$\boldsymbol{R}_{LLS}(t, z) = \exp[-n\lambda(z)t]. \tag{3.1.59}$$

Similarly, in case of a homogeneous and aging series system Corollary 3.1.1 takes the following form.

Corollary 3.1.2. *If, in a homogeneous and aging series system following the LLS dependency rule, the components have piecewise exponential reliability functions* (3.1.34) *with the coordinates* (3.1.55), *then the system's mean lifetime in the reliability state subset* $\{u, u+1, \ldots, z\}$, $u = 1, 2, \ldots, z$, *is given by*

$$\mu_{LLS}(u) = \frac{1}{n\lambda(u+1)} + \sum_{j=1}^{n} \frac{1}{n} \cdot \left[\frac{1}{\lambda(u) \sum_{i=1}^{n} \frac{1}{q(u, E_j, E_i)}} - \frac{1}{n(\lambda(u+1) - \lambda(u)) + \lambda(u) \sum_{i=1}^{n} \frac{1}{q(u, E_j, E_i)}} \right],$$

$$u = 1, 2, \ldots, z-1, \tag{3.1.60}$$

$$\mu_{LLS}(z) = \frac{1}{n\lambda(z)}, \tag{3.1.61}$$

and the standard deviation of the system lifetime is

$$\sigma_{LLS}(u) = \sqrt{n_{LLS}(u) - [\mu_{LLS}(u)]^2}, \quad u = 1, 2, \ldots, z-1, \tag{3.1.62}$$

where

$$n_{LLS}(u) = \frac{2}{[n\lambda(u+1)]^2} + 2\sum_{j=1}^{n} \frac{1}{n} \cdot \left[\frac{1}{\left[\lambda(u) \sum_{i=1}^{n} \frac{1}{q(u, E_j, E_i)}\right]^2} - \frac{1}{\left[n(\lambda(u+1) - \lambda(u)) + \lambda(u) \sum_{i=1}^{n} \frac{1}{q(u, E_j, E_i)}\right]^2} \right],$$

$$u = 1, 2, \ldots, z-1, \tag{3.1.63}$$

and

$$\sigma_{LLS}(z) = \frac{1}{n\lambda(z)}. \tag{3.1.64}$$

The proof of Corollary 3.1.2 is analogical to the proof of Corollary 3.1.1.

Using Proposition 3.1.4 in case of homogeneous multistate series system, we determine another, practically very important, reliability characteristic of the system, namely, the system intensities of departures from the reliability state subsets (Blokus-Roszkowska and Kołowrocki, 2015c).

Corollary 3.1.3. *If, in a homogeneous and aging series system following the LLS dependency rule, the components have piecewise exponential reliability functions* (3.1.34) *with the coordinates* (3.1.55), *then the intensities of system's departures from the reliability state subsets* $\{u, u+1, \ldots, z\}$, $u=1, 2, \ldots, z$, *are given by*

$$\boldsymbol{\lambda}_{LLS}(t,u)=\left\{n\lambda(u+1)+\sum_{j=1}^{n}\frac{1}{n}\cdot\left[\lambda(u)\sum_{i=1}^{n}\frac{1}{q(u,E_j,E_i)}\right.\right.$$

$$\cdot\exp\left[n\lambda(u+1)t-\lambda(u)\sum_{i=1}^{n}\frac{1}{q(u,E_j,E_i)}t\right]$$

$$-\left(n\lambda(u+1)-n\lambda(u)+\lambda(u)\sum_{i=1}^{n}\frac{1}{q(u,E_j,E_i)}\right)$$

$$\left.\left.\cdot\exp\left[n\lambda(u)t-\lambda(u)\sum_{i=1}^{n}\frac{1}{q(u,E_j,E_i)}t\right]\right]\right\}\Big/ \tag{3.1.65}$$

$$\left\{1+\sum_{j=1}^{n}\frac{1}{n}\cdot\left[\exp\left[n\lambda(u+1)t-\lambda(u)\sum_{i=1}^{n}\frac{1}{q(u,E_j,E_i)}t\right]\right.\right.$$

$$\left.\left.-\exp\left[n\lambda(u)t-\lambda(u)\sum_{i=1}^{n}\frac{1}{q(u,E_j,E_i)}t\right]\right]\right\},$$

$u=1,2,\ldots,z-1,$

$$\boldsymbol{\lambda}_{LLS}(t,z)=n\lambda(z). \tag{3.1.66}$$

Proof. From the definition of the intensity of a system departure from the reliability state subset $\{u, u+1, \ldots, z\}$, $u=1, 2, \ldots, z$, we have

$$\boldsymbol{\lambda}_{LLS}(t,u)=\frac{\boldsymbol{f}_{LLS}(t,u)}{\boldsymbol{R}_{LLS}(t,u)},\quad u=1,2,\ldots,z, \tag{3.1.67}$$

where $\boldsymbol{R}_{LLS}(t,u)$, $u=1, 2, \ldots, z$, is the reliability function coordinate of a system with components dependent according to the LLS rule and $\boldsymbol{f}_{LLS}(t,u)$ is the density function coordinate of a system lifetime in the reliability state subset $\{u, u+1, \ldots, z\}$, $u=1, 2, \ldots, z$. In case of a homogeneous multistate series system with components having piecewise exponential reliability functions (3.1.34) with the coordinates (3.1.55), the coordinates of the system reliability function take form (3.1.58), (3.1.59).

Considering a well-known relationship between the density function and the reliability function

$$f_{LLS}(t, u) = \frac{d}{dt}(-\boldsymbol{R}_{LLS}(t, u)), \quad u=1,2,\ldots,z, \tag{3.1.68}$$

and using the system reliability function coordinate given by (3.1.58), we obtain the density function coordinate of a system lifetime in the subset $\{u, u+1, \ldots, z\}$

$$\begin{aligned} f_{LLS}(t, u) &= n\lambda(u+1)\exp[-n(\lambda(u+1)t] \\ &+\frac{1}{n}\sum_{j=1}^{n}\left[\lambda(u)\sum_{i=1}^{n}\frac{1}{q(u, E_j, E_i)}\cdot\exp\left[-\lambda(u)\sum_{i=1}^{n}\frac{1}{q(u, E_j, E_i)}t\right]\right. \\ &-\left(n\lambda(u+1) - n\lambda(u) + \lambda(u)\sum_{i=1}^{n}\frac{1}{q(u, E_j, E_i)}\right) \\ &\left.\cdot\exp\left[-n\lambda(u+1)t + n\lambda(u)t - \lambda(u)\sum_{i=1}^{n}\frac{1}{q(u, E_j, E_i)}t\right]\right], \end{aligned} \tag{3.1.69}$$

for $u=1, 2, \ldots, z-1$ and respectively, using (3.1.59), for $u=z$ we obtain the density function coordinate given by

$$f_{LLS}(t, z) = n\lambda(z)\cdot\exp[-n\lambda(z)t]. \tag{3.1.70}$$

Finally, substituting (3.1.69)–(3.1.79) and (3.1.58), (3.1.59) into (3.1.67), we get (3.1.65), (3.1.66), which completes the proof. □

Next, we assume that components E_i, $i=1, \ldots, n$, have piecewise Rayleigh reliability functions (3.1.6) with the coordinates

$$R_i(t, u) = \exp\left[-\lambda_i(u)t^2\right], \quad t\geq 0, \quad u=1,2,\ldots,z, \tag{3.1.71}$$

where $\lambda_i(u)$, $i=1, \ldots, n$, $\lambda_i(u)\geq 0$, are components' intensities of departure from the reliability state subset $\{u, u+1, \ldots, z\}$, $u=1, 2, \ldots, z$.

According to the relationship between the lifetime mean value in the reliability state subset and the intensity of departure from this subset

$$E[T_i(u)] = \frac{1}{2}\sqrt{\frac{\pi}{\lambda_i(u)}}, \quad u=1,2,\ldots,z,$$

we get the formula for the conditional intensities $\lambda_{i/j}(v)$, $i=1, \ldots, n$, $j=1, \ldots, n$, of the component E_i departure from the subset $\{v, v+1, \ldots, z\}$, $v=u, u-1, \ldots, 1$, $u=1, 2, \ldots, z-1$, after the departure of the component E_j. Namely, from (3.1.1), assuming the LLS dependency rule we obtain

$$\lambda_{i/j}(v) = \frac{\lambda_i(v)}{\left[q(v, E_j, E_i)\right]^2}. \tag{3.1.72}$$

Thus, considering (3.1.3)–(3.1.5) and (3.1.71), (3.1.72), the components E_i, $i=1, \ldots, n$, after the departure of E_j, $j=1, \ldots, n$, from that subset $\{u, u+1, \ldots, z\}$, $u=1, 2, \ldots, z$, have the reliability functions (3.1.3) with the coordinates

$$R_{i/j}(t, v) = \exp\left[-\frac{\lambda_i(v)}{\left[q(v, E_j, E_i)\right]^2} t^2\right], \quad v=u, u-1, \ldots, 1, \\ u=1, 2, \ldots, z-1, \tag{3.1.73}$$

$$R_{i/j}(t, v) = \exp\left[-\lambda_i(v) t^2\right], \quad v=u+1, \ldots, z, \quad u=1, 2, \ldots, z-1. \tag{3.1.74}$$

Subsequently, the distribution function $\widetilde{F}_j(t, u+1)$ corresponding to E_j, $j=1, \ldots, n$, defined by (3.1.10), is given by

$$\widetilde{F}_j(t, u+1) = 1 - \exp\left[-\left(\lambda_j(u+1) - \lambda_j(u)\right)t^2\right], \quad u=1, 2, \ldots, z-1, \quad t \geq 0,$$

and the corresponding density function is

$$\widetilde{f}_j(t, u+1) = 2\left(\lambda_j(u+1) - \lambda_j(u)\right)t \cdot \exp\left[-\left(\lambda_j(u+1) - \lambda_j(u)\right)t^2\right], \\ u=1, 2, \ldots, z-1, \quad t \geq 0. \tag{3.1.75}$$

Considering (3.1.73)–(3.1.75), in case the system components have piecewise Rayleigh distribution, from Proposition 3.1.1 we obtain the reliability function of a series system following the LLS dependency rule. More specifically, substituting the reliability function coordinates of components (3.1.71), the density function coordinates (3.1.75), and the reliability function coordinates of components after the departure of component E_j (3.1.73), (3.1.74), into (3.1.7)–(3.1.9), we obtain the thesis of Proposition 3.1.5.

Proposition 3.1.5. *If, in an aging series system following the LLS dependency rule, the components have piecewise Rayleigh reliability functions with the coordinates* (3.1.71), *then the system's reliability function is*

$$\boldsymbol{R}_{LLS}(t,\ \cdot) = [1, \boldsymbol{R}_{LLS}(t, 1), \ldots, \boldsymbol{R}_{LLS}(t, z)], \quad t \geq 0, \tag{3.1.76}$$

with the coordinates

$$\boldsymbol{R}_{LLS}(t, u) = \exp\left[-\sum_{i=1}^{n} \lambda_i(u+1)t^2\right] + \sum_{j=1}^{n} \frac{\lambda_j(u+1) - \lambda_j(u)}{\sum_{i=1}^{n}(\lambda_i(u+1) - \lambda_i(u))}$$

$$\cdot\left[\exp\left[-\sum_{i=1}^{n} \frac{\lambda_i(u)}{\left[q\left(u, E_j, E_i\right)\right]^2} t^2\right] - \exp\left[-\sum_{i=1}^{n}\left(\lambda_i(u+1) - \lambda_i(u) + \frac{\lambda_i(u)}{\left[q\left(u, E_j, E_i\right)\right]^2}\right) t^2\right]\right],$$

$u = 1, 2, \ldots, z-1,$

(3.1.77)

$$\boldsymbol{R}_{LLS}(t, z) = \exp\left[-\sum_{i=1}^{n} \lambda_i(z) t^2\right]. \tag{3.1.78}$$

In case of a homogeneous multistate series system with components having piecewise Rayleigh reliability functions (3.1.34) *with the coordinates*

$$R(t, u) = \exp\left[-\lambda(u)t^2\right], \quad t \geq 0, \quad \lambda(u) \geq 0, \quad u = 1, 2, \ldots, z, \tag{3.1.79}$$

Proposition 3.1.5 takes the form presented here.

Proposition 3.1.6. *If, in a homogeneous and aging series system following the LLS dependency rule, the components have piecewise Rayleigh reliability functions* (3.1.34) *with the coordinates* (3.1.79), *then the system's reliability function given by the vector*

$$\boldsymbol{R}_{LLS}(t,\ \cdot) = [1, \boldsymbol{R}_{LLS}(t, 1), \ldots, \boldsymbol{R}_{LLS}(t, z)], \quad t \geq 0, \tag{3.1.80}$$

with the coordinates

$$\boldsymbol{R}_{LLS}(t, u) = \exp\left[-n\lambda(u+1)t^2\right] + \frac{1}{n}\sum_{j=1}^{n}\left[\exp\left[-\lambda(u)\sum_{i=1}^{n}\frac{1}{\left[q\left(u, E_j, E_i\right)\right]^2} t^2\right]\right.$$

$$\left. - \exp\left[-\left(n\lambda(u+1) - n\lambda(u) + \lambda(u)\sum_{i=1}^{n}\frac{1}{\left[q\left(u, E_j, E_i\right)\right]^2}\right) t^2\right]\right],$$

$$u = 1, 2, \ldots, z-1, \tag{3.1.81}$$

$$\boldsymbol{R}_{LLS}(t, z) = \exp\left[-n\lambda(z)t^2\right]. \tag{3.1.82}$$

For an aging series system with components having piecewise Rayleigh reliability functions, to determine other reliability characteristics approximate methods should be used. Namely, then the system's mean lifetime in the reliability state subset $\{u, u+1, \ldots, z\}$, $u=1, 2, \ldots, z$, *is calculated by the formula* (2.1.16) *from Chapter* 2, *and the standard deviation of the system lifetime from* (2.1.17), (2.1.18) *and using the trapezoid rule for approximating integrals. The same procedure should be adopted in case of a series system with components having piecewise Erlang distribution.*

In case, when components of a series system have piecewise Erlang-l reliability functions (3.1.34) *with the coordinates*

$$R(t,u)=\sum_{\omega=0}^{l-1}\frac{[\lambda(u)t]^{\omega}}{\omega!}\exp[-\lambda(u)t],\quad t\geq 0,\quad \lambda(u)\geq 0,\quad u=1,2,\ldots,z, \tag{3.1.83}$$

the distribution function $\widetilde{F}(t,u+1)$, *defined by* (3.1.38), *is given by* (*Blokus and Kołowrocki,* 2019)

$$\widetilde{F}(t,u+1)=1-\frac{\sum_{\omega=0}^{l-1}\frac{[\lambda(u+1)t]^{\omega}}{\omega!}}{\sum_{\omega=0}^{l-1}\frac{[\lambda(u)t]^{\omega}}{\omega!}}\exp[-(\lambda(u+1)-\lambda(u))t],\quad u=1,2,\ldots,z-1,\quad t\geq 0, \tag{3.1.84}$$

and the corresponding density function is

$$\widetilde{f}(t,u+1)=\frac{d}{dt}\left(1-\frac{\sum_{\omega=0}^{l-1}\frac{[\lambda(u+1)t]^{\omega}}{\omega!}}{\sum_{\omega=0}^{l-1}\frac{[\lambda(u)t]^{\omega}}{\omega!}}\exp[-(\lambda(u+1)-\lambda(u))t]\right)$$

$$=\left[\frac{\sum_{\omega=0}^{l-1}\frac{[\lambda(u+1)t]^{\omega}}{\omega!}\cdot\sum_{\omega=0}^{l-2}\frac{[\lambda(u)]^{\omega+1}t^{\omega}}{\omega!}-\sum_{\omega=0}^{l-2}\frac{[\lambda(u+1)]^{\omega+1}t^{\omega}}{\omega!}\cdot\sum_{\omega=0}^{l-1}\frac{[\lambda(u)t]^{\omega}}{\omega!}}{\left[\sum_{\omega=0}^{l-1}\frac{[\lambda(u)t]^{\omega}}{\omega!}\right]^{2}}\right.$$

$$\left.+\frac{\sum_{\omega=0}^{l-1}\frac{[\lambda(u+1)t]^{\omega}}{\omega!}(\lambda(u+1)-\lambda(u))}{\sum_{\omega=0}^{l-1}\frac{[\lambda(u)t]^{\omega}}{\omega!}}\right]$$

$$\cdot\exp[-(\lambda(u+1)-\lambda(u))t],\quad u=1,2,\ldots,z-1,\quad t\geq 0. \tag{3.1.85}$$

The conditional intensities $\lambda_{i/j}(v)$, $i=1, 2, \ldots, n, j=1, 2, \ldots, n$, *of the component* E_i *departure from the subset* $\{v, v+1, \ldots, z\}$, $v=u, u-1, \ldots, 1$, $u=1, 2, \ldots, z-1$, *after the exit of* E_j, *by* (3.1.1) *are*

$$\lambda_{i/j}(v) = \frac{\lambda(v)}{q(v, E_j, E_i)}. \tag{3.1.86}$$

Further, by considering (3.1.83), (3.1.86), *the components* E_i, $i=1, 2, \ldots, n$, *after the departure of* E_j, $j=1, 2, \ldots, n$, *from the subset* $\{u, u+1, \ldots, z\}$, $u=1, 2, \ldots, z$, *have the reliability functions given by* (3.1.3) *with the coordinates*

$$R_{i/j}(t, v) = \sum_{\omega=0}^{l-1} \frac{\left[\frac{\lambda(v)}{q(v, E_j, E_i)} t\right]^{\omega}}{\omega!} \exp\left[-\frac{\lambda(v)}{q(v, E_j, E_i)} t\right], \tag{3.1.87}$$

$$v = u, u-1, \ldots, 1, \quad u = 1, 2, \ldots, z-1,$$

$$R_{i/j}(t, v) = \sum_{\omega=0}^{l-1} \frac{[\lambda(v)t]^{\omega}}{\omega!} \exp[-\lambda(v)t], \quad v = u+1, \ldots, z, \tag{3.1.88}$$
$$u = 1, 2, \ldots, z-1.$$

From (3.1.87), (3.1.88) and Proposition 3.1.2, we obtain the next theorem (Blokus and Kołowrocki, 2019).

Proposition 3.1.7. *If, in a homogeneous and aging series system with the LLS dependency rule, the components have piecewise Erlang reliability functions with the coordinates* (3.1.83), *then the system's reliability function is*

$$\boldsymbol{R}_{LLS}(t, \cdot) = [1, \boldsymbol{R}_{LLS}(t, 1), \ldots, \boldsymbol{R}_{LLS}(t, z)], \quad t \geq 0, \tag{3.1.89}$$

where

$$\boldsymbol{R}_{LLS}(t, u) = \left[\sum_{\omega=0}^{l-1} \frac{[\lambda(u+1)t]^{\omega}}{\omega!}\right]^{n} \cdot \exp[-n\lambda(u+1)t] + \sum_{j=1}^{n} \int_0^t \tilde{f}(a, u+1)$$
$$\left[\sum_{\omega=0}^{l-1} \frac{[\lambda(u+1)a]^{\omega}}{\omega!}\right]^{n-1} \cdot \sum_{\omega=0}^{l-1} \frac{[\lambda(u)a]^{\omega}}{\omega!} \cdot \exp[-((n-1)\lambda(u+1) + \lambda(u))a]$$
$$\cdot \left[\prod_{i=1}^{n} \frac{\sum_{\omega=0}^{l-1} \frac{\left[\frac{\lambda(u)}{q(u, E_j, E_i)} t\right]^{\omega}}{\omega!}}{\sum_{\omega=0}^{l-1} \frac{[\lambda(u)a]^{\omega}}{\omega!}} \cdot \exp\left[-\frac{\lambda(u)}{q(u, E_j, E_i)} t + \lambda(u)a\right]\right] da,$$
$$u = 1, 2, \ldots, z-1, \tag{3.1.90}$$

and after substituting the density function coordinate $\widetilde{f}(a, u+1)$ *given by* (3.1.85) *into* (3.1.90), *the coordinates take form*

$$\boldsymbol{R}_{LLS}(t,u) = \left[\sum_{\omega=0}^{l-1}\frac{[\lambda(u+1)t]^{\omega}}{\omega!}\right]^{n} \cdot \exp[-n\lambda(u+1)t] + \sum_{j=1}^{n}\int_{0}^{t}\left[\sum_{\omega=0}^{l-1}\frac{[\lambda(u+1)a]^{\omega}}{\omega!}\right]^{n-1} \cdot \sum_{\omega=0}^{l-1}\frac{[\lambda(u)a]^{\omega}}{\omega!}$$

$$\left[\frac{\sum_{\omega=0}^{l-1}\frac{[\lambda(u+1)a]^{\omega}}{\omega!}\cdot\sum_{\omega=0}^{l-2}\frac{[\lambda(u)]^{\omega+1}a^{\omega}}{\omega!} - \sum_{\omega=0}^{l-2}\frac{[\lambda(u+1)]^{\omega+1}a^{\omega}}{\omega!}\cdot\sum_{\omega=0}^{l-1}\frac{[\lambda(u)a]^{\omega}}{\omega!}}{\left[\sum_{\omega=0}^{l-1}\frac{[\lambda(u)a]^{\omega}}{\omega!}\right]^{2}} + \frac{\sum_{\omega=0}^{l-1}\frac{[\lambda(u+1)a]^{\omega}}{\omega!}(\lambda(u+1)-\lambda(u))}{\sum_{\omega=0}^{l-1}\frac{[\lambda(u)a]^{\omega}}{\omega!}}\right] \cdot \exp[-n\lambda(u+1)a]$$

$$\cdot\left[\prod_{i=1}^{n}\frac{\sum_{\omega=0}^{l-1}\frac{\left[\frac{\lambda(u)}{q(u,E_j,E_i)}t\right]^{\omega}}{\omega!}}{\sum_{\omega=0}^{l-1}\frac{[\lambda(u)a]^{\omega}}{\omega!}} \cdot \exp\left[-\frac{\lambda(u)}{q(u,E_j,E_i)}t + \lambda(u)a\right]\right] da, \quad u = 1,2,\ldots,z-1, \tag{3.1.91}$$

$$\boldsymbol{R}_{LLS}(t,z) = \left[\sum_{\omega=0}^{l-1}\frac{[\lambda(z)t]^{\omega}}{\omega!}\right]^{n} \cdot \exp[-n\lambda(z)t]. \tag{3.1.92}$$

3.1.3 Reliability of multistate series-parallel system with dependent components of its subsystems

In this section, we consider a multistate series-parallel system, with a scheme given in Fig. 3.2. A system is composed of k multistate series subsystems linked parallel and working independently with dependent components. In each series subsystem, we assume local load-sharing model of components dependency described in Section 3.1.1. In the ith series subsystem S_i, $i=1, 2, \ldots, k$, there are l_i components dependent according to the LLS rule. We denote by E_{ij}, $i=1, 2, \ldots, k, j=1, 2, \ldots, l_i$, $k, l_i \in N$, components of a system and assume that all components E_{ij} have the reliability state set $\{0, 1, \ldots, z\}$. Subsequently, $T_{ij}(u)$, $i=1, 2, \ldots, k, j=1, 2, \ldots, l_i$, are random variables representing lifetimes of components E_{ij} in the reliability state subset $\{u, u+1, \ldots, z\}$, $u=1, 2, \ldots, z$, while they were in the reliability state z at the moment $t=0$.

Similarly as in Section 3.1.1, we assume that after changing the reliability state subset by one of components E_{ig_i}, $g_i=1, 2, \ldots, l_i$, in the ith series subsystem S_i, $i=1, 2, \ldots, k$, to the worse state subset, the lifetimes of remaining components in this subsystem in the reliability state subsets decrease. We denote by $E[T_{i,j}(u)]$ and $E[T_{i,j/g_i}(u)]$, $i=1, 2, \ldots, k, j=1, 2, \ldots, l_i$, $g_i=1, 2, \ldots, l_i$ for $u=1, 2, \ldots, z$, the mean values of components' lifetimes, respectively, before and after departure of the component E_{ig_i}, $g_i=1, 2, \ldots, l_i$, from the reliability state subset $\{u, u+1, \ldots, z\}$, $u=1, 2, \ldots, z$. The reliability parameters of remaining components E_{ij}, $i=1, 2, \ldots, k, j=1, 2, \ldots, l_i$, $j \neq g_i$, in this subsystem change depending on their exposure to the component E_{ig_i} reliability state change, expressed by the coefficients $q(v, E_{ig_i}, E_{ij})$ for $v=u, u-1, \ldots, 1$ and $u=1, 2, \ldots, z-1$. Further, it is assumed that the mean values of these components lifetimes in the reliability state subset $\{v, v+1, \ldots, z\}$, $v=u, u-1, \ldots, 1$, $u=1, 2, \ldots, z-1$, decrease according to the following formula:

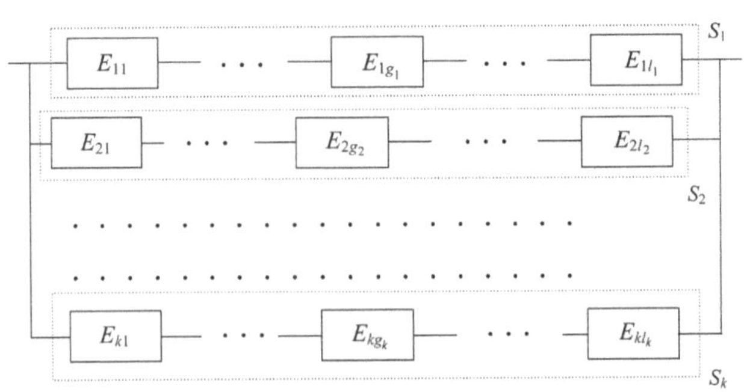

Fig. 3.2 The scheme of a series-parallel system.

$$E\left[T_{i,j/g_i}(v)\right] = q\left(v, E_{ig_i}, E_{ij}\right) \cdot E\left[T_{i,j}(v)\right], \quad i = 1,2,\ldots,k,$$
$$j = 1,2,\ldots,l_i, \quad g_i = 1,2,\ldots,l_i, \tag{3.1.93}$$

where the coefficients fulfill following conditions

$$0 < q\left(v, E_{ig_i}, E_{ij}\right) \le 1, \quad i = 1,2,\ldots,k, \quad j = 1,2,\ldots,l_i, \quad g_i = 1,2,\ldots,l_i, \quad j \ne g_i,$$

and

$$q\left(v, E_{ig_i}, E_{ig_i}\right) = 1, \quad g_i = 1,2,\ldots,l_i.$$

By combining the results for a multistate series system assuming its components' dependency according to the LLS rule and the reliability function of a parallel system with independent components given by (2.3.7), (2.3.8) in Chapter 2 and in Kołowrocki (2014), we obtain the Proposition 3.1.8.

Proposition 3.1.8. *If, in an aging series-parallel system, the subsystems are working independently and components of these series subsystems are dependent according to the LLS rule and have reliability functions* (3.1.6), *then the system's reliability function is*

$$\overline{\boldsymbol{R}}_{LLS}(t, \cdot) = [1, \overline{\boldsymbol{R}}_{LLS}(t, 1), \ldots, \overline{\boldsymbol{R}}_{LLS}(t, z)], \quad t \ge 0, \tag{3.1.94}$$

with the coordinates

$$\overline{\boldsymbol{R}}_{LLS}(t, u) = 1 - \prod_{i=1}^{k} \left[1 - \boldsymbol{R}_{LLS}^{(i)}(t, u)\right], \quad u = 1,2,\ldots,z, \tag{3.1.95}$$

where $\boldsymbol{R}_{LLS}^{(i)}(t, u)$, $u = 1, 2, \ldots, z$, *is the reliability function coordinate, given by* (3.1.8), (3.1.9), *of series subsystem* S_i, $i = 1, 2, \ldots, k$, *following the LLS dependency rule.*

Further, we assume that components E_{ij}, $i = 1, 2, \ldots, k$, $j = 1, 2, \ldots, l_i$, *have piecewise exponential reliability functions given by*

$$R_{ij}(t, \cdot) = \left[1, R_{ij}(t, 1), \ldots, R_{ij}(t, z)\right], \quad t \ge 0, \tag{3.1.96}$$

with the coordinates

$$R_{ij}(t, u) = \exp\left[-\lambda_{ij}(u)t\right], \quad t \ge 0, \quad u = 1,2,\ldots,z, \tag{3.1.97}$$

where $\lambda_{ij}(u)$, $i = 1, 2, \ldots, k$, $j = 1, 2, \ldots, l_i$, $\lambda_{ij}(u) \ge 0$, *are* E_{ij} *components' intensities of departure from the reliability state subset* $\{u, u+1, \ldots, z\}$, $u = 1, 2, \ldots, z$.

If components have piecewise exponential reliability functions given by (3.1.96), (3.1.97), *then after departure of the component* E_{ig_i}, $g_i = 1, 2, \ldots, l_i$, *from the subset* $\{u, u+1, \ldots, z\}$, $u = 1, 2, \ldots, z$, *in the ith subsystem* S_i, $i = 1, 2, \ldots, k$, *the conditional intensities* $\lambda_{i,\,j/g_i}(v)$, $i=1, 2, \ldots, k, j=1, 2, \ldots, l_i, g_i=1, 2, \ldots, l_i$, *of departure from the reliability state subset* $\{v, v+1, \ldots, z\}$, $v=u, u-1, \ldots, 1$, $u=1, 2, \ldots, z-1$, *of remaining components* E_{ij}, $j=1, 2, \ldots, l_i$, $j \neq g_i$, *in this subsystem, by* (3.1.93), *are given by*

$$\lambda_{i,j/g_i}(v) = \frac{\lambda_{ij}(v)}{q(v, E_{ig_i}, E_{ij})}, \quad i=1,2,\ldots,k, \quad j=1,2,\ldots,l_i, \qquad (3.1.98)$$
$$g_i = 1,2,\ldots,l_i.$$

Next, applying (3.1.94), (3.1.95), in case the system components have piecewise exponential reliability functions from Proposition 3.1.4 we obtain the following result.

Proposition 3.1.9. *If, in an aging series-parallel system, the subsystems are working independently and components of these series subsystems are dependent according to the LLS rule and have piecewise exponential reliability functions* (3.1.96), (3.1.97), *then the system's reliability function is*

$$\overline{\boldsymbol{R}}_{LLS}(t, \cdot) = [1, \overline{\boldsymbol{R}}_{LLS}(t, 1), \ldots, \overline{\boldsymbol{R}}_{LLS}(t, z)], \quad t \geq 0, \qquad (3.1.99)$$

with the coordinates

$$\overline{\boldsymbol{R}}_{LLS}(t, u) = 1 - \prod_{i=1}^{k}\left[1 - \exp\left[-\sum_{j=1}^{l_i}\lambda_{ij}(u+1)t\right]\right.$$
$$-\sum_{g_i=1}^{l_i}\frac{\lambda_{ig_i}(u+1) - \lambda_{ig_i}(u)}{\sum_{j=1}^{l_i}\left(\lambda_{ij}(u+1) - \lambda_{ij}(u)\right)} \cdot \left[\exp\left[-\sum_{j=1}^{l_i}\frac{\lambda_{ij}(u)}{q(u, E_{ig_i}, E_{ij})}t\right]\right. \qquad (3.1.100)$$
$$\left.\left.-\exp\left[-\sum_{j=1}^{l_i}\left(\lambda_{ij}(u+1) - \lambda_{ij}(u) + \frac{\lambda_{ij}(u)}{q(u, E_{ig_i}, E_{ij})}\right)t\right]\right]\right],$$
$$u = 1,2,\ldots,z-1,$$

$$\overline{\boldsymbol{R}}_{LLS}(t, z) = 1 - \prod_{i=1}^{k}\left[1 - \exp\left[-\sum_{j=1}^{l_i}\lambda_{ij}(z)t\right]\right], \qquad (3.1.101)$$

where k *is the number of series subsystems and* l_i, $i=1, 2, \ldots, k$, *is the number of components in ith subsystem.*

As a special case a multistate regular series-parallel system, with a scheme given in Fig. 3.3, is analyzed. In that case, we assume that there are *k* series subsystems and in each subsystem there are *l* dependent according to the LLS rule components.

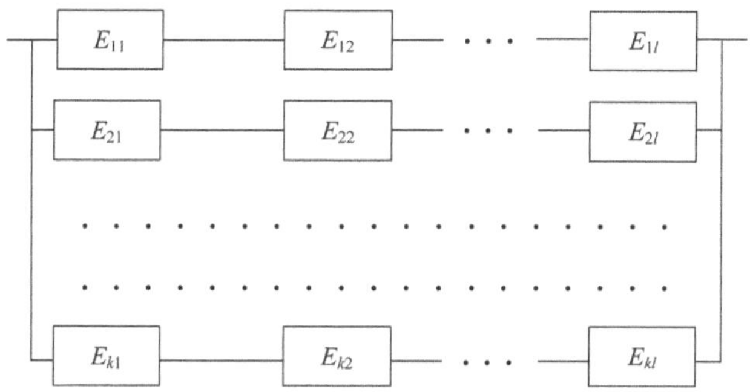

Fig. 3.3 The scheme of a regular series-parallel system.

In that case, by combining the results for a multistate series system assuming its components' dependency and the results for a homogeneous multistate parallel system with independent components given by (2.3.10), (2.3.11) in Chapter 2 and in Kołowrocki (2014), we immediately get the reliability function of a multistate series-parallel system with dependent components of its subsystems.

Proposition 3.1.10. *If in a homogeneous and aging regular series-parallel system, the subsystems are working independently and components of these series subsystems are dependent according to the LLS rule and have reliability functions* (3.1.34), *then the system's reliability function is*

$$\overline{\boldsymbol{R}}_{LLS}(t,\ \cdot\ ) = [1, \overline{\boldsymbol{R}}_{LLS}(t,1), \ldots, \overline{\boldsymbol{R}}_{LLS}(t,z)], \quad t \geq 0, \tag{3.1.102}$$

with the coordinates

$$\overline{\boldsymbol{R}}_{LLS}(t,u) = 1 - [1 - \boldsymbol{R}_{LLS}(t,u)]^{k}, \quad u = 1,2,\ldots,z, \tag{3.1.103}$$

where the reliability function coordinates of series subsystems following the LLS dependency rule $\boldsymbol{R}_{LLS}(t,u)$, $u = 1, 2, \ldots, z$, *are given by* (3.1.36), (3.1.37).

Next, applying (3.1.102), (3.1.103), in case the system components have piecewise exponential reliability functions we obtain immediately following result.

Proposition 3.1.11. *If in a homogeneous and aging regular series-parallel system, the subsystems are working independently and components of these series subsystems are dependent according to the LLS rule and have piecewise exponential reliability functions with the coordinates* (3.1.55), *then the system's reliability function is*

$$\overline{\boldsymbol{R}}_{LLS}(t,\ \cdot\ ) = [1, \overline{\boldsymbol{R}}_{LLS}(t,1), \ldots, \overline{\boldsymbol{R}}_{LLS}(t,z)], \quad t \geq 0, \tag{3.1.104}$$

with the coordinates

$$\overline{\boldsymbol{R}}_{LLS}(t,u) = 1 - \left[1 - \exp[-l\lambda(u+1)t] - \frac{1}{l}\sum_{j=1}^{l}\left[\exp\left[-\lambda(u)\sum_{i=1}^{l}\frac{1}{q\left(u, E_j, E_i\right)}t\right] - \exp\left[-\left(l\lambda(u+1) - l\lambda(u) + \lambda(u)\sum_{i=1}^{l}\frac{1}{q\left(u, E_j, E_i\right)}\right)t\right]\right]\right]^k,$$

$$u = 1, 2, \ldots, z-1, \tag{3.1.105}$$

$$\overline{\boldsymbol{R}}_{LLS}(t,z) = 1 - [1 - \exp[-l\lambda(z)t]]^k, \tag{3.1.106}$$

where k is the number of series subsystems and l is the number of components in subsystems.

3.1.4 Reliability of multistate series-"*m* out of *k*" system with dependent components of its subsystems

Similarly as in Section 3.1.3, we consider a multistate series-"*m* out of *k*" system, with a scheme given in Fig. 3.4, as an "*m* out of *k*" system composed of series subsystems with dependent components. We assume that k is a number of series subsystems S_i, $i=1, 2, \ldots, k$, and l_i, $i=1, 2, \ldots, k$, is a number of components in ith subsystem. Subsystems are working independently, whereas in each series subsystem we assume local load-sharing model of components dependency described in Sections 3.1.1 and 3.1.3.

The reliability function of a multistate series-"*m* out of *k*" system with dependent components of its subsystems is determined by combining the results for a multistate series system following the LLS rule and the results for an "*m* out of *n*" system with independent components, given by (2.3.14)–(2.3.16) in Chapter 2 and in Kołowrocki (2014).

Proposition 3.1.12. *If, in an aging series-"m out of k" system, the subsystems are working independently and components of these series subsystems are dependent according to the LLS rule and have reliability functions* (3.1.6), *then the system's reliability function is given by the vector*

$$\overline{\boldsymbol{R}}_{LLS}^{(m)}(t, \cdot) = \left[1, \overline{\boldsymbol{R}}_{LLS}^{(m)}(t, 1), \ldots, \overline{\boldsymbol{R}}_{LLS}^{(m)}(t, z)\right], \quad t \geq 0, \tag{3.1.107}$$

with the coordinates

$$\overline{\boldsymbol{R}}_{LLS}^{(m)}(t, u) = \sum_{\substack{r_1, r_2, \ldots, r_k = 0 \\ m \leq r_1 + \cdots + r_k \leq k}}^{1} \prod_{i=1}^{k} \left[\boldsymbol{R}_{LLS}^{(i)}(t, u)\right]^{r_i} \left[1 - \boldsymbol{R}_{LLS}^{(i)}(t, u)\right]^{1-r_i}, \quad u = 1, 2, \ldots, z, \tag{3.1.108}$$

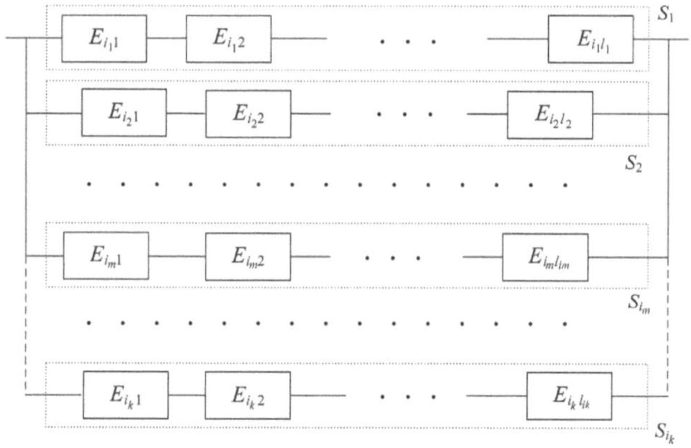

Fig. 3.4 The scheme of a series-"*m* out of *k*" system.

or with the coordinates

$$\overline{\boldsymbol{R}}_{LLS}^{(m)}(t,u)=1-\sum_{\substack{r_1,r_2,\ldots,r_k=0\\ 0\le r_1+\cdots+r_k\le m-1}}^{1}\prod_{i=1}^{k}\left[\boldsymbol{R}_{LLS}^{(i)}(t,u)\right]^{r_i}\left[1-\boldsymbol{R}_{LLS}^{(i)}(t,u)\right]^{1-r_i},$$
$$u=1,2,\ldots,z, \tag{3.1.109}$$

where $\mathbf{R}_{LLS}^{(i)}(t,u)$, $u=1, 2, \ldots, z$, *is the reliability function coordinate, given by* (3.1.8), (3.1.9), *of series subsystem* S_i, $i=1, 2, \ldots, k$, *following the LLS dependency rule.*

If we assume that components E_{ij}, $i=1, 2, \ldots, k$ *and* $j=1, 2, \ldots, l_i$, *have piecewise exponential reliability functions given by* (3.1.96), (3.1.97), *then the conditional intensities* $\lambda_{i,\,j/g_i}(v)$, $i=1, 2, \ldots, k$, $j=1, 2, \ldots, l_i$, $g_i=1, 2, \ldots, l_i$, *of the component* E_{ij} *departure from the subset* $\{v, v+1, \ldots, z\}$ *for* $v=u, u-1, \ldots, 1$ *and* $u=1, 2, \ldots, z-1$, *after the departure of* E_{ig_i} *from the subset* $\{u, u+1, \ldots, z\}$, $u=1, 2, \ldots, z$, *are given by* (3.1.98). *In that case, applying* (3.1.107), (3.1.108), *or* (3.1.107), (3.1.109), *respectively, we obtain immediately following result.*

Proposition 3.1.13. *If, in an aging series-"m out of k" system, the subsystems are working independently and components of these series subsystems are dependent according to the LLS rule and have piecewise exponential reliability functions (3.1.96), (3.1.97), then the system's reliability function is given by the vector*

$$\overline{\boldsymbol{R}}_{LLS}^{(m)}(t,\cdot)=\left[1,\overline{\boldsymbol{R}}_{LLS}^{(m)}(t,1),\ldots,\overline{\boldsymbol{R}}_{LLS}^{(m)}(t,z)\right],\quad t\ge 0, \tag{3.1.110}$$

with the coordinates

$$\overline{\boldsymbol{R}}_{LLS}^{(m)}(t,u)=\sum_{\substack{r_1,r_2,\ldots,r_k=0\\ m\le r_1+\cdots+r_k\le k}}^{1}\prod_{i=1}^{k}\left[\exp\left[-\sum_{j=1}^{l_i}\lambda_{ij}(u+1)t\right]\right.$$

$$+\sum_{g_i=1}^{l_i}\frac{\lambda_{ig_i}(u+1)-\lambda_{ig_i}(u)}{\sum_{j=1}^{l_i}\left(\lambda_{ij}(u+1)-\lambda_{ij}(u)\right)}\cdot\left[\exp\left[-\sum_{j=1}^{l_i}\frac{\lambda_{ij}(u)}{q(u,E_{ig_i},E_{ij})}t\right]\right.$$

$$\left.\left.-\exp\left[-\sum_{j=1}^{l_i}\left(\lambda_{ij}(u+1)-\lambda_{ij}(u)+\frac{\lambda_{ij}(u)}{q(u,E_{ig_i},E_{ij})}\right)t\right]\right]\right]^{r_i}$$

$$\cdot\left[1-\exp\left[-\sum_{j=1}^{l_i}\lambda_{ij}(u+1)t\right]-\sum_{g_i=1}^{l_i}\frac{\lambda_{ig_i}(u+1)-\lambda_{ig_i}(u)}{\sum_{j=1}^{l_i}\left(\lambda_{ij}(u+1)-\lambda_{ij}(u)\right)}\right.$$

$$\cdot\left[\exp\left[-\sum_{j=1}^{l_i}\frac{\lambda_{ij}(u)}{q\left(u, E_{ig_i}, E_{ij}\right)}t\right]\right.$$

$$\left.\left.\left.-\exp\left[-\sum_{j=1}^{l_i}\left(\lambda_{ij}(u+1)-\lambda_{ij}(u)+\frac{\lambda_{ij}(u)}{q\left(u, E_{ig_i}, E_{ij}\right)}\right)t\right]\right]\right]^{1-r_i},$$

$$u = 1, 2, \ldots, z-1, \tag{3.1.111}$$

$$\overline{\boldsymbol{R}}_{LLS}^{(m)}(t, z) = \sum_{\substack{r_1, r_2, \ldots, r_k = 0\\ m \le r_1 + \cdots + r_k \le k}}^{1} \prod_{i=1}^{k}\left[\exp\left[-\sum_{j=1}^{l_i}\lambda_{ij}(z)t\right]\right]^{r_i} \left[1-\exp\left[-\sum_{j=1}^{l_i}\lambda_{ij}(z)t\right]\right]^{1-r_i} \tag{3.1.112}$$

or with the coordinates

$$\overline{\boldsymbol{R}}_{LLS}^{(m)}(t, u) = 1 - \sum_{\substack{r_1, r_2, \ldots, r_k = 0\\ 0 \le r_1 + \cdots + r_k \le m-1}}^{1} \prod_{i=1}^{k}\left[\exp\left[-\sum_{j=1}^{l_i}\lambda_{ij}(u+1)t\right]\right.$$

$$+\sum_{g_i=1}^{l_i}\frac{\lambda_{ig_i}(u+1)-\lambda_{ig_i}(u)}{\sum_{j=1}^{l_i}\left(\lambda_{ij}(u+1)-\lambda_{ij}(u)\right)}\cdot\left[\exp\left[-\sum_{j=1}^{l_i}\frac{\lambda_{ij}(u)}{q\left(u, E_{ig_i}, E_{ij}\right)}t\right]\right.$$

$$\left.\left.\left.-\exp\left[-\sum_{j=1}^{l_i}\left(\lambda_{ij}(u+1)-\lambda_{ij}(u)+\frac{\lambda_{ij}(u)}{q\left(u, E_{ig_i}, E_{ij}\right)}\right)t\right]\right]\right]^{r_i}$$

$$\left[1-\exp\left[-\sum_{j=1}^{l_i}\lambda_{ij}(u+1)t\right]-\sum_{g_i=1}^{l_i}\frac{\lambda_{ig_i}(u+1)-\lambda_{ig_i}(u)}{\sum_{j=1}^{l_i}\left(\lambda_{ij}(u+1)-\lambda_{ij}(u)\right)}\right. \tag{3.1.113}$$

$$\cdot\left[\exp\left[-\sum_{j=1}^{l_i}\frac{\lambda_{ij}(u)}{q\left(u, E_{ig_i}, E_{ij}\right)}t\right]\right.$$

$$\left.\left.\left.-\exp\left[-\sum_{j=1}^{l_i}\left(\lambda_{ij}(u+1)-\lambda_{ij}(u)+\frac{\lambda_{ij}(u)}{q\left(u, E_{ig_i}, E_{ij}\right)}\right)t\right]\right]\right]^{1-r_i}$$

$$u = 1, 2, \ldots, z-1,$$

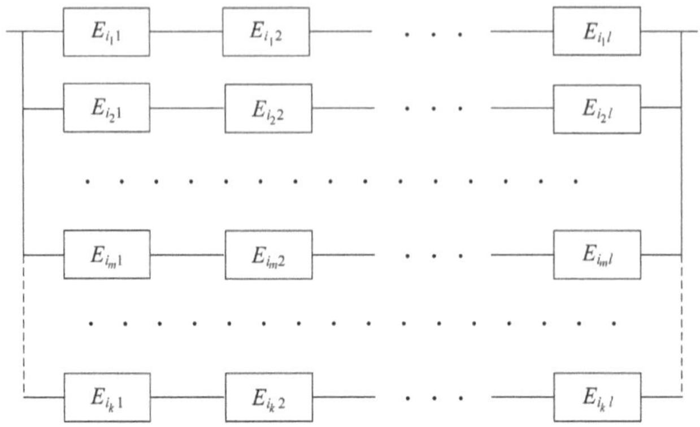

Fig. 3.5 The scheme of a regular series-"*m* out of *k*" system.

$$\overline{\boldsymbol{R}}_{LLS}^{(m)}(t, z) = 1 - \sum_{\substack{r_1, r_2, \ldots, r_k = 0 \\ 0 \le r_1 + \cdots + r_k \le m-1}}^{1} \prod_{i=1}^{k} \left[\exp\left[-\sum_{j=1}^{l_i} \lambda_{ij}(z) t \right] \right]^{r_i} \left[1 - \exp\left[-\sum_{j=1}^{l_i} \lambda_{ij}(z) t \right] \right]^{1-r_i}, \tag{3.1.114}$$

where k is the number of series subsystems and l_i, $i = 1, 2, \ldots, k$, *is the number of components in ith subsystem.*

As a special case, a multistate regular series-"*m* out of *k*" system, with a scheme given in Fig. 3.5, is discussed. We assume that *k* is a number of series subsystems working independently and *l* is a number of dependent components of these series subsystems. In each series subsystem, we assume LLS model of components dependency.

In that case, the reliability function of a homogeneous and aging regular series-"*m* out of *k*" system with dependent components of its subsystems is proposed in the form of Proposition 3.1.14, by using the reliability function of a homogeneous multistate "*m* out of *n*" system (2.3.17)–(2.3.19) from Chapter 2.

Proposition 3.1.14. *If in a homogeneous and aging regular series-"m out of k" system, the subsystems are working independently and components of these series subsystems are dependent according to the LLS rule and have reliability functions* (3.1.34), *then the system's reliability function is given by the vector*

$$\overline{\boldsymbol{R}}_{LLS}^{(m)}(t, \cdot) = \left[1, \overline{\boldsymbol{R}}_{LLS}^{(m)}(t, 1), \ldots, \overline{\boldsymbol{R}}_{LLS}^{(m)}(t, z) \right], \quad t \ge 0, \tag{3.1.115}$$

with the coordinates

$$\overline{\boldsymbol{R}}_{LLS}^{(m)}(t,u) = \sum_{r=m}^{k} \binom{k}{r} [\boldsymbol{R}_{LLS}(t,u)]^r [1 - \boldsymbol{R}_{LLS}(t,u)]^{k-r}, \tag{3.1.116}$$

$$u = 1, 2, \ldots, z,$$

or with the coordinates

$$\overline{\boldsymbol{R}}_{LLS}^{(m)}(t,u) = 1 - \sum_{r=0}^{m-1} \binom{k}{r} [\boldsymbol{R}_{LLS}(t,u)]^r [1 - \boldsymbol{R}_{LLS}(t,u)]^{k-r}, \tag{3.1.117}$$

$$u = 1, 2, \ldots, z,$$

where the reliability function coordinates of series subsystems following the LLS dependency rule $\boldsymbol{R}_{LLS}(t,u)$, $u = 1, 2, \ldots, z$, *are given by* (3.1.36), (3.1.37).

Next, in case the system components have piecewise exponential reliability functions with the coordinates (3.1.55), *considering reliability function for a homogeneous multistate series system given by* (3.1.58), (3.1.59) *and applying* (3.1.115), (3.1.116), *or* (3.1.115), (3.1.117), *respectively, from Proposition* 3.1.14, *we obtain immediately following result.*

Proposition 3.1.15. *If in a homogeneous and aging regular series-"m out of k" system, the subsystems are working independently and components of these series subsystems are dependent according to the LLS rule and have piecewise exponential reliability functions with the coordinates (3.1.55), then the system's reliability function is given by the vector*

$$\overline{\boldsymbol{R}}_{LLS}^{(m)}(t,\cdot) = \left[1, \overline{\boldsymbol{R}}_{LLS}^{(m)}(t,1), \ldots, \overline{\boldsymbol{R}}_{LLS}^{(m)}(t,z)\right], \quad t \geq 0, \tag{3.1.118}$$

with the coordinates

$$\overline{\boldsymbol{R}}_{LLS}^{(m)}(t,u)=\sum_{r=m}^{k}\binom{k}{r}\left[\exp[-l\lambda(u+1)t]]+\frac{1}{l}\sum_{j=1}^{l}\left\{\exp\left[-\lambda(u)\sum_{i=1}^{l}\frac{1}{q\left(u,E_j,E_i\right)}t\right]-\exp\left[-\left(l\lambda(u+1)-l\lambda(u)+\lambda(u)\sum_{i=1}^{l}\frac{1}{q\left(u,E_j,E_i\right)}\right)t\right]\right\}\right]^{r}$$

$$\cdot\left[1-\exp[-l\lambda(u+1)t]]-\frac{1}{l}\sum_{j=1}^{l}\left\{\exp\left[-\lambda(u)\sum_{i=1}^{l}\frac{1}{q\left(u,E_j,E_i\right)}t\right]-\exp\left[-\left(l\lambda(u+1)-l\lambda(u)+\lambda(u)\sum_{i=1}^{l}\frac{1}{q\left(u,E_j,E_i\right)}\right)t\right]\right\}\right]^{k-r},$$

$$u=1,2,\ldots,z-1, \tag{3.1.119}$$

$$\overline{\boldsymbol{R}}_{LLS}^{(m)}(t,z)=\sum_{r=m}^{k}\binom{k}{r}\,[\exp[-l\lambda(z)t]]^{r}[1-\exp[-l\lambda(z)t]]^{k-r}, \tag{3.1.120}$$

or with the coordinates

$$\overline{\boldsymbol{R}}_{LLS}^{(m)}(t,u)=1-\sum_{r=0}^{m-1}\binom{k}{r}\left[\exp[-l\lambda(u+1)t]]+\frac{1}{l}\sum_{j=1}^{l}\left\{\exp\left[-\lambda(u)\sum_{i=1}^{l}\frac{1}{q\left(u,E_j,E_i\right)}t\right]-\exp\left[-\left(l\lambda(u+1)-l\lambda(u)+\lambda(u)\sum_{i=1}^{l}\frac{1}{q\left(u,E_j,E_i\right)}\right)t\right]\right\}\right]^{r}$$

$$\cdot\left[1-\exp[-l\lambda(u+1)t]]-\frac{1}{l}\sum_{j=1}^{l}\left\{\exp\left[-\lambda(u)\sum_{i=1}^{l}\frac{1}{q\left(u,E_j,E_i\right)}t\right]-\exp\left[-\left(l\lambda(u+1)-l\lambda(u)+\lambda(u)\sum_{i=1}^{l}\frac{1}{q\left(u,E_j,E_i\right)}\right)t\right]\right\}\right]^{k-r},$$

$$u=1,2,\ldots,z-1, \tag{3.1.121}$$

$$\overline{\boldsymbol{R}}_{LLS}^{(m)}(t,z)=1-\sum_{r=0}^{m-1}\binom{k}{r}\,[\exp[-l\lambda(z)t]]^{r}[1-\exp[-l\lambda(z)t]]^{k-r}. \tag{3.1.122}$$

For multistate series-parallel and series-"m out of k" systems, their mean lifetimes in the reliability state subsets should be estimated using approximate methods of calculating integrals and the formula (2.1.16), and similarly the standard deviation of their lifetimes by (2.1.17), (2.1.18) from Chapter 2.

3.2 Equal load-sharing model of system components dependency

The ELS dependency model of components for multistate aging parallel and "m out of n" systems is described in this section. Regarding components' lifetimes in reliability state subsets for multistate systems following the ELS dependency rule, it is assumed that the influence of components departures from the reliability state subsets on other system components may differ depending on system and components' features, expressed by a component stress proportionality correction coefficient. Under such assumptions the system reliability function, the mean lifetime, and the standard deviation of system lifetime in the reliability state subset for multistate piecewise exponential parallel and "m out of n" systems are determined. Next, the ELS dependency model is extended for multistate aging parallel-series and "m out of l"-series systems, assuming dependence according to the ELS rule among components in their subsystems. For those systems, it is assumed that after departure of some components of a subsystem from the reliability state subset, the lifetimes of remaining components in this subsystem decrease equally depending on the number of these components that have left the reliability state subset and on the component stress proportionality correction coefficient.

3.2.1 Approach description

In a system with components dependent according to the ELS rule, we assume that after decreasing the reliability state by one of a parallel or an "m out of n" system components the increased load can be shared equally among remaining components (Crowder et al., 1991; Daniels, 1945; Pradhan et al., 2010; Smith, 1982, 1983). Subsequently, inside interactions among remaining components may cause the decrease of these components lifetimes in the reliability state subset equally. More exactly, we assume that if anyone of system components gets out of the reliability state subset $\{u, u+1, \ldots, z\}$, $u=1, 2, \ldots, z$, then the reliability of remaining ones is getting worse so that the system component lifetimes and their mean values in reliability state subset $\{u, u+1, \ldots, z\}$, $u=1, 2, \ldots, z$, decrease according to the formulas

$$T_i'(u) = c(u)\frac{n-1}{n} \cdot T_i(u),$$
$$E\left[T_i'(u)\right] = c(u)\frac{n-1}{n} \cdot E[T_i(u)], \quad u = 1, \ldots, z, \tag{3.2.1}$$

where $T_i'(u)$, $i=1, \ldots, n$, denotes the lifetime in the reliability state subset $\{u, u+1, \ldots, z\}$, $u=1, 2, \ldots, z$, of component E_i, $i=1, 2, \ldots, n$, after the departure of one of system components from this subset and $c(u)$ is the component stress proportionality correction coefficient for each u, $u=1, 2, \ldots, z$ (Blokus-Roszkowska and Kołowrocki, 2015a). This model of equal load sharing (ELS) is often applied to parallel or "m out of n" systems and has been analyzed in Blokus (2006), Blokus-Roszkowska (2007a), and Blokus-Roszkowska and Kołowrocki (2014). A brief overview of load-sharing and dependency models in the system reliability modeling is presented in the book by K.B. Misra "Reliability Analysis and Prediction" (1992). However, the two short sections "Shared Load Models" and "Dependency Models" in that book apply to two-state modeling of systems.

Here, we assume that component lifetimes decrease uniformly depending on the number of components that have left the reliability state subset. Additionally, these changes are influenced by a component stress proportionality correction coefficient, concerned with particular components' features. The value of this coefficient can be estimated on the basis of behavior and dynamics of changes in component reliability state or it can be assumed a priori. However, in both cases, it should be verified by the actual reliability data analysis and experts' judgment.

Generalizing, if ω components, $\omega=0, 1, 2, \ldots, n-1$, of a system are out of the reliability state subset $\{u, u+1, \ldots, z\}$, $u=1, 2, \ldots, z$, then the system remaining component lifetimes $T_i'(u)$ and their mean values in the reliability state subset $\{u, u+1, \ldots, z\}$, $u=1, 2, \ldots, z$, are given by

$$T_i'(u) = c(u)\frac{n-\omega}{n} \cdot T_i(u),$$

$$E\left[T_i'(u)\right] = c(u)\frac{n-\omega}{n} \cdot E[T_i(u)], \quad u = 1, 2, \ldots, z, \quad \omega = 0, 1, \ldots, n-1. \tag{3.2.2}$$

Hence, for case when considered system is homogeneous with components having piecewise exponential reliability functions with the coordinates (3.1.55), according to the relationship between the lifetime mean value in this reliability state subset and intensity of departure from this reliability state subset of the form

$$E[T_i(u)] = \frac{1}{\lambda(u)}, \quad i = 1, 2, \ldots, n, \quad u = 1, 2, \ldots, z,$$

we get following formula for intensities of departure from this reliability state subset of remaining components

$$\lambda^{(\omega)}(u) = \frac{n}{n - \omega} \frac{\lambda(u)}{c(u)}, \quad u = 1, 2, \ldots, z, \quad \omega = 0, 1, \ldots, n - 1. \qquad (3.2.3)$$

3.2.2 Reliability of multistate parallel system with dependent components

With the ELS approach to inside dependencies of parallel systems with homogeneous components, we find analytical solutions of their reliability characteristics.

Proposition 3.2.1. *If, in a homogeneous and aging parallel system following the ELS dependency rule, the components have piecewise exponential reliability functions with the coordinates* (3.1.55), *then the system's reliability function is*

$$\boldsymbol{R}_{ELS}(t, \cdot) = [1, \boldsymbol{R}_{ELS}(t, 1), \ldots, \boldsymbol{R}_{ELS}(t, z)], \quad t \geq 0, \qquad (3.2.4)$$

with the coordinates

$$\boldsymbol{R}_{ELS}(t, u) = \sum_{\omega=0}^{n-1} \frac{\left[\frac{n\lambda(u)}{c(u)} t\right]^{\omega}}{\omega!} \exp\left[-\frac{n\lambda(u)}{c(u)} t\right], \quad u = 1, 2, \ldots, z. \qquad (3.2.5)$$

Corollary 3.2.1. *If, in a homogeneous and aging parallel system following the ELS dependency rule, the components have piecewise exponential reliability functions with the coordinates (3.1.55), then the system's lifetime in the reliability state subset* $\{u, u+1, \ldots, z\}$, $u = 1, 2, \ldots, z$, *exhibits an Erlang distribution with the shape parameter n and the intensity parameter* $n\lambda(u)/c(u)$ *and the system's distribution function is given by the vector*

$$\boldsymbol{F}_{ELS}(t, \cdot) = [0, \boldsymbol{F}_{ELS}(t, 1), \ldots, \boldsymbol{F}_{ELS}(t, z)], \quad t \geq 0, \qquad (3.2.6)$$

with the coordinates

$$\boldsymbol{F}_{ELS}(t, u) = 1 - \sum_{\omega=0}^{n-1} \frac{\left[\frac{n\lambda(u)}{c(u)} t\right]^{\omega}}{\omega!} \exp\left[-\frac{n\lambda(u)}{c(u)} t\right], \quad u = 1, 2, \ldots, z. \qquad (3.2.7)$$

Using the result of Corollary 3.2.1, we determine the reliability characteristics, the mean values and standard deviations of the system lifetimes in the reliability state subsets, for a parallel system following the ELS dependency rule.

Corollary 3.2.2. *If, in a homogeneous and aging parallel system following the ELS dependency rule, the components have piecewise exponential reliability functions with the coordinates* (3.1.55), *then the system's mean lifetime in the reliability state subset* $\{u, u+1, \ldots, z\}$, $u=1, 2, \ldots, z$, *is given by*

$$\mu_{ELS}(u) = \frac{c(u)}{\lambda(u)}, \quad u = 1, 2, \ldots, z, \tag{3.2.8}$$

and the standard deviation of the system lifetime is

$$\sigma_{ELS}(u) = \frac{c(u)}{\sqrt{n}\lambda(u)}, \quad u = 1, 2, \ldots, z. \tag{3.2.9}$$

3.2.3 Reliability of multistate "*m* out of *n*" system with dependent components

The definition of a multistate "*m* out of *n*" system means that it is in the reliability state subset $\{u, u+1, \ldots, z\}$, $u=1, 2, \ldots, z$, if and only if at least m out of its n components are in this reliability state subset. Thus, if ω components, $\omega=0, 1, 2, \ldots, n-m$, of a system with the ELS rule leave the state subset $\{u, u+1, \ldots, z\}$, $u=1, 2, \ldots, z$, then remaining components' lifetimes $T_i'(u)$, $i=1, 2, \ldots, n$, and their mean values in the reliability state subset $\{u, u+1, \ldots, z\}$, $u=1, 2, \ldots, z$, decrease according to the following rule (Blokus-Roszkowska, 2007a; Blokus-Roszkowska and Kołowrocki, 2015a)

$$\begin{aligned} T_i'(u) &= c(u)\frac{n-\omega}{n} \cdot T_i(u), \\ E[T_i'(u)] &= c(u)\frac{n-\omega}{n} \cdot E[T_i(u)], \quad u = 1, 2, \ldots, z, \quad \omega = 0, 1, \ldots, n-m. \end{aligned} \tag{3.2.10}$$

In case components have identical piecewise exponential reliability functions with the coordinates (3.1.55), the intensities of departure from the reliability state subset $\{u, u+1, \ldots, z\}$, $u=1, 2, \ldots, z$, of remaining components are given by

$$\lambda^{(\omega)}(u) = \frac{n}{n-\omega}\frac{\lambda(u)}{c(u)}, \quad u = 1, 2, \ldots, z, \quad \omega = 0, 1, \ldots, n-m. \tag{3.2.11}$$

A similar result to the result presented in the form of Proposition 3.2.2, concerned with reliability of an "*m* out of *n*" system with dependent components, is formulated and proved in Blokus-Roszkowska (2007a). Birolini (1994) proposed a similar approach to reliability analysis of parallel redundancy structures with constant failure rates and load sharing. Considering a k out of n standby redundancy, Birolini (1994) obtained the reliability function similarly as in Blokus-Roszkowska (2007a) and Blokus-Roszkowska and Kołowrocki (2015a).

Proposition 3.2.2. *If, in a homogeneous and aging "m out of n" system following the ELS dependency rule, the components have piecewise exponential reliability functions with the coordinates* (3.1.55), *then the system's reliability function is*

$$\boldsymbol{R}_{ELS}(t,\ \cdot) = [1, \boldsymbol{R}_{ELS}(t,1), \ldots, \boldsymbol{R}_{ELS}(t,z)], \quad t \geq 0, \tag{3.2.12}$$

with the coordinates

$$\boldsymbol{R}_{ELS}(t,u) = \sum_{\omega=0}^{n-m} \frac{\left[\frac{n\lambda(u)}{c(u)}t\right]^{\omega}}{\omega!} \exp\left[-\frac{n\lambda(u)}{c(u)}t\right], \quad u = 1,2,\ldots,z. \tag{3.2.13}$$

The proof of analytical solution for reliability function of a homogeneous and aging "m out of n" system with components following the ELS dependency rule is given in Blokus-Roszkowska and Kołowrocki (2015a). In the proof, by defining a number of components that got out of the reliability state subset up to the moment t, $t \geq 0$ as the Markov process and using stochastic process theory (Aven and Jensen, 1999; Grabski, 2014; Limnios and Oprisan, 2005; Rausand and Høyland, 2004), the probability of particular states of this process was determined, and thus the system's reliability function was determined.

Corollary 3.2.3. *If, in a homogeneous and aging "m out of n" system following the ELS dependency rule, the components have piecewise exponential reliability functions with the coordinates* (3.1.55), *then the system's lifetime in the reliability state subset* $\{u, u+1, \ldots, z\}$, $u = 1, 2, \ldots, z$, *exhibits an Erlang distribution with the shape parameter* $n - m + 1$ *and the intensity parameter* $n\lambda(u)/c(u)$ *and the system's distribution function is given by the vector*

$$\boldsymbol{F}_{ELS}(t,\ \cdot) = [0, \boldsymbol{F}_{ELS}(t,1), \ldots, \boldsymbol{F}_{ELS}(t,z)], \quad t \geq 0, \tag{3.2.14}$$

with the coordinates

$$\boldsymbol{F}_{ELS}(t,u) = 1 - \sum_{\omega=0}^{n-m} \frac{\left[\frac{n\lambda(u)}{c(u)}t\right]^{\omega}}{\omega!} \exp\left[-\frac{n\lambda(u)}{c(u)}t\right], \quad u = 1,2,\ldots,z. \tag{3.2.15}$$

Next, using the result of Corollary 3.2.3, we determine the mean values and standard deviations of lifetimes in the reliability state subsets for an "m out of *n*" system following the ELS dependency rule.**Corollary 3.2.4.** *If, in a homogeneous and aging "m out of n" system following the ELS dependency rule, the components have piecewise exponential reliability functions with the coordinates (3.1.55), then the system's mean lifetime in the reliability state subset* $\{u, u+1, \ldots, z\}$, $u=1, 2, \ldots, z$, *is given by*

$$\mu_{ELS}(u) = \frac{(n-m+1)\cdot c(u)}{n\lambda(u)}, \quad u=1,2,\ldots,z, \tag{3.2.16}$$

and the standard deviation of the system lifetime is

$$\sigma_{ELS}(u) = \frac{\sqrt{n-m+1}\cdot c(u)}{n\lambda(u)}, \quad u=1,2,\ldots,z. \tag{3.2.17}$$

3.2.4 Reliability of multistate parallel-series system with dependent components of its subsystems

In this section, in reliability analysis of a multistate aging parallel-series system composed of parallel subsystems, it is assumed that components of these subsystems are dependent. It means that degradation of a component in a parallel subsystem can have influence on other components degradation in this subsystem and cause their lifetimes shortening. Similarly, Mohammad et al. (2012) pointed out that upon a failure of a component in a subsystem, the load on the surviving components increases, causing their failure rates to increase, too. The authors assume that load on the subsystem can vary with the phases, to which a mission of the system is divided. The system mission is succeed if all system phases are completed, and the system is failed in a phase if any one of its subsystems fails in that phase. The load for a given subsystem during a specific phase is equally shared among all its surviving components. A recursive algorithm for the reliability evaluation of phased-mission systems, assuming equal load sharing for their components, is proposed in Mohammad et al. (2012) by using a modularization technique. An expression for conditional reliability of subsystems and a recursive formula for calculating the subsystem reliability across the phases are given.

Maatouk et al. (2011) also present reliability evaluation of multistate parallel-series or series-parallel systems with dependent components. Authors consider the propagation failure effect and load sharing among parallel components. It is assumed in Maatouk et al. (2011) that components of a system are sharing the same load uniformly, and their failure rates can affect subsystems reliability. Moreover, authors take into account load sharing

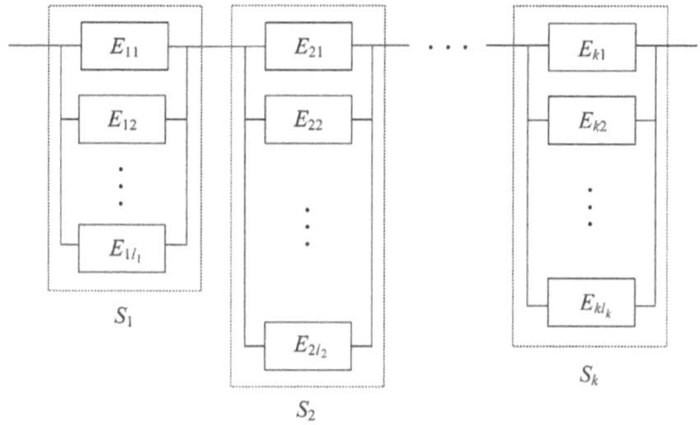

Fig. 3.6 The scheme of a parallel-series system reliability structure.

among parallel-connected subsystems. The approach presented by Maatouk et al. (2011) is based on the hybridization of Markov process and the universal generating function (UGF), introduced by Ushakov (1986, 1994) and described by Levitin (2004, 2005).

We consider here a multistate parallel-series system with a scheme presented in Fig. 3.6. We assume that k is a number of parallel subsystems working independently linked in series. In the ith subsystem S_i, $i=1, 2, \ldots, k$, there are l_i components dependent according to the ELS rule. We denote by E_{ij}, $i=1, 2, \ldots, k$, $j=1, 2, \ldots, l_i$, k, $l_i \in N$, components of a system and assume that all components E_{ij} have the reliability state set $\{0, 1, \ldots, z\}$. Subsequently, $T_{ij}(u)$, $i=1, 2, \ldots, k$, $j=1, 2, \ldots, l_i$, are random variables representing lifetimes of components E_{ij} in the reliability state subset $\{u, u+1, \ldots, z\}$, $u=1, 2, \ldots, z$, while they were in the reliability state z at the moment $t=0$.

In the ELS dependency model for a parallel-series system, we assume that after leaving the reliability state subset by some of components in a subsystem, the lifetimes of remaining components in this subsystem decrease equally depending on the number of these components that have left the reliability state subset. Additionally, these changes are influenced by the component stress proportionality correction coefficient $c_i(u)$, $i=1, 2, \ldots, k$, $u=1, 2, \ldots, z$, concerned with features of ith subsystem and its components. We assume, similarly as in formula (3.2.2) for a multistate parallel system, that if ω_i components, $\omega_i=0, 1, 2, \ldots, l_i-1$, in ith subsystem, $i=1, 2, \ldots, k$, are out of the reliability state subset $\{u, u+1, \ldots, z\}$, $u=1, 2, \ldots, z$, then the mean values of lifetimes $T'_{ij}(u)$, $i=1, 2, \ldots, k$, $j=1, 2, \ldots, l_i$, in the reliability state subset $\{u, u+1, \ldots, z\}$ of this subsystem remaining components are given by (Blokus-Roszkowska, 2007a,b; Blokus-Roszkowska and Kołowrocki, 2014)

$$E\left[T'_{ij}(u)\right] = c_i(u)\frac{l_i-\omega_i}{l_i}E\left[T_{ij}(u)\right], \quad \omega_i = 0, 1, 2, \ldots, l_i-1,$$
$$u = 1, 2, \ldots, z, \tag{3.2.18}$$

where $c_i(u)$ for $i=1, 2, \ldots, k$ and $u=1, 2, \ldots, z$, are component stress proportionality correction coefficients.

We assume that in ith parallel subsystem S_i, $i=1, 2, \ldots, k$, components are dependent according to the ELS rule and have identical piecewise exponential reliability functions

$$R_{ij}(t, \cdot) = [1, R_{ij}(t, 1), \ldots, R_{ij}(t, z)], \quad t \geq 0, \quad i=1,2,\ldots,k, \quad j=1,2,\ldots,l_i, \tag{3.2.19}$$

with the coordinates

$$R_{ij}(t, u) = \exp[-\lambda_i(u)t], \quad t \geq 0, \quad u=1,2,\ldots,z, \tag{3.2.20}$$

where $\lambda_i(u)$, $i=1, 2, \ldots, k$, $\lambda_i(u) \geq 0$, are intensities of departure from the reliability state subset $\{u, u+1, \ldots, z\}$, $u=1, 2, \ldots, z$, of components in subsystem S_i.

In that case, after the departure of ω_i components, $\omega_i=0, 1, 2, \ldots, l_i-1, i=1, 2, \ldots, k$, from this subset in subsystem S_i we get following formula for the intensities of departure from this subset of remaining components in subsystem S_i

$$\lambda_i^{(\omega_i)}(u) = \frac{l_i}{l_i - \omega_i} \frac{\lambda_i(u)}{c_i(u)}, \quad \omega_i = 0,1,2,\ldots,l_i-1, \quad i=1,2,\ldots,k, \quad u=1,2,\ldots,z. \tag{3.2.21}$$

Considering results, for a parallel system with components dependent according to the ELS rule, given in Proposition 3.2.1 and by combining these results with the reliability function of a series system with independent components, we obtain the formula for the reliability function of a parallel-series system.

Proposition 3.2.3. *If, in an aging parallel-series system, its subsystems are working independently and the components of these parallel subsystems follow the ELS dependency rule and have piecewise exponential reliability functions* (3.2.19), (3.2.20), *then the system's reliability function is given by the vector*

$$\boldsymbol{R}_{ELS}(t, \cdot) = [1, \boldsymbol{R}_{ELS}(t, 1), \ldots, \boldsymbol{R}_{ELS}(t, z)], \quad t \geq 0, \tag{3.2.22}$$

with the coordinates

$$\boldsymbol{R}_{ELS}(t, u) = \prod_{i=1}^{k} \left[\sum_{\omega_i=0}^{l_i-1} \frac{\left[\frac{l_i\lambda_i(u)}{c_i(u)} t\right]^{\omega_i}}{\omega_i!} \exp\left[-\frac{l_i\lambda_i(u)}{c_i(u)} t\right] \right], \quad u=1,2,\ldots,z. \tag{3.2.23}$$

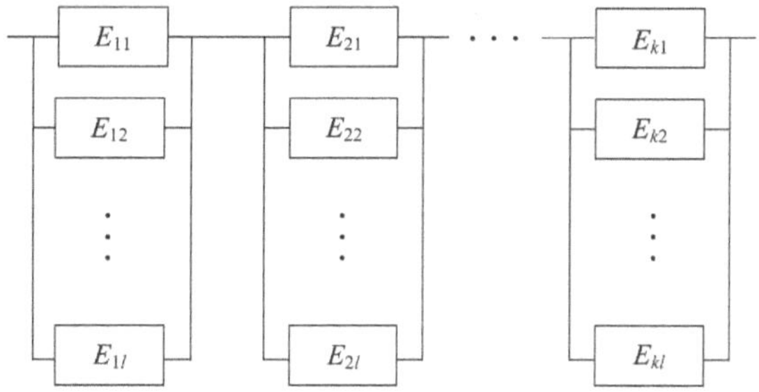

Fig. 3.7 The scheme of a regular parallel-series system.

where k is the number of parallel subsystems and l_i, $i=1, 2, \ldots, k$, *is the number of components in ith subsystem.*

In a special case, we analyze a multistate regular parallel-series system with a scheme presented in Fig. 3.7. We assume that k is a number of parallel subsystems working independently linked in series and l is a number of dependent components of these parallel subsystems.

In such case, under assumption that the system is homogeneous with components having piecewise exponential reliability functions with the coordinates (3.1.55), after the departure from the reliability state subset $\{u, u+1, \ldots, z\}$, $u=1, 2, \ldots, z$, of ω components, $\omega=0, 1, 2, \ldots, l-1$, in a parallel subsystem, the intensities of departure from this subset of the subsystem's remaining components are given by

$$\lambda^{(\omega)}(u) = \frac{l}{l-\omega}\frac{\lambda(u)}{c(u)}, \quad u=1,2,\ldots,z, \quad \omega=0,1,2,\ldots,l-1, \tag{3.2.24}$$

where $c(u)$, $u=1, 2, \ldots, z$, are component stress proportionality correction coefficients.

Next, we get the result being a particular case of Proposition 3.2.3.

Proposition 3.2.4. *If, in a homogeneous and aging regular parallel-series system, its subsystems are working independently and the components of these parallel subsystems follow the ELS dependency rule and have piecewise exponential reliability functions with the coordinates* (3.1.55), *then the system's reliability function is given by the vector*

$$\boldsymbol{R}_{ELS}(t, \cdot) = [1, \boldsymbol{R}_{ELS}(t, 1), \ldots, \boldsymbol{R}_{ELS}(t, z)], \quad t \ge 0, \tag{3.2.25}$$

with the coordinates

$$\boldsymbol{R}_{ELS}(t, u) = \left[\sum_{\omega=0}^{l-1} \frac{\left[\frac{l\lambda(u)}{c(u)}t\right]^{\omega}}{\omega!} \exp\left[-\frac{l\lambda(u)}{c(u)}t\right]\right]^{k}, \quad u=1,2,\ldots,z. \tag{3.2.26}$$

3.2.5 Reliability of multistate "*m* out of *l*"-series system with dependent components of its subsystems

In this section, we consider a multistate "m_i out of l_i"-series system composed of k linked in series "m_i out of l_i," $i=1, 2, \ldots, k$, subsystems. The scheme of such system is presented in Fig. 3.8. We assume that components in each "m_i out of l_i," $i=1, 2, \ldots, k$, subsystem are dependent according to the ELS rule and subsystems are working independently.

In the ith subsystem S_i, $i=1, 2, \ldots, k$, there are l_i components dependent according to the ELS rule, described in Section 3.2.1. We assume, similarly as for a parallel-series system, that if ω_i components, $\omega_i=0, 1, 2, \ldots, l_i-m_i$, in ith subsystem S_i, $i=1, 2, \ldots, k$, are out of the reliability state subset $\{u, u+1, \ldots, z\}$, $u=1, 2, \ldots, z$, the mean values of lifetimes $T'_{ij}(u)$, $i=1, 2, \ldots, k$, $j=1, 2, \ldots, l_i$, in the reliability state subset $\{u, u+1, \ldots, z\}$ of this subsystem remaining components are given by (Blokus-Roszkowska and Kołowrocki, 2014, 2015a)

$$E\left[T'_{ij}(u)\right] = c_i(u)\frac{l_i-\omega_i}{l_i}E\left[T_{ij}(u)\right], \quad \omega_i = 0,1,2,\ldots,l_i-m_i,$$
$$u = 1,2,\ldots,z, \tag{3.2.27}$$

where $c_i(u)$ for $i=1, 2, \ldots, k$ and $u=1, 2, \ldots, z$, are component stress proportionality correction coefficients.

We assume that in ith subsystem S_i, $i=1, 2, \ldots, k$, components are dependent according to ELS rule and have identical piecewise exponential

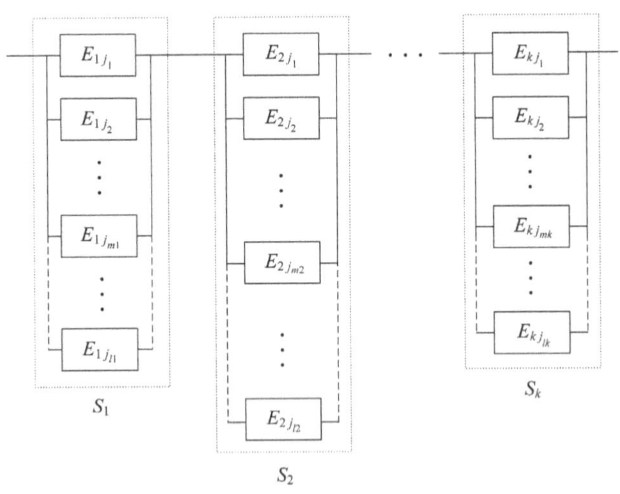

Fig. 3.8 The scheme of an "m_i out of l_i"-series system reliability structure.

reliability functions (3.2.19), (3.2.20). In that case, the intensities of departure from the reliability state subset $\{u, u+1, \ldots, z\}$, $u=1, 2, \ldots, z$, of remaining components in the subsystem S_i, $i=1, 2, \ldots, k$, are given by

$$\lambda_i^{(\omega)}(u) = \frac{l_i}{l_i - \omega_i} \frac{\lambda_i(u)}{c_i(u)}, \quad i=1,2,\ldots,k, \quad \omega_i = 0,1,2,\ldots,l_i - m_i,$$
$$u=1,2,\ldots,z. \tag{3.2.28}$$

Proposition 3.2.3 slight extension yields the following result.

Proposition 3.2.5. *If, in an aging "m_i out of l_i"-series system, its subsystems are working independently and the components of these "m_i out of l_i" subsystems follow the ELS dependency rule and have piecewise exponential reliability functions (3.2.19), (3.2.20), then the system's reliability function is given by the vector*

$$\boldsymbol{R}_{ELS}(t, \cdot) = [1, \boldsymbol{R}_{ELS}(t, 1), \ldots, \boldsymbol{R}_{ELS}(t, z)], \quad t \geq 0, \tag{3.2.29}$$

with the coordinates

$$\boldsymbol{R}_{ELS}(t, u) = \prod_{i=1}^{k} \left[\sum_{\omega_i=0}^{l_i - m_i} \frac{\left[\frac{l_i \lambda_i(u)}{c_i(u)} t \right]^{\omega_i}}{\omega_i!} \exp\left[-\frac{l_i \lambda_i(u)}{c_i(u)} t \right] \right],$$
$$u=1,2,\ldots,z, \tag{3.2.30}$$

where k is the number of "m_i out of l_i" subsystems and l_i, $i=1, 2, \ldots, k$, is the number of components in ith subsystem.

As a special case, we consider a multistate regular "m out of l"-series system with a scheme presented in Fig. 3.9. We assume that k is a number

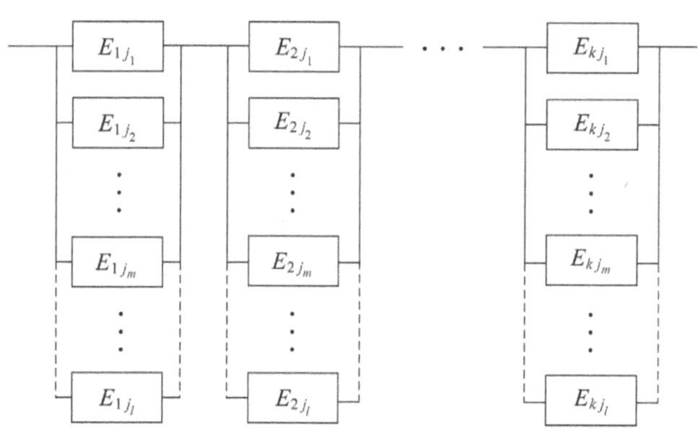

Fig. 3.9 The scheme of a regular "*m* out of *l*"-series system.

of "m out of l" subsystems working independently linked in series and l is a number of dependent components of these "m out of l" subsystems.

Further, we assume that a system is homogeneous with components having piecewise exponential reliability functions with the coordinates (3.1.55), and the intensities of departure from the reliability state subset $\{u, u+1, \ldots, z\}$, $u=1, 2, \ldots, z$, of the subsystem's remaining components are given by

$$\lambda^{(\omega)}(u) = \frac{l}{l-\omega}\frac{\lambda(u)}{c(u)}, \quad u=1,2,\ldots,z, \quad \omega=0,1,2,\ldots,l-m. \qquad (3.2.31)$$

Proposition 3.2.4 slight extension yields the following result.

Proposition 3.2.6. *If, in a homogeneous and aging regular "m out of l"-series system, its subsystems are working independently and the components of these "m out of l" subsystems follow the ELS dependency rule and have piecewise exponential reliability functions with the coordinates* (3.1.55), *then the system's reliability function is given by the vector*

$$\boldsymbol{R}_{ELS}(t, \cdot) = [1, \boldsymbol{R}_{ELS}(t, 1), \ldots, \boldsymbol{R}_{ELS}(t, z)], \quad t \geq 0, \qquad (3.2.32)$$

with the coordinates

$$\boldsymbol{R}_{ELS}(t, u) = \left[\sum_{\omega=0}^{l-m} \frac{\left[\frac{l\lambda(u)}{c(u)} t \right]^{\omega}}{\omega!} \exp\left[-\frac{l\lambda(u)}{c(u)} t \right] \right]^{k}, \quad u=1,2,\ldots,z. \qquad (3.2.33)$$

Due to the form of the reliability functions of multistate parallel-series and "m out of l"-series systems, their mean lifetimes in the reliability state subsets are estimated from the formula (2.1.16), and the standard deviation of their lifetimes by (2.1.17), (2.1.18), using approximate methods of calculating integrals.

3.3 Mixed load-sharing model of components and subsystems dependency

Linking the LLS and ELS dependency models, the mixed load-sharing (MLS) model for multistate aging parallel-series and "m out of l"-series systems is proposed. For the MLS dependency model, the reliability functions of aging parallel-series and "m out of l"-series systems are determined, assuming that their components follow piecewise exponential reliability functions. Moreover, the results are transferred to a special case of a homogeneous and regular parallel-series system and a homogeneous and regular "m out of l"-series system.

3.3.1 Approach description

In reliability analysis of various system structures, we link the results for previously described models of dependency among their components and subsystems. For instance, obtained results for a parallel-series and "m out of l"-series system with independent subsystems and their components dependent according to the equal load sharing rule, are presented in Sections 3.2.4 and 3.2.5, respectively. Reliability analysis of such systems has also been presented in the references (Blokus-Roszkowska and Kołowrocki, 2014, 2015a). The results for a series-parallel and series-"m out of k" system composed of independent subsystems with their components dependent according to the local load-sharing rule are given in Sections 3.1.3 and 3.1.4, and in Blokus-Roszkowska (2016).

In more complex models of dependency, apart from the dependency of subsystems' departures from the reliability states subsets we take into account the dependencies among components in subsystems. This way we proceed with parallel-series and "m out of l"-series systems assuming dependence among their parallel, respectively "m out of l," subsystems according to the local load-sharing rule and dependence among their components in subsystems according to the equal load-sharing rule (Blokus and Kołowrocki, 2019). Further, such dependency model will be called a mixed load-sharing (MLS) model.

3.3.2 Reliability of multistate parallel-series system with dependent components and subsystems

In this section, the MLS dependency model is applied to a multistate parallel-series system assuming dependence among subsystems and among components in these subsystems. We consider a multistate parallel-series system composed of k parallel subsystems S_i, $i=1, 2, \ldots, k$, connected in series, illustrated in Fig. 3.10. Further, by E_{ij}, $i=1, 2, \ldots, k$, $j=1, 2, \ldots, l_i$, we denote the jth component being in the ith subsystem S_i, and we assume that components in the ith subsystem have identical piecewise exponential reliability functions, given by (3.2.19), (3.2.20).

In the ith parallel subsystem S_i, $i=1, 2, \ldots, k$, we consider dependency of its l_i components according to the ELS model, presented in Section 3.2. In that case, after departure from the reliability state subset $\{u, u+1, \ldots, z\}$,

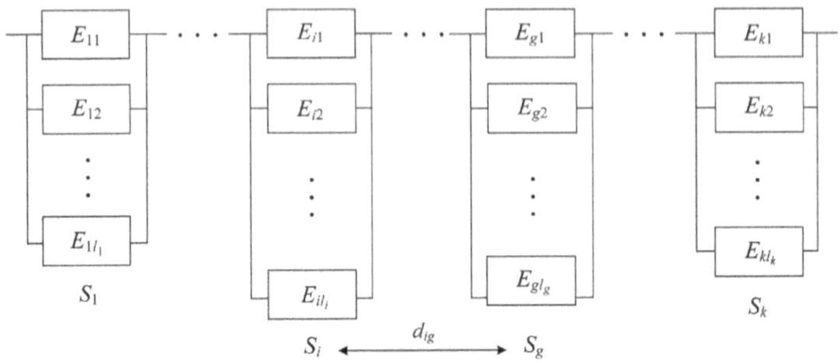

Fig. 3.10 The scheme of a parallel-series system.

$u=1, 2, \ldots, z$, by ω_i components, $\omega_i=0, 1, 2, \ldots, l_i-1$, of the subsystem, the intensities of departure from this reliability state subset of the subsystem's remaining components are given by (3.2.21).

Further, we assume the LLS dependency model, presented in Section 3.1, among these subsystems linked in series. Subsequently, we assume that after leaving the reliability state subset $\{u, u+1, \ldots, z\}$, $u=1, 2, \ldots, z$, by the subsystem S_g, $g=1, 2, \ldots, k$, the reliability parameters of components in other subsystems S_i, $i=1, 2, \ldots, k$, $i \neq g$, vary depending on their exposure to the subsystem S_g reliability state change. However, within a single subsystem, the changes of the reliability parameters for all of its components are on the same level according to the ELS rule.

We denote by $T_{i,j}(v)$ and $T_{i/g,j}(v)$, $i=1, 2, \ldots, k$, $g=1, 2, \ldots, k$, $j=1, 2, \ldots, l$, the lifetimes in the reliability state subset $\{v, v+1, \ldots, z\}$, $v=u, u-1, \ldots, 1$, $u=1, 2, \ldots, z-1$, of ith subsystem's components E_{ij}, $j=1, 2, \ldots, l$, respectively, before and after departure of the subsystem S_g, $g=1, \ldots, k$, from the subset $\{u, u+1, \ldots, z\}$, $u=1, 2, \ldots, z$. The impact of subsystem S_g reliability state change on lifetimes in the reliability state subsets of components in other subsystems S_i, $i=1, 2, \ldots, k$, $i \neq g$, is expressed by the coefficients $q(v, S_g, S_i)$ for $v=u, u-1, \ldots, 1$ and $u=1, 2, \ldots, z-1$. It is assumed that $0 < q(v, S_g, S_i) \leq 1$, $i=1, 2, \ldots, k$, $g=1, 2, \ldots, k$, $i \neq g$, and $q(v, S_g, S_g)=1$, $g=1, 2, \ldots, k$.

Under these notations and assumptions, it follows from (3.1.1), in the LLS dependency model used among subsystems, that the mean values of components lifetimes in the reliability state subset $\{v, v+1, \ldots, z\}$, $v=u$, $u-1, \ldots, 1$, $u=1, 2, \ldots, z-1$, decrease according to the formula (Blokus and Kołowrocki, 2019)

$$E\left[T_{i/g,j}(v)\right] = q\left(v, S_g, S_i\right) \cdot E\left[T_{i,j}(v)\right], \quad i=1,2,\ldots,k,$$
$$g=1,2,\ldots,k, \quad j=1,2,\ldots,l_i. \tag{3.3.1}$$

In particular case, the coefficients $q(v, S_g, S_i)$ are functions of a distance d_{ig} between subsystems S_g and S_i, $i, g=1, 2, \ldots, k$, expressed by $q(v, d_{ig})$, and $q(v, 0)=1$. Interpretation of the distance index in the MLS model is illustrated in Fig. 3.10. Subsequently, formula (3.3.1) takes the following form

$$E\left[T_{i/g,j}(v)\right] = q\left(v, d_{ig}\right) \cdot E\left[T_{i,j}(v)\right], \quad i=1,2,\ldots,k,$$
$$g=1,2,\ldots,k, \quad j=1,2,\ldots,l. \tag{3.3.2}$$

Considering Corollary 3.2.1 concerned with Erlang distribution of system lifetime in the reliability state subset $\{u, u+1, \ldots, z\}$, $u=1, 2, \ldots, z$, in case components of parallel system are dependent according to the ELS rule, and by combining this result with the reliability function of a series system with components dependent according to the LLS rule and having piecewise Erlang reliability functions, determined in Section 3.1.2, we obtain the reliability function of a homogeneous multistate regular parallel-series system with MLS dependency model. Applying Proposition 3.1.7 for a series system composed of k subsystems S_i, $i=1, 2, \ldots, k$, and using fact that ith subsystem has Erlang reliability functions with the shape parameter l_i and with the intensity parameter $l_i\lambda_i(u)/c_i(u)$, $u=1, 2, \ldots, z$, we immediately get Proposition 3.3.7.

Proposition 3.3.1. *If, in an aging parallel-series system with the MLS dependency rule, its subsystems* S_i, $i=1, 2, \ldots, k$, *are dependent according to the LLS rule and components of these parallel subsystems follow the ELS dependency rule and have piecewise exponential reliability functions* (3.2.19), (3.2.20), *then the system's reliability function is given by the vector*

$$\boldsymbol{R}_{MLS}(t, \cdot) = \left[1, \boldsymbol{R}_{MLS}(t, 1), \ldots, \boldsymbol{R}_{MLS}(t, z)\right], t \geq 0, \tag{3.3.3}$$

with the coordinates

$$\boldsymbol{R}_{MLS}(t,u)=\prod_{i=1}^{k}\left[\sum_{\omega=0}^{l_i-1}\frac{\left[\frac{l_i\lambda_i(u+1)}{c_i(u+1)}t\right]^{\omega}}{\omega!}\right]\cdot\exp\left[-\frac{l_i\lambda_i(u+1)}{c_i(u+1)}t\right]$$

$$+\sum_{g=1}^{k}\int_0^t\widetilde{f}_g(a,u+1)\cdot\prod_{\substack{i=1\\i\neq g}}^{k}\left[\sum_{\omega=0}^{l_i-1}\frac{\left[\frac{l_i\lambda_i(u+1)}{c_i(u+1)}a\right]^{\omega}}{\omega!}\right]\cdot\sum_{\omega=0}^{l_g-1}\frac{\left[\frac{l_g\lambda_g(u)}{c_g(u)}a\right]^{\omega}}{\omega!}$$

$$\cdot\exp\left[-\left(\sum_{\substack{i=1\\i\neq g}}^{k}\frac{l_i\lambda_i(u+1)}{c_i(u+1)}+\frac{l_g\lambda_g(u)}{c_g(u)}\right)a\right]$$

$$\cdot\left[\prod_{i=1}^{k}\frac{\sum_{\omega=0}^{l_i-1}\frac{\left[\frac{l_i\lambda_i(u)}{c_i(u)q(u,S_g,S_i)}t\right]^{\omega}}{\omega!}}{\sum_{\omega=0}^{l_i-1}\frac{\left[\frac{l_i\lambda_i(u)}{c_i(u)}a\right]^{\omega}}{\omega!}}\cdot\exp\left[-\frac{l_i\lambda_i(u)}{c_i(u)q(u,S_g,S_i)}t+\frac{l_i\lambda_i(u)}{c_i(u)}a\right]\right]da,$$

$$u=1,2,\ldots,z-1, \tag{3.3.4}$$

where $\widetilde{f}_g(a, u+1)$ *is given by*

$$\widetilde{f}_g(a,u+1)=\left[\frac{\displaystyle\sum_{\omega=0}^{l_g-1}\frac{\left[\frac{l_g\lambda_g(u+1)}{c_g(u+1)}a\right]^{\omega}}{\omega!}\cdot\sum_{\omega=0}^{l_g-2}\frac{\left[\frac{l_g\lambda_g(u)}{c_g(u)}\right]^{\omega+1}a^{\omega}}{\omega!}-\sum_{\omega=0}^{l_g-2}\frac{\left[\frac{l_g\lambda_g(u+1)}{c_g(u+1)}\right]^{\omega+1}a^{\omega}}{\omega!}\cdot\sum_{\omega=0}^{l_g-1}\frac{\left[\frac{l_g\lambda_g(u)}{c_g(u)}a\right]^{\omega}}{\omega!}}{\left[\displaystyle\sum_{\omega=0}^{l_g-1}\frac{\left[\frac{l_g\lambda_g(u)}{c_g(u)}a\right]^{\omega}}{\omega!}\right]^{2}}\right.$$

$$\left.+\frac{\displaystyle\sum_{\omega=0}^{l_g-1}\frac{\left[\frac{l_g\lambda_g(u+1)}{c_g(u+1)}a\right]^{\omega}}{\omega!}\left(\frac{l_g\lambda_g(u+1)}{c_g(u+1)}-\frac{l_g\lambda_g(u)}{c_g(u)}\right)}{\displaystyle\sum_{\omega=0}^{l_g-1}\frac{\left[\frac{l_g\lambda_g(u)}{c_g(u)}a\right]^{\omega}}{\omega!}}\right]\cdot\exp\left[-\left(\frac{l_g\lambda_g(u+1)}{c_g(u+1)}-\frac{l_g\lambda_g(u)}{c_g(u)}\right)a\right],\quad u=1,2,\ldots,z-1, \tag{3.3.5}$$

and

$$\boldsymbol{R}_{MLS}(t,z)=\prod_{i=1}^{k}\left[\sum_{\omega=0}^{l_i-1}\frac{\left[\frac{l_i\lambda_i(z)}{c_i(z)}t\right]^{\omega}}{\omega!}\right]^{k}\cdot\exp\left[-\sum_{i=1}^{k}\frac{l_i\lambda_i(z)}{c_i(z)}t\right]. \tag{3.3.6}$$

Some results for a multistate regular parallel-series system with dependent components and subsystems are presented in Blokus-Roszkowska and Kolowrocki (2017).

In a special case, we apply the MLS model of components and subsystems dependency to reliability analysis of a multistate regular parallel-series system. We consider a multistate regular parallel-series system composed of k parallel subsystems S_i, $i=1, 2, \ldots, k$, linked in series, as illustrated in Fig. 3.11. Further, we assume all system components have identical piecewise exponential reliability functions with the coordinates (3.1.55).

In each of parallel subsystems, we consider dependency of its l components according to the ELS model, presented in Section 3.2. In that case, after departure from the reliability state subset $\{u, u+1, \ldots, z\}$, $u=1, 2, \ldots, z$, by ω components, $\omega=0, 1, 2, \ldots, l-1$, of a subsystem, the intensities of departure from this reliability state subset of the subsystem's remaining components are given by (3.2.24). From Corollary 3.2.1, we conclude that the lifetime in the reliability state subset $\{u, u+1, \ldots, z\}$, $u=1, 2, \ldots, z$, of each parallel subsystem, assuming that its components are dependent according to the ELS rule, has Erlang distribution with the shape parameter l and the intensity parameter $l\lambda(u)/c(u)$, $u=1, 2, \ldots, z$.

Next, the LLS dependency model is assumed among subsystems linked in series. We assume that after leaving the reliability state subset $\{u, u+1, \ldots, z\}$, $u=1, 2, \ldots, z$, by the subsystem S_g, $g=1, 2, \ldots, k$, the reliability parameters of components in other subsystems S_i, $i=1, 2, \ldots, k$, $i \neq g$, vary depending on their exposure to the subsystem S_g reliability state change. And, this impact is expressed by the coefficients $q(v, S_g, S_i)$, $v=u, u-1, \ldots, 1$, $u=1, 2, \ldots, z-1$. Subsequently, the mean values of components lifetimes in the reliability state subset decrease according to the formula (3.3.1).

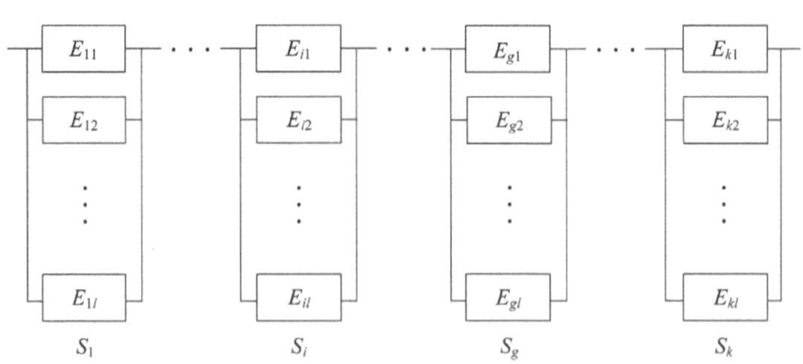

Fig. 3.11 The scheme of a regular parallel-series system.

By combining the results for a parallel system following the ELS rule with the reliability function of a series system with components dependent according to the LLS rule and having piecewise Erlang reliability functions, presented in Section 3.1.2, we obtain the reliability function of a homogeneous multistate regular parallel-series system with the MLS dependency model.

Proposition 3.3.2. *If, in a homogeneous and aging regular parallel-series system with the MLS dependency rule, the subsystems are dependent according to the LLS rule and components of these parallel subsystems follow the ELS dependency rule and have piecewise exponential reliability functions with the coordinates* (3.1.55), *then the system's reliability function is*

$$\boldsymbol{R}_{MLS}(t,\cdot)=[1,\boldsymbol{R}_{MLS}(t,1),\ldots,\boldsymbol{R}_{MLS}(t,z)],\quad t\geq 0, \tag{3.3.7}$$

where

$$\begin{aligned}
\boldsymbol{R}_{MLS}(t,u)={}&\left[\sum_{\omega=0}^{l-1}\frac{\left[\frac{l\lambda(u+1)}{c(u+1)}t\right]^{\omega}}{\omega!}\right]^{k}\cdot\exp\left[-\frac{kl\lambda(u+1)}{c(u+1)}t\right]\\
&+\sum_{g=1}^{k}\int_{0}^{t}\tilde{f}(a,u+1)\left[\sum_{\omega=0}^{l-1}\frac{\left[\frac{l\lambda(u+1)}{c(u+1)}a\right]^{\omega}}{\omega!}\right]^{k-1}\cdot\sum_{\omega=0}^{l-1}\frac{\left[\frac{l\lambda(u)}{c(u)}a\right]^{\omega}}{\omega!}\\
&\cdot\exp\left[-\left((k-1)\frac{l\lambda(u+1)}{c(u+1)}+\frac{l\lambda(u)}{c(u)}\right)a\right]\\
&\cdot\left[\prod_{i=1}^{k}\frac{\sum_{\omega=0}^{l-1}\frac{\left[\frac{l\lambda(u)}{c(u)q(u,S_g,S_i)}t\right]^{\omega}}{\omega!}}{\sum_{\omega=0}^{l-1}\frac{\left[\frac{l\lambda(u)}{c(u)}a\right]^{\omega}}{\omega!}}\cdot\exp\left[-\frac{l\lambda(u)}{c(u)q(u,S_g,S_i)}t+\frac{l\lambda(u)}{c(u)}a\right]\right]da,
\end{aligned}$$

$$u=1,2,\ldots,z-1, \tag{3.3.8}$$

where $\widetilde{f}(a, u+1)$ *is given by*

$$\widetilde{f}(a, u+1) = \left[\frac{\sum_{\omega=0}^{l-1} \frac{\left[\frac{l\lambda(u+1)}{c(u+1)}a\right]^{\omega}}{\omega!} \cdot \sum_{\omega=0}^{l-2} \frac{\left[\frac{l\lambda(u)}{c(u)}\right]^{\omega+1} a^{\omega}}{\omega!} - \sum_{\omega=0}^{l-2} \frac{\left[\frac{l\lambda(u+1)}{c(u+1)}\right]^{\omega+1} a^{\omega}}{\omega!} \cdot \sum_{\omega=0}^{l-1} \frac{\left[\frac{l\lambda(u)}{c(u)}a\right]^{\omega}}{\omega!}}{\left[\sum_{\omega=0}^{l-1} \frac{\left[\frac{l\lambda(u)}{c(u)}a\right]^{\omega}}{\omega!}\right]^{2}} + \frac{\sum_{\omega=0}^{l-1} \frac{\left[\frac{l\lambda(u+1)}{c(u+1)}a\right]^{\omega}}{\omega!} \left(\frac{l\lambda(u+1)}{c(u+1)} - \frac{l\lambda(u)}{c(u)}\right)}{\sum_{\omega=0}^{l-1} \frac{\left[\frac{l\lambda(u)}{c(u)}a\right]^{\omega}}{\omega!}} \right] \cdot \exp\left[-\left(\frac{l\lambda(u+1)}{c(u+1)} - \frac{l\lambda(u)}{c(u)}\right)a\right], \quad u = 1, 2, \ldots, z-1, \tag{3.3.9}$$

and

$$\boldsymbol{R}_{MLS}(t, z) = \left[\sum_{\omega=0}^{l-1} \frac{\left[\frac{l\lambda(z)}{c(z)}t\right]^{\omega}}{\omega!}\right]^{k} \cdot \exp\left[-k \cdot \frac{l\lambda(z)}{c(z)}t\right]. \tag{3.3.10}$$

3.3.3 Reliability of multistate "*m* out of *l*"-series system with dependent components and subsystems

Next, we apply the MLS model of components and subsystems dependency to the reliability analysis of a multistate "m_i out of l_i"-series system. We consider a multistate "m_i out of l_i"-series system composed of k "m_i out of l_i" subsystems S_i, $i=1, 2, \ldots, k$, linked in series, illustrated in Fig. 3.12. Similarly as for a parallel-series system, by E_{ij}, $i=1, 2, \ldots, k$, $j=1, 2, \ldots, l$, we denote jth component being in ith subsystem S_i, and we assume all components in ith subsystem have identical piecewise exponential reliability functions (3.2.19), (3.2.20).

In ith "m_i out of l_i" subsystem S_i we consider, similarly as in previous section, dependency of its l_i components according to the ELS model. In that case, after departure from the reliability state subset $\{u, u+1, \ldots, z\}$, $u=1, 2, \ldots, z$, by ω_i components, $\omega_i=0, 1, 2, \ldots, l_i - m_i$, of the subsystem S_i, $i=1, 2, \ldots, k$, the intensities of departure from this reliability state subset of remaining components in this subsystem are given by (3.2.28).

Moreover, the LLS dependency model is assumed among these subsystems linked in series. We assume that after the exit of the subsystem S_g, $g=1, 2, \ldots, k$, from the reliability state subset $\{u, u+1, \ldots, z\}$, $u=1, 2, \ldots, z$, the reliability parameters of the components of other subsystems S_i, $i=1, 2, \ldots, k$, $i \neq g$, vary depending on their exposure to the subsystem S_g reliability state change. Subsequently, the mean values of these components lifetimes in the reliability state subset $\{v, v+1, \ldots, z\}$, $v=u, u-1, \ldots, 1$, $u=1, 2, \ldots, z$, change according to (3.3.1), where the strength of this impact is expressed by the coefficients $q(v, S_g, S_i)$, $v=u, u-1, \ldots, 1$, $u=1, 2, \ldots, z-1$.

Considering Corollary 3.2.3, concerned with Erlang distribution of system lifetime in the reliability state subset $\{u, u+1, \ldots, z\}$, $u=1, 2, \ldots, z$, in

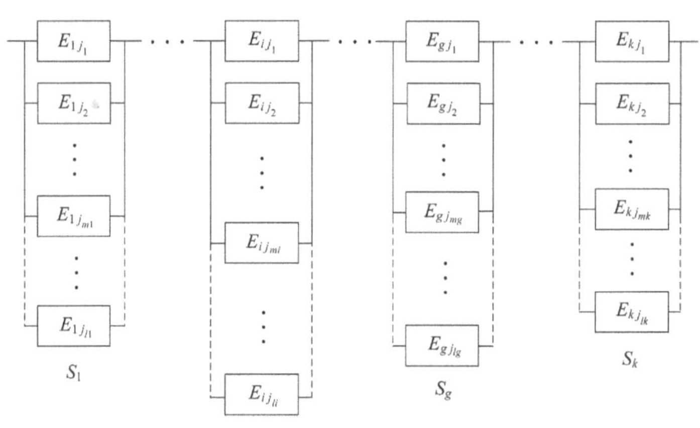

Fig. 3.12 The scheme of an "m_i out of l_i"-series system.

case components of "m out of l" system are dependent according to the ELS rule, and by combining this result with the reliability function of a series system with components dependent according to LLS rule and having piecewise Erlang reliability functions, we determine the reliability function of a multistate "m_i out of l_i"-series system with MLS dependency model. Applying Proposition 3.1.7 for series system composed of k subsystems S_i, $i=1, 2, \ldots, k$, and using fact that ith subsystem has Erlang reliability functions with the shape parameter $l_i - m_i + 1$ and with the intensity parameter $l_i\lambda_i(u)/c_i(u)$, $u=1, 2, \ldots, z$, we immediately the following result.

Proposition 3.3.3. *If, in an aging "m_i out of l_i"-series system with the MLS dependency rule, its subsystems S_i, $i=1, 2, \ldots, k$, are dependent according to the LLS rule and components of these parallel subsystems follow the ELS dependency rule and have piecewise exponential reliability functions* (3.2.19), (3.2.20), *then the system's reliability function is given by the vector*

$$\boldsymbol{R}_{MLS}(t, \cdot) = [1, \boldsymbol{R}_{MLS}(t, 1), \ldots, \boldsymbol{R}_{MLS}(t, z)], \quad t \geq 0, \tag{3.3.11}$$

with the coordinates

$$\begin{aligned}
\boldsymbol{R}_{MLS}(t, u) = & \prod_{i=1}^{k} \left[\sum_{\omega=0}^{l_i - m_i} \frac{\left[\frac{l_i\lambda_i(u+1)}{c_i(u+1)}\right] t^{\omega}}{\omega!} \right] \cdot \exp\left[-\frac{l_i\lambda_i(u+1)}{c_i(u+1)} t \right] \\
& + \sum_{g=1}^{k} \int_0^t \tilde{f}_g(a, u+1) \cdot \prod_{\substack{i=1 \\ i \neq g}}^{k} \left[\sum_{\omega=0}^{l_i - m_i} \frac{\left[\frac{l_i\lambda_i(u+1)}{c_i(u+1)} a\right]^{\omega}}{\omega!} \right] \cdot \sum_{\omega=0}^{l_g - m_g} \frac{\left[\frac{l_g\lambda_g(u)}{c_g(u)} a\right]^{\omega}}{\omega!} \\
& \cdot \exp\left[-\left(\sum_{\substack{i=1 \\ i \neq g}}^{k} \frac{l_i\lambda_i(u+1)}{c_i(u+1)} + \frac{l_g\lambda_g(u)}{c_g(u)} \right) a \right] \\
& \cdot \left[\prod_{i=1}^{k} \frac{\sum_{\omega=0}^{l_i - m_i} \frac{\left[\frac{l_i\lambda_i(u)}{c_i(u)q(u, S_g, S_i)} t\right]^{\omega}}{\omega!}}{\sum_{\omega=0}^{l_i - m_i} \frac{\left[\frac{l_i\lambda_i(u)}{c_i(u)} a\right]^{\omega}}{\omega!}} \cdot \exp\left[-\frac{l_i\lambda_i(u)}{c_i(u)q(u, S_g, S_i)} t + \frac{l_i\lambda_i(u)}{c_i(u)} a \right] \right] da, \\
& u = 1, 2, \ldots, z-1,
\end{aligned} \tag{3.3.12}$$

where $\widetilde{f}_g(a, u+1)$ *is given by*

$$\widetilde{f}_g(a,u+1) = \left[\frac{ \sum_{\omega=0}^{l_g-m_g} \frac{\left[\frac{l_g\lambda_g(u+1)}{c_g(u+1)}a\right]^{\omega}}{\omega!} \cdot \sum_{\omega=0}^{l_g-m_g-1} \frac{\left[\frac{l_g\lambda_g(u)}{c_g(u)}\right]^{\omega+1} a^{\omega}}{\omega!} - \sum_{\omega=0}^{l_g-m_g-1} \frac{\left[\frac{l_g\lambda_g(u+1)}{c_g(u+1)}\right]^{\omega+1} a^{\omega}}{\omega!} \cdot \sum_{\omega=0}^{l_g-m_g} \frac{\left[\frac{l_g\lambda_g(u)}{c_g(u)}a\right]^{\omega}}{\omega!} }{ \left[\sum_{\omega=0}^{l_g-m_g} \frac{\left[\frac{l_g\lambda_g(u)}{c_g(u)}a\right]^{\omega}}{\omega!} \right]^2 } \right.$$

$$\left. + \frac{ \sum_{\omega=0}^{l_g-m_g} \frac{\left[\frac{l_g\lambda_g(u+1)}{c_g(u+1)}a\right]^{\omega}}{\omega!} \left(\frac{l_g\lambda_g(u+1)}{c_g(u+1)} - \frac{l_g\lambda_g(u)}{c_g(u)} \right) }{ \sum_{\omega=0}^{l_g-m_g} \frac{\left[\frac{l_g\lambda_g(u)}{c_g(u)}a\right]^{\omega}}{\omega!} } \right] \cdot \exp\left[-\left(\frac{l_g\lambda_g(u+1)}{c_g(u+1)} - \frac{l_g\lambda_g(u)}{c_g(u)} \right) a \right], \tag{3.3.13}$$

$$u = 1, 2, \ldots, z-1,$$

and

$$\boldsymbol{R}_{MLS}(t,z) = \prod_{i=1}^{k} \left[\sum_{\omega=0}^{l_i-m_i} \frac{\left[\frac{l_i\lambda_i(z)}{c_i(z)}t\right]^{\omega}}{\omega!} \right]^k \cdot \exp\left[-\sum_{i=1}^{k} \frac{l_i\lambda_i(z)}{c_i(z)} t \right]. \tag{3.3.14}$$

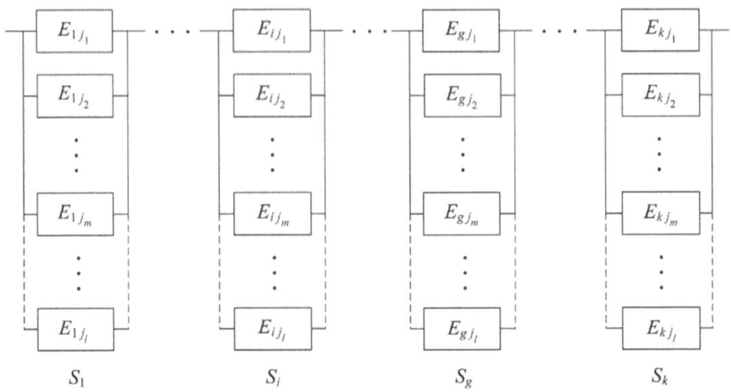

Fig. 3.13 The scheme of a regular "*m* out of *l*"-series system.

The results for a multistate regular "m out of l"-series system with dependent components and subsystems are presented in Blokus and Kołowrocki (2019) *and Blokus-Roszkowska and Kolowrocki* (2017).

Next, we analyze reliability of a multistate regular "m out of l"-series system, presented in Fig. 3.13, assuming the MLS model of components and subsystems dependency. We assume all system components have identical piecewise exponential reliability functions with the coordinates (3.1.55).

In each of the "m out of l" subsystem we consider dependency of its l components according to the ELS rule. In that case, after departure from the reliability state subset $\{u, u+1, \ldots, z\}$, $u=1, 2, \ldots, z$, by ω components, $\omega=0, 1, \ldots, l-m$, of a subsystem, the intensities of departure from this reliability state subset of the subsystem remaining components are given by (3.2.31). From Corollary 3.2.3, we conclude that the lifetime in the subset $\{u, u+1, \ldots, z\}$, $u=1, 2, \ldots, z$, of each "m out of l" subsystem, assuming that its components follow the ELS dependency rule, has Erlang distribution with the shape parameter $l-m+1$ and the intensity parameter $l\lambda(u)/c(u)$.

In turn, the LLS dependency model is assumed among these subsystems linked in series. Subsequently, the mean values of these components lifetimes in the reliability state subset $\{v, v+1, \ldots, z\}$, $v=u, u-1, \ldots, 1$, $u=1, 2, \ldots, z$, decrease according to (3.3.1).

Under these assumptions, by combining results of Corollary 3.2.3 for an aging "m out of n" system following the ELS dependency rule with results of Proposition 3.1.7 for a series system following the LLS rule, we obtain the reliability function of a homogeneous aging regular "m out of l"-series system with MLS dependency model (Blokus and Kołowrocki, 2019).

Proposition 3.3.4. *If, in a homogeneous and aging regular "m out of l"-series system with the MLS dependency rule, the subsystems are dependent according to the LLS rule and components of these "m out of l" subsystems follow the ELS dependency rule and have piecewise exponential reliability functions with the coordinates* (3.1.55), *then the system's reliability function is*

$$\boldsymbol{R}_{MLS}(t,\ \cdot\ ) = [1, \boldsymbol{R}_{MLS}(t,1), \ldots, \boldsymbol{R}_{MLS}(t,z)], \quad t \geq 0, \tag{3.3.15}$$

with the coordinates

$$\begin{aligned}
\boldsymbol{R}_{MLS}(t,u) = {} & \left[\sum_{\omega=0}^{l-m} \frac{\left[\frac{l\lambda(u+1)}{c(u+1)}t\right]^{\omega}}{\omega!}\right]^{k} \cdot \exp\left[-\frac{kl\lambda(u+1)}{c(u+1)}t\right] \\
& + \sum_{g=1}^{k} \int_0^t \tilde{f}(a,u+1) \left[\sum_{\omega=0}^{l-m} \frac{\left[\frac{l\lambda(u+1)}{c(u+1)}a\right]^{\omega}}{\omega!}\right]^{k-1} \\
& \cdot \sum_{\omega=0}^{l-m} \frac{\left[\frac{l\lambda(u)}{c(u)}a\right]^{\omega}}{\omega!} \cdot \exp\left[-\left((k-1)\frac{l\lambda(u+1)}{c(u+1)} + \frac{l\lambda(u)}{c(u)}\right)a\right] \\
& \cdot \left[\prod_{i=1}^{k} \frac{\sum_{\omega=0}^{l-m} \frac{\left[\frac{l\lambda(u)}{c(u)q(u,S_g,S_i)}t\right]^{\omega}}{\omega!}}{\sum_{\omega=0}^{l-m} \frac{\left[\frac{l\lambda(u)}{c(u)}a\right]^{\omega}}{\omega!}} \cdot \exp\left[-\frac{l\lambda(u)}{c(u)q(u,S_g,S_i)}t + \frac{l\lambda(u)}{c(u)}a\right]\right] da,
\end{aligned}$$

$u = 1, 2, \ldots, z-1,$

(3.3.16)

where $\widetilde{f}(a, u+1)$ *is given by*

$$\widetilde{f}(a, u+1) = \left[\frac{\sum_{\omega=0}^{l-m} \frac{\left[\frac{l\lambda(u+1)}{c(u+1)} a\right]^{\omega}}{\omega!} \cdot \sum_{\omega=0}^{l-m-1} \frac{\left[\frac{l\lambda(u)}{c(u)}\right]^{\omega+1} a^{\omega}}{\omega!} - \sum_{\omega=0}^{l-m-1} \frac{\left[\frac{l\lambda(u+1)}{c(u+1)}\right]^{\omega+1} a^{\omega}}{\omega!} \cdot \sum_{\omega=0}^{l-m} \frac{\left[\frac{l\lambda(u)}{c(u)} a\right]^{\omega}}{\omega!}}{\left[\sum_{\omega=0}^{l-m} \frac{\left[\frac{l\lambda(u)}{c(u)} a\right]^{\omega}}{\omega!}\right]^{2}} + \frac{\sum_{\omega=0}^{l-m} \frac{\left[\frac{l\lambda(u+1)}{c(u+1)} a\right]^{\omega}}{\omega!} \left(\frac{l\lambda(u+1)}{c(u+1)} - \frac{l\lambda(u)}{c(u)}\right)}{\sum_{\omega=0}^{l-m} \frac{\left[\frac{l\lambda(u)}{c(u)} a\right]^{\omega}}{\omega!}} \right] \cdot \exp\left[-\left(\frac{l\lambda(u+1)}{c(u+1)} - \frac{l\lambda(u)}{c(u)}\right) t\right], \tag{3.3.17}$$

$$u = 1, 2, \ldots, z-1,$$

and

$$\boldsymbol{R}_{MLS}(t, z) = \left[\sum_{\omega=0}^{l-m} \frac{\left[\frac{l\lambda(z)}{c(z)} t\right]^{\omega}}{\omega!}\right]^{k} \cdot \exp\left[-k \cdot \frac{l\lambda(z)}{c(z)} t\right]. \tag{3.3.18}$$

Due to the reliability functions of parallel-series and "*m* out of *l*"-series systems following the MLS dependency rule, given by (3.3.3)–(3.3.6) and (3.3.7)–(3.3.10), respectively, are very complicated, other reliability characteristics are estimated using approximate methods. The system's mean lifetime in the reliability state subset is calculated by the formula (2.1.16), and the standard deviation of the system lifetime by (2.1.17), (2.1.18) from Chapter 2.

3.4 Applications

Obtained theoretical results for a multistate aging series system with dependent components are illustrated by their application to the reliability evaluation of an exemplary system. The MLS model for a multistate regular "m out of l"-series system is applied to reliability evaluation of a shipyard rope elevator, in case its components have piecewise exponential reliability functions assuming dependence of departures rates from the subsets of their reliability states. Moreover, a discussion of results obtained for different models of dependency and without taking into account dependencies is presented. Finally, the maintenance of this elevator as a repairable multistate aging system with negligible renovation time is analyzed.

3.4.1 Reliability of exemplary system

We consider an exemplary system S as a homogeneous multistate aging series system composed of three components E_i, $i = 1, 2, 3$, with reliability structure presented in Fig. 3.14 (Blokus-Roszkowska and Kołowrocki, 2015c).

We assume that the system S is a 5-state system and we arbitrarily distinguish the following five reliability states of the system and its components:

- state 4—the system operation is fully effective,
- state 3—the system operation is less effective because of aging,
- state 2—the system operation is less effective because of aging and dangerous for the environment,
- state 1—the system operation is less effective because of aging and more dangerous for the environment,
- state 0—the system is destroyed.

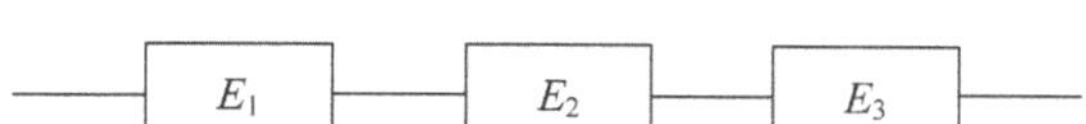

Fig. 3.14 The scheme of the exemplary system S reliability structure.

To have the assumption on aging satisfied, we assume that the transitions between the reliability states of components and the system only from better to worse ones are possible. Moreover, we assume that the critical reliability state of the system and its components is $r=2$.

The components' lifetimes in the reliability states are expressed in years and they have the identical reliability functions (Blokus-Roszkowska and Kołowrocki, 2015c)

$$R(t,\ \cdot\,)=[1, R(t,1), R(t,2), R(t,3), R(t,4)],\quad t\geq 0, \tag{3.4.1}$$

with the coordinates that, by the assumption, are piecewise exponential

$$\begin{aligned}R(t,1)&=\exp[-0.25t],\ R(t,2)=\exp[-0.5t],\ R(t,3)\\&=\exp[-0.75t],\ R(t,4)=\exp[-1t].\end{aligned} \tag{3.4.2}$$

The coefficients $q(\upsilon, E_j, E_i)$ for $\upsilon=u, u-1, \ldots, 1$ and $u=1, 2, 3$, in the LLS dependency model, are functions of component E_j, $j=1, 2, 3$, impact on other components E_i, $i=1, 2, 3$, $i\neq j$, lifetimes in the reliability state subsets, and take values regarding the system and its components properties. In presented example, we assume that these coefficients are given by the formula (Blokus-Roszkowska and Kołowrocki, 2015c)

$$q(\upsilon, E_j, E_i)=1-[q(\upsilon)]^{d_{ij}},\quad i=1,2,3,\quad j=1,2,3,\quad i\neq j, \tag{3.4.3}$$

where $q(\upsilon)=\frac{1}{2^{\upsilon}}$, $\upsilon=u, u-1, \ldots, 1$, $u=1, 2, 3$, and $d_{ij}=|i-j|$ denotes the distance between components E_i and E_j, $i, j=1, 2, 3$, $i\neq j$. By assumption $q(\upsilon, E_j, E_j)=1$, $j=1, 2, 3$.

Hence, we get for the reliability state $\upsilon=1$:

$$\begin{aligned}&q(1,E_1,E_1)=q(1,E_2,E_2)=q(1,E_3,E_3)=1,\\&q(1,E_1,E_2)=q(1,E_2,E_1)=q(1,E_2,E_3)=q(1,E_3,E_2)=0.5,\\&q(1,E_1,E_3)=q(1,E_3,E_1)=0.75,\end{aligned} \tag{3.4.4}$$

for the reliability state $\upsilon=2$:

$$\begin{aligned}&q(2,E_1,E_1)=q(2,E_2,E_2)=q(2,E_3,E_3)=1,\\&q(2,E_1,E_2)=q(2,E_2,E_1)=q(2,E_2,E_3)=q(2,E_3,E_2)=0.75,\\&q(2,E_1,E_3)=q(2,E_3,E_1)=0.94,\end{aligned} \tag{3.4.5}$$

and for the reliability state $\upsilon=3$:

$$\begin{aligned}&q(3,E_1,E_1)=q(3,E_2,E_2)=q(3,E_3,E_3)=1,\\&q(3,E_1,E_2)=q(3,E_2,E_1)=q(3,E_2,E_3)=q(3,E_3,E_2)=0.88,\\&q(3,E_1,E_3)=q(3,E_3,E_1)=0.98.\end{aligned} \tag{3.4.6}$$

From Proposition 3.1.4, according to (3.1.58), (3.1.59) and by (3.4.2), (3.4.4)–(3.4.6), the system reliability function is given by the vector

$$\boldsymbol{R}_{LLS}(t,\cdot) = [1, \boldsymbol{R}_{LLS}(t,1), \boldsymbol{R}_{LLS}(t,2), \boldsymbol{R}_{LLS}(t,3), \boldsymbol{R}_{LLS}(t,4)], \quad t \geq 0, \qquad (3.4.7)$$

with the coordinates

$$\boldsymbol{R}_{LLS}(t,1) = \exp[-1.5t] + \frac{1}{3}[\exp[-1.25t] - \exp[-2t]] + \frac{2}{3}[\exp[-1.0833t] - \exp[-1.8333t]], \qquad (3.4.8)$$

$$\boldsymbol{R}_{LLS}(t,2) = \exp[-2.25t] + \frac{1}{3}[\exp[-1.8333t] - \exp[-2.5833t]] + \frac{2}{3}[\exp[-1.6986t] - \exp[-2.4486t]], \qquad (3.4.9)$$

$$\boldsymbol{R}_{LLS}(t,3) = \exp[-3t] + \frac{1}{3}[\exp[-2.4545t] - \exp[-3.2045t]] + \frac{2}{3}[\exp[-2.3676t] - \exp[-3.1176t]], \qquad (3.4.10)$$

$$\boldsymbol{R}_{LLS}(t,4) = \exp[-3t]. \qquad (3.4.11)$$

The reliability function coordinates of the exemplary series system are illustrated in Fig. 3.15.

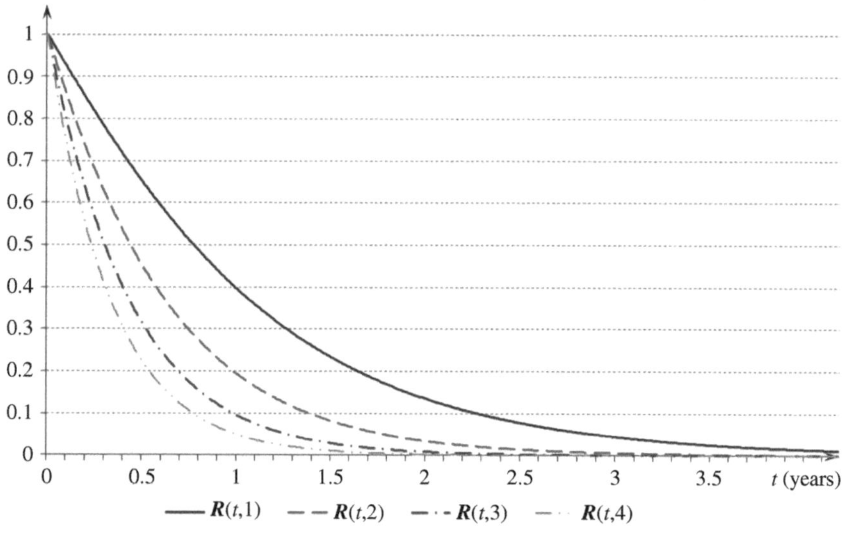

Fig. 3.15 The graphs of the exemplary system reliability function coordinates.

From Corollary 3.1.2, according to (3.1.60), (3.1.61) and substituting the coefficients (3.4.4)–(3.4.6) and using (3.4.2), the mean lifetimes of the exemplary system in the reliability state subsets {1,2,3,4}, {2,3,4}, {3,4}, {4}, in years, are:

$$\mu_{LLS}(1) \cong 1.0184, \quad \mu_{LLS}(2) \cong 0.6647,$$
$$\mu_{LLS}(3) \cong 0.4254, \quad \mu_{LLS}(4) \cong 0.3333. \tag{3.4.12}$$

Similarly, form Corollary 3.1.2, applying (3.1.62)–(3.1.64), we determine the standard deviations of the exemplary system lifetimes and their values in years are:

$$\sigma_{LLS}(1) \cong 0.9226, \quad \sigma_{LLS}(2) \cong 0.5398,$$
$$\sigma_{LLS}(3) \cong 0.4332, \quad \sigma_{LLS}(4) \cong 0.3333. \tag{3.4.13}$$

The mean values of the exemplary system lifetimes in the particular states 1, 2, 3, 4, by (2.1.20) and using (3.4.12), are:

$$\bar{\mu}_{LLS}(1) \cong 0.3537, \quad \bar{\mu}_{LLS}(2) \cong 0.2393,$$
$$\bar{\mu}_{LLS}(3) \cong 0.0921, \quad \bar{\mu}_{LLS}(4) \cong 0.3333. \tag{3.4.14}$$

As we assumed that the critical reliability state is $r=2$, then under the definition introduced in Chapter 2, applying (2.1.21), we obtain the risk function of the exemplary system

$$\begin{aligned} \boldsymbol{r}_{LLS}(t) &= 1 - \boldsymbol{R}_{LLS}(t, 2) \\ &= 1 - \exp[-2.25t] - \frac{1}{3}[\exp[-1.8333t] - \exp[-2.5833t]] \\ &\quad - \frac{2}{3}[\exp[-1.6986t] - \exp[-2.4486t]], \quad t \geq 0. \end{aligned} \tag{3.4.15}$$

Hence, the moment of exceeding an acceptable level $\delta=0.1$, by (2.1.22), is

$$\tau = r^{-1}(\delta) \cong 0.07 \text{ years} \cong 613 \text{ h}. \tag{3.4.16}$$

The exemplary system risk function is illustrated in Fig. 3.16.

From Corollary 3.1.3, according to (3.1.65), (3.1.66), the intensities of departure from the subsets {1,2,3,4}, {2,3,4}, {3,4}, {4}, respectively, are:

$$\begin{aligned} \boldsymbol{\lambda}_{LLS}(t, 1) &= \{1.5 + 0.7222\exp[0.4167t] - 1.2222\exp[-0.3333t] \\ &\quad + 0.4167\exp[0.25t] - 0.6667\exp[-0.5t]\}/\{1 + 0.6667\exp[0.4167t] \\ &\quad - 0.6667\exp[-0.3333t] + 0.3333\exp[0.25t] - 0.3333\exp[-0.5t]\}, \\ &\quad t \geq 0, \end{aligned} \tag{3.4.17}$$

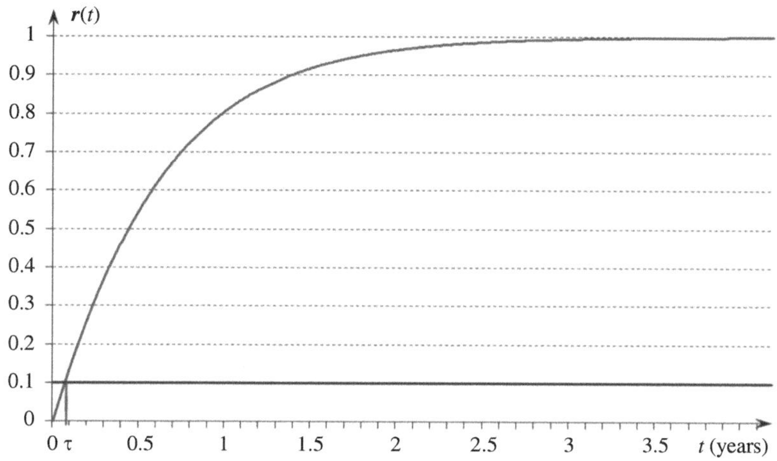

Fig. 3.16 The graph of the risk function ***r***(*t*) of the exemplary system.

$$\lambda_{LLS}(t,2) = \{2.25 + 1.1324\exp[0.5514t] - 1.6324\exp[-0.1986t] + 0.6111\exp[0.4167t] - 0.8611\exp[-0.3333t]\}/\{1 + 0.6667\exp[0.5514t] - 0.6667\exp[-0.1986t] + 0.3333\exp[0.4167t] - 0.3333\exp[-0.3333t]\},\ t \geq 0, \tag{3.4.18}$$

$$\lambda_{LLS}(t,3) = \{3 + 1.5784\exp[0.6324t] - 2.0784\exp[-0.1176t] + 0.8182\exp[0.5455t] - 1.0682\exp[-0.2045t]\}/\{1 + 0.6667\exp[0.6324t] - 0.6667\exp[-0.1176t] + 0.3333\exp[0.5455t] - 0.3333\exp[-0.2045t]\},\ t \geq 0, \tag{3.4.19}$$

$$\lambda_{LLS}(t,4) = 3, \quad t \geq 0, \tag{3.4.20}$$

and their graphs are illustrated in Fig. 3.17.

3.4.2 Reliability of shipyard rope elevator

Aging and fatigue processes of system components and the problem of assessing the impact of this aging on functioning and reliability of the entire system is the subject of many works in many different areas (Biondini and Frangopol, 2016; Liu et al., 2019; Woch et al., 2018; Zieja et al., 2018).

As a practical application of this chapter results, regarding the reliability analysis for multistate aging systems with dependent components, we consider a shipyard rope elevator. The shipyard rope elevator has been described and analyzed assuming that its components are independent in Chapter 2.

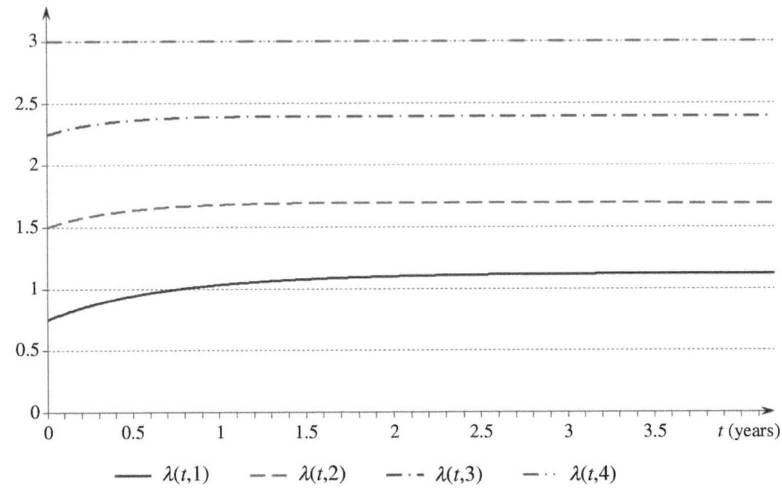

Fig. 3.17 The graphs of the exemplary system intensities.

The shipyard rope elevator has been analyzed assuming the ELS model of dependency among strands in the ropes only and without assumptions of dependence among the ropes in Blokus-Roszkowska (2007b). In reliability analysis of the shipyard rope elevator presented in Blokus-Roszkowska and Kołowrocki (2014), variable operation conditions of the system were additionally taken into account. Another example of a shipyard transport system analyzed under variable operation conditions is the ground ship-rope transporter (Blokus-Roszkowska and Kołowrocki, 2008). Its reliability assuming the time-varying its reliability structure and components' reliability parameters has been performed in Blokus-Roszkowska and Kołowrocki (2015a). In that reference, the dependencies among its components following the ELS rule were also taken into account.

It is assumed, as in Chapter 2, that the components E_{ij}, $i=1, 2, \ldots, 10$, $j=1, 2, \ldots, 22$, have reliability functions with the following piecewise exponential reliability function coordinates:

$$R(t, 1)=\exp[-\lambda(1)t], \quad R(t, 2)=\exp[-\lambda(2)t],$$
$$R(t, 3)=\exp[-\lambda(3)t], \quad t \geq 0. \tag{3.4.21}$$

In (3.4.21), the intensities of leaving the reliability state subsets $\{1,2,3\}$, $\{2,3\}$, $\{3\}$ are expressed in years^{-1} and are as follows:

$$\lambda(1)=0.1613, \quad \lambda(2)=0.2041, \quad \lambda(3)=0.2326. \tag{3.4.22}$$

Taking into account the reliability structure of the shipyard rope elevator, described in Chapter 2 and presented in Fig. 2.12, we propose the MLS dependency model, presented in Section 3.3. Thereby, we assume the dependence according to the LLS rule among subsystems, i.e., the ropes, and the ELS dependency model among strands in the ropes.

Each rope consists of 22 identical strands following the ELS dependency rule, forming a "5 out of 22" reliability structure. The dependence among multistate components means that after leaving the reliability state subset $\{u, u+1, \ldots, 3\}$, $u=1, 2, 3$, by ω, $\omega=0, 1, 2, \ldots, 17$, strands, the lifetimes of other strands of a rope in this subset decrease equally depending on the number ω. The component stress proportionality correction coefficients $c(u)$, by assumption, are equal to 0.9 for $u=1, 2, 3$. Hence, the intensities of getting out of the subsets $\{1,2,3\}$, $\{2,3\}$, $\{3\}$ of other strands in ropes, according to (3.2.31) and using values given in (3.4.22), are as follows (Blokus and Kołowrocki, 2019):

$$\lambda^{(\omega)}(1)=\frac{22}{22-\omega}\cdot\frac{\lambda(1)}{c(1)}=\frac{3.9429}{22-\omega},\quad \omega=0,1,2,\ldots,17, \tag{3.4.23}$$

$$\lambda^{(\omega)}(2)=\frac{22}{22-\omega}\cdot\frac{\lambda(2)}{c(2)}=\frac{4.9891}{22-\omega},\quad \omega=0,1,2,\ldots,17, \tag{3.4.24}$$

$$\lambda^{(\omega)}(3)=\frac{22}{22-\omega}\cdot\frac{\lambda(3)}{c(3)}=\frac{5.6858}{22-\omega},\quad \omega=0,1,2,\ldots,17. \tag{3.4.25}$$

Further, we assume that subsystems, i.e., the ropes are dependent according to the LLS rule. The coefficients of impact of the subsystem S_g, $g=1, \ldots, 10$, that has left the subset $\{u, u+1, \ldots, 3\}$, $u=1, 2, 3$, on other subsystems S_i, $i=1, \ldots, 10$, $i\neq g$, are expressed as a function of distance $d_{ig}=|i-g|$ between the subsystems S_g and S_i. They are arbitrarily defined as in Blokus and Kołowrocki (2019) by

$$\begin{gathered} q(\upsilon, S_g, S_i)=q(\upsilon, d_{ig})=1-\exp\left[-d_{ig}\right],\quad i,g=1,2,\ldots,10, \\ i\neq g,\quad \upsilon=u,\quad u-1,\ldots,1,\quad u=1,2, \end{gathered} \tag{3.4.26}$$

and

$$\begin{gathered} q(\upsilon, S_g, S_g)=q(\upsilon, 0)=1, g=1,2,\ldots,10, \\ \upsilon=u, u-1,\ldots,1, u=1,2. \end{gathered} \tag{3.4.27}$$

This implies that if the subsystem S_g, $g=1, \ldots, 10$, gets out of the subset $\{u, u+1, \ldots, 3\}$, $u=1, 2, 3$, then the mean values of lifetimes in the subset

$\{\upsilon, \upsilon+1, \ldots, 3\}$, $\upsilon=u, u-1, \ldots, 1$, $u=1, 2$, of components $E_{i,j}$, $i=1, \ldots, 10, j=1, \ldots, 22$, of subsystem S_i, decrease by formula (3.3.2) according to

$$E\left[T_{i/g,j}(\upsilon)\right] = q\left(\upsilon, d_{ig}\right) \cdot E\left[T_{i,j}(\upsilon)\right], \; i,g=1,2,\ldots,10, \quad j=1,2,\ldots,22. \tag{3.4.28}$$

From Proposition 3.3.4, by applying formulas (3.3.15)–(3.3.18) and substituting the intensities (3.4.22) and the coefficients of impact (3.4.26), (3.4.27), the reliability function of shipyard rope elevator (Blokus and Kołowrocki, 2019) is as follows:

$$\boldsymbol{R}_{MLS}(t, \cdot) = [1, \boldsymbol{R}_{MLS}(t,1), \boldsymbol{R}_{MLS}(t,2), \boldsymbol{R}_{MLS}(t,3)], \quad t \geq 0, \tag{3.4.29}$$

where

$$\begin{aligned}
\boldsymbol{R}_{MLS}(t,1) = {} & \left[\sum_{\omega=0}^{17} \frac{[4.9891t]^{\omega}}{\omega!}\right]^{10} \cdot \exp[-49.8911t] \\
& + \sum_{g=1}^{10} \int_0^t \left[\sum_{\omega=0}^{17} \frac{[4.9891a]^{\omega}}{\omega!}\right]^{9} \cdot \sum_{\omega=0}^{17} \frac{[3.9429a]^{\omega}}{\omega!} \\
& \left[\frac{\sum_{\omega=0}^{17} \frac{[4.9891a]^{\omega}}{\omega!} \cdot \sum_{\omega=0}^{16} \frac{[3.9429]^{\omega+1} a^{\omega}}{\omega!} - \sum_{\omega=0}^{16} \frac{[4.9891]^{\omega+1} a^{\omega}}{\omega!} \cdot \sum_{\omega=0}^{17} \frac{[3.9429a]^{\omega}}{\omega!}}{\left[\sum_{\omega=0}^{17} \frac{[3.9429a]^{\omega}}{\omega!}\right]^2} \right. \\
& \left. + \frac{1.0462 \cdot \sum_{\omega=0}^{17} \frac{[4.9891a]^{\omega}}{\omega!}}{\sum_{\omega=0}^{17} \frac{[3.9429a]^{\omega}}{\omega!}} \right] \cdot \exp[-49.8911a] \\
& \cdot \left[\prod_{i=1}^{10} \frac{\sum_{\omega=0}^{17} \frac{\left[\frac{3.9429}{1-\exp[-d_{ig}]} t\right]^{\omega}}{\omega!}}{\sum_{\omega=0}^{17} \frac{[3.9429a]^{\omega}}{\omega!}} \cdot \exp\left[-\frac{3.9429}{1-\exp[-d_{ig}]} t + 3.9429a \right] \right] da,
\end{aligned} \tag{3.4.30}$$

$$
\begin{aligned}
\boldsymbol{R}_{MLS}(t,2) = {} & \left[\sum_{\omega=0}^{17}\frac{[5.6858t]^{\omega}}{\omega!}\right]^{10}\cdot\exp[-56.8578t] \\
& + \sum_{g=1}^{10}\int_0^t\left[\sum_{\omega=0}^{17}\frac{[5.6858a]^{\omega}}{\omega!}\right]^{9}\cdot\sum_{\omega=0}^{17}\frac{[4.9891a]^{\omega}}{\omega!} \\
& \left[\frac{\sum_{\omega=0}^{17}\frac{[5.6858a]^{\omega}}{\omega!}\cdot\sum_{\omega=0}^{16}\frac{[4.9891]^{\omega+1}a^{\omega}}{\omega!}-\sum_{\omega=0}^{16}\frac{[5.6858]^{\omega+1}a^{\omega}}{\omega!}\cdot\sum_{\omega=0}^{17}\frac{[4.9891a]^{\omega}}{\omega!}}{\left[\sum_{\omega=0}^{17}\frac{[4.9891a]^{\omega}}{\omega!}\right]^{2}}\right. \\
& \left.+\frac{0.6967\cdot\sum_{\omega=0}^{17}\frac{[5.6858a]^{\omega}}{\omega!}}{\sum_{\omega=0}^{17}\frac{[4.9891a]^{\omega}}{\omega!}}\right]\cdot\exp[-56.8578a] \\
& \cdot\left[\prod_{i=1}^{10}\frac{\sum_{\omega=0}^{17}\frac{\left[\frac{4.9891}{1-\exp[-d_{ig}]}t\right]^{\omega}}{\omega!}}{\sum_{\omega=0}^{17}\frac{[4.9891a]^{\omega}}{\omega!}}\cdot\exp\left[-\frac{4.9891}{1-\exp[-d_{ig}]}t+4.9891a\right]\right]da,
\end{aligned}
\tag{3.4.31}
$$

$$
\boldsymbol{R}_{MLS}(t,3) = \left[\sum_{\omega=0}^{17}\frac{[5.6858t]^{\omega}}{\omega!}\right]^{10}\cdot\exp[-56.8578t]. \tag{3.4.32}
$$

Applying formula (2.1.16), we determine the expected value of system lifetime in the reliability state subset $\{1,2,3\}$

$$
\mu_{MLS}(1) = \int_0^{\infty}\boldsymbol{R}_{MLS}(t,1)\,dt \cong 2.514, \tag{3.4.33}
$$

and its standard deviation, by formulas (2.1.17), (2.1.18), is

$$\sigma_{MLS}(1) = \sqrt{2\int_0^\infty t \cdot \boldsymbol{R}_{MLS}(t,1)dt - [\mu_{MLS}(1)]^2} \cong 0.270, \qquad (3.4.34)$$

where $\boldsymbol{R}_{MLS}(t,1)$ is expressed by (3.4.30).

Due to the complicated formula of reliability function coordinate $\boldsymbol{R}_{MLS}(t,1)$, a computer program written in Java with the SSJ V2.1.3 library has been used to calculate the formulas (3.4.33), (3.4.34).

Similarly, using the reliability function coordinate $\boldsymbol{R}_{MLS}(t,2)$ given by (3.4.31), the expected value and standard deviation of system lifetime in subset $\{2,3\}$, respectively, are

$$\mu_{MLS}(2) = \int_0^\infty \boldsymbol{R}_{MLS}(t,2)dt \cong 2.149, \qquad (3.4.35)$$

$$\sigma_{MLS}(2) = \sqrt{2\int_0^\infty t \cdot \boldsymbol{R}_{MLS}(t,2)dt - [\mu_{MLS}(2)]^2} \cong 0.290. \qquad (3.4.36)$$

The expected value and standard deviation of system lifetime in subset $\{3\}$, obtained on the basis of reliability function coordinate $\boldsymbol{R}_{MLS}(t,3)$ given by (3.4.32), respectively, are

$$\mu_{MLS}(3) = \int_0^\infty \boldsymbol{R}_{MLS}(t,3)dt \cong 2.124, \qquad (3.4.37)$$

$$\sigma_{MLS}(3) = \sqrt{2\int_0^\infty t \cdot \boldsymbol{R}_{MLS}(t,3)dt - [\mu_{MLS}(3)]^2} \cong 0.327. \qquad (3.4.38)$$

Additionally, according to (2.1.20), we determine the mean lifetime of rope system in individual reliability states 1, 2, 3. Their values calculated for $c(1) = c(2) = c(3) = 0.9$, using results obtained from (3.4.33), (3.4.35), and (3.4.37), in years are:

$$\bar{\mu}_{MLS}(1) = \mu_{MLS}(1) - \mu_{MLS}(2) = 0.365, \qquad (3.4.39)$$

$$\bar{\mu}_{MLS}(2) = \mu_{MLS}(2) - \mu_{MLS}(3) = 0.025, \qquad (3.4.40)$$

Table 3.1 The expected values and standard deviations of lifetimes in the reliability state subsets {1,2,3}, {2,3}, {3} for shipyard rope elevator in three cases of dependency models and without dependency.

	$\mu(1)$	$\sigma(1)$	$\mu(2)$	$\sigma(2)$	$\mu(3)$	$\sigma(3)$
MLS	2.514	0.270	2.149	0.290	2.124	0.327
ELS $c(u)=0.9$	3.063	0.471	2.420	0.372	2.124	0.327
ELS $c(u)=1.0$	3.403	0.524	2.689	0.414	2.360	0.363
Independent	6.441	1.040	5.091	0.808	4.467	0.704

$$\overline{\mu}_{MLS}(3)=\mu_{MLS}(3)=2.124. \quad (3.4.41)$$

The expected values and standard deviations of the shipyard rope elevator lifetimes in the reliability state subsets {1,2,3}, {2,3}, {3}, are given in Table 3.1 for different models of dependency. Namely, assuming the MLS model of dependency among components and subsystems, assuming only the ELS dependency model among components in subsystems for the values of component stress proportionality correction coefficients $c(u)=0.9$ and $c(u)=1.0$, $u=1, 2, 3$, and finally without dependency among components and subsystems.

For the assumed critical state $r=2$, we determine, according to the Definition 2.1.3 and formula (2.1.21), the risk function of shipyard rope elevator:

$$\boldsymbol{r}_{MLS}(t)=1-\boldsymbol{R}_{MLS}(t,2), \quad (3.4.42)$$

where the reliability function coordinate $\boldsymbol{R}_{MLS}(t,2)$ is given by (3.4.31).

Using the defined in (3.4.42) risk function, we determine the moment of exceeding the assumed risk level δ. For the coefficient $c(2)=0.9$ and the level $\delta=0.05$, this moment by formula (2.1.22) is:

$$\tau=\boldsymbol{r}_{MLS}^{-1}(\delta)\cong 1.75\,\text{years}\cong 1\,\text{year}\,274\,\text{days}, \quad (3.4.43)$$

where $\boldsymbol{r}_{MLS}^{-1}(t)$, if exists, is the inverse function of the risk function $\boldsymbol{r}_{MLS}(t)$.

The graphs of reliability function's coordinates of the shipyard rope transportation system are presented in Fig. 3.18 in case of the MLS dependency model. Additionally, Figs. 3.19 and 3.20 show graphs taking into account dependencies among components in subsystems according to the ELS rule and without dependency among the subsystems. The results for the values of component stress proportionality correction coefficients $c(1)=c(2)=c(3)=0.9$ are shown in Fig. 3.19, and for the values

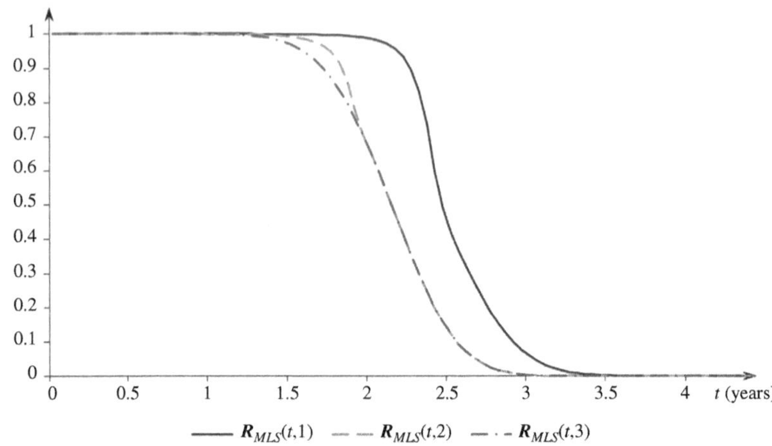

Fig. 3.18 The graphs of reliability function's coordinates of shipyard rope elevator in case of the MLS dependency model.

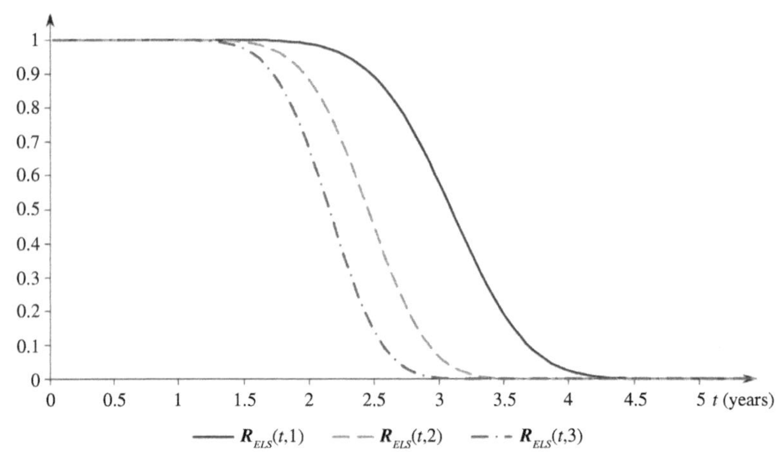

Fig. 3.19 The graphs of reliability function's coordinates of shipyard rope elevator in case of the ELS dependency model with coefficients $c(1)=c(2)=c(3)=0.9$.

$c(1)=c(2)=c(3)=1.0$, respectively, in Fig. 3.20. In case of the MLS dependency model, the results both for values of coefficients 0.9 and 1.0 are very similar, and for this reason they were presented in Fig. 3.18 only for $c(1)=c(2)=c(3)=0.9$. These results, obtained for different dependency models, are compared with the reliability function's coordinates determined for shipyard rope transportation system assuming that all components and subsystems are independent (Fig. 3.21). Similar results are presented in Blokus and Kołowrocki (2019).

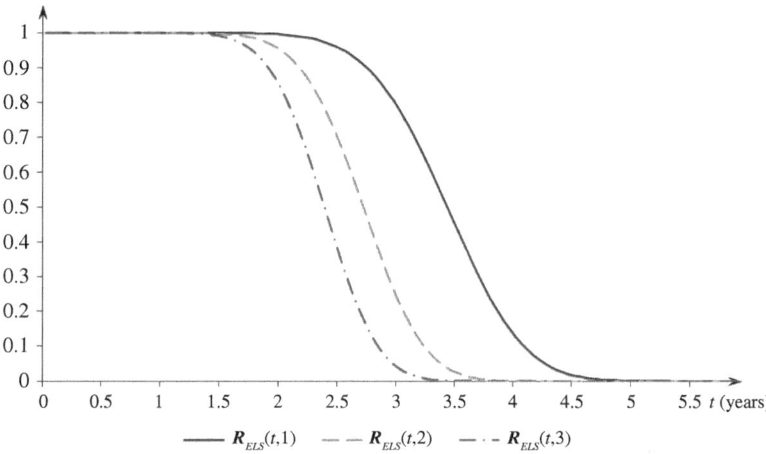

Fig. 3.20 The graphs of reliability function's coordinates of shipyard rope elevator in case of the ELS dependency model with coefficients $c(1)=c(2)=c(3)=1.0$.

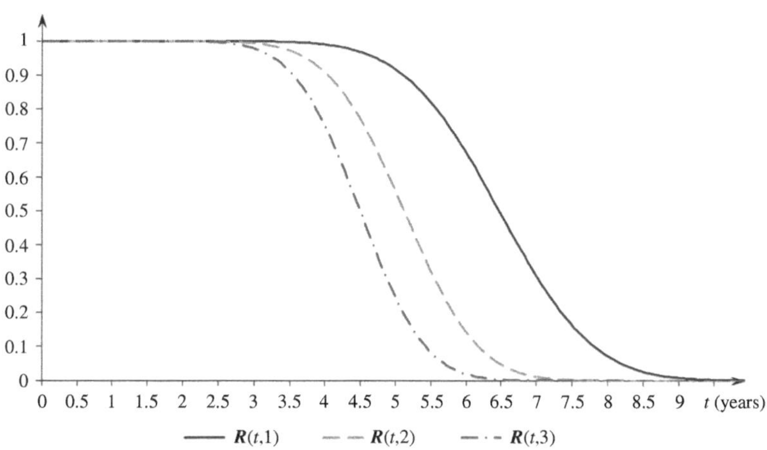

Fig. 3.21 The graphs of reliability function's coordinates of shipyard rope elevator in case of independent components and subsystems.

Comparing the mean lifetimes in reliability state subsets, presented in Table 3.1, and the graphs of reliability function coordinates Figs. 3.18–3.21 for various dependency models of the shipyard rope elevator and in case of independence between both subsystems and components in subsystems, we can conclude that the dependence among components in subsystems is of key importance. To be specific, the mean lifetimes in reliability state subsets {1,2,3}, {2,3}, {3} assuming dependence among components in subsystems according to the ELS rule at a correction coefficient $c(u)=1.0$, $u=1, 2, 3$, are shorter by about 47%, and at the coefficient of $c(u)=0.9$, $u=1, 2, 3$, these lifetimes are shorter by about 52% relative to the results for the system without taking into account the dependencies. Assuming additionally the dependence according to the LLS rule among subsystems, i.e., for the entire system with the MLS dependency model, this difference increases, but not so much. The mean lifetime of shipyard rope elevator following the MLS dependency model in the state subset $\{1,2,3\}$ is shorter by about 61%, the mean lifetime in the state subset $\{2,3\}$ is shorter by about 58%, and the mean lifetime in the state 3 by 52%, compared to these mean lifetimes for shipyard rope elevator with independent components and subsystems. In general, it can be concluded that deterioration or failure of some system components significantly reduces the lifetimes of other system components and, as a consequence, the entire system reliability deteriorates. A similar conclusion was drawn in Blokus and Kołowrocki (2019).

References

Aven, T., Jensen, U., 1999. Stochastic Models in Reliability. Springer-Verlag, New York.

Biondini, F., Frangopol, D.M., 2016. Life-cycle performance of deteriorating structural systems under uncertainty. J. Struct. Eng. 142(9), F4016001.

Birolini, A., 1994. Quality and Reliability of Technical Systems. Theory—Practice—Management. Springer Verlag, Berlin Heidelberg.

Blokus, A., 2006. Reliability analysis of large systems with dependent components. Int. J. Reliab. Qual. Saf. Eng. 13 (1), 1–14.

Blokus, A., Kołowrocki, K., 2019. Reliability and maintenance strategy for systems with aging-dependent components. Qual. Reliab. Eng. Int. 35, 2709–2731. https://doi.org/10.1002/qre.2552.

Blokus-Roszkowska, A., 2007a. Reliability Analysis of Homogenous Large Systems With Component Dependent Failures (Ph.D. Thesis). Gdynia Maritime University—Polish Academy of Sciences, System Research Institute, Warsaw (in Polish).

Blokus-Roszkowska, A., 2007b. On component failures' dependency influence on system's lifetime. Int. J. Reliab. Qual. Saf. Eng. 14 (6), 1–19 (Special Issue: System Reliability and Safety).

Blokus-Roszkowska, A., 2016. Reliability analysis of the bulk cargo loading system including dependent components. In: Simos, T., Tsitouras, C. (Eds.), Proceedings of the International Conference of Numerical Analysis and Applied Mathematics 2015 (ICNAAM 2015). AIP Conf. Proc., vol. 1738, AIP Publishing, pp. 440002-1 – 440002-4.

Blokus-Roszkowska, A., Kołowrocki, K., 2007. Reliability and risk analysis of large systems with ageing components. In: Aven, T., Vinnem, J.E. (Eds.), Risk, Reliability and Societal Safety, Three Volume Set: Proceedings of the European Safety and Reliability Conference 2007 (ESREL 2007). CRC Press/Balkema, Stavanger, pp. 995–1003.

Blokus-Roszkowska, A., Kołowrocki, K., 2008. Modelling environment and infrastructure of shipyard transportation systems and processes. Int. J. Mater. Struct. Reliab. 6 (2), 153–166.

Blokus-Roszkowska, A., Kołowrocki, K., 2014. Reliability analysis of complex shipyard transportation system with dependent components. J. Pol. Saf. Reliab. Assoc. Summer Saf. Reliab. Semin. 5 (1), 21–31.
Blokus-Roszkowska, A., Kołowrocki, K., 2015a. Reliability analysis of ship-rope transporter with dependent components. In: Nowakowski, T. et al., (Eds.), Safety and Reliability: Methodology and Applications—Proceedings of the European Safety and Reliability Conference, ESREL 2014. Taylor & Francis Group, London, pp. 255–263.
Blokus-Roszkowska, A., Kołowrocki, K., 2015b. Reliability analysis of multistate series systems with dependent components. J. Pol. Saf. Reliab. Assoc. Summer Saf. Reliab. Semin. 6 (1), 31–36.
Blokus-Roszkowska, A., Kołowrocki, K., 2015c. Reliability of the exemplary multistate series system with dependent components. J. Pol. Saf. Reliab. Assoc. Summer Saf. Reliab. Semin. 6 (1), 37–46.
Blokus-Roszkowska, A., Kolowrocki, K., 2017. Modelling safety of multistate systems with dependent components and subsystems. J. Pol. Saf. Reliab. Assoc. Summer Saf. Reliab. Semin. 8 (3), 23–41.
Crowder, M.J., Kimber, A.C., Smith, R.L., Sweeting, T.J., 1991. Statistical Analysis of Reliability Data. Chapman & Hall/CRC. first ed., 2017, eBook.
Daniels, H.E., 1945. The statistical theory of the strength of bundles of threads I. Proc. R. Soc. Lond. Ser. A 183, 404–435.
Grabski, F., 2014. Semi-Markov Processes: Applications in System Reliability and Maintenance. Elsevier, London.
Harlow, D.G., Phoenix, S.L., 1978. The chain-of-bundles probability model for the strength of fibrous materials. J. Compos. Mater. 12, 195–214.
Harlow, D.G., Phoenix, S.L., 1982. Probability distribution for the strength of fibrous materials under local load sharing. Adv. Appl. Probab. 14, 68–94.
Kołowrocki, K., 2014. Reliability of Large and Complex Systems, second ed. Elsevier, London.
Kołowrocki, K., Soszyńska-Budny, J., 2011. Reliability and Safety of Complex Technical Systems and Processes: Modeling—Identification—Prediction—Optimization, first ed. Springer-Verlag, London.
Lai, C.-D., Xie, M., 2006. Stochastic Ageing and Dependence for Reliability. Springer, New York.
Levitin, G., 2004. A universal generating function approach for the analysis of multi-state systems with dependent elements. Reliab. Eng. Syst. Saf. 84, 285–292.
Levitin, G., 2005. The universal generating function in reliability analysis and optimization. In: Reliability Engineering, first ed. Springer, London.
Limnios, N., Oprisan, G., 2005. Semi-Markov Processes and Reliability. Birkhauser, Boston.
Liu, Y., Frangopol, D.M., Cheng, M., 2019. Risk-informed structural repair decision making for service life extension of aging naval ships. Mar. Struct. 64, 305–321.
Maatouk, I., Châtelet, E., Chebbo, N., 2011. Reliability of multi-states systems with load sharing and propagation failure dependence. In: Proceedings of ICQRMS 2011, International Conference on Quality, Reliability, Risk, Maintenance, and Safety Engineering. IEEE, China, ISBN: 978-1-4577-1232-6.
Misra, K.B., 1992. Reliability Analysis and Prediction. Elsevier.
Mohammad, R., Amari, S.V., Kalam, A., 2012. Reliability evaluation of phased-mission systems with load-sharing components. IEEE Trans. Reliab. Maintain. Symp. 1–6. https://doi.org/10.1109/RAMS.2012.6175468.
Phoenix, S.L., Smith, R.L., 1983. A comparison of probabilistic techniques for the strength of fibrous materials under local load sharing among fibres. Int. J. Solids Struct. 19 (6), 479–496.

Pradhan, S., Hansen, A., Chakrabarti, B.K., 2010. Failure processes in elastic fiber bundles. Rev. Mod. Phys. 82, 499–555.
Rausand, M., Høyland, A., 2004. System Reliability Theory: Models, Statistical Methods, and Applications, second ed. John Wiley & Sons, Inc., New Jersey.
Smith, R.L., 1982. The asymptotic distribution of the strength of a series-parallel system with equal load sharing. Ann. Probab. 10, 137–171.
Smith, R.L., 1983. Limit theorems and approximations for the reliability of load sharing systems. Adv. Appl. Probab. 15, 304–330.
Ushakov, I.A., 1986. Universal generating function. Sov. J. Comput. Syst. Sci. 24 (5), 118–129.
Ushakov, I. (Ed.), 1994. Handbook of Reliability Engineering. Wiley, New York.
Woch, M., Zieja, M., Tomaszewska, J., 2018. The impact of input parameters on the reliability of aircraft. In: 2018 3rd International Conference on System Reliability and Safety (ICSRS), Barcelona, Spain, pp. 210–214.
Xue, J., 1985. On multi-state system analysis. IEEE Trans. Reliab. 34, 329–337.
Xue, J., Yang, K., 1995a. Dynamic reliability analysis of coherent multi-state systems. IEEE Trans. Reliab. 44 (4), 683–688.
Xue, J., Yang, K., 1995b. Symmetric relations in multi-state systems. IEEE Trans. Reliab. 44 (4), 689–693.
Zieja, M., Ważny, M., Stępień, S., 2018. Outline of a method for estimating the durability of components or device assemblies while maintaining the required reliability level. Eksploat. Niezawod.—Maint. Reliab. 20 (2), 260–266.

CHAPTER FOUR

Availability analysis of aging-dependent systems under imperfect repair

Basic assumptions and definitions related to renewal stream and renewal process are introduced. Availability characteristics of a repairable system with negligible renovation time and in case when the renewal time is non-negligible are determined, under the assumption of imperfect repair. Namely, the expected values of the times until the successive times that the reliability critical state is exceeded, and the expected values of the numbers of times that the reliability critical state is exceeded up to the fixed time point, are given. The procedure for determining the expected value of the number of times the system exceeds the reliability critical state up to the fixed time point are proposed for a system with negligible renewal time and in case the renewal time is significant and cannot be omitted. For a repairable system with non-negligible renewal time the expected values of time until the successive system renewals, the expected values of the number of system renewals until the fixed moment and the availability coefficient of the system at the fixed moment are determined.

Moreover, the availability analysis of a multistate series system with components following the LLS (local load sharing) dependency rule is performed and the availability of a series-parallel system and a series-"m out of k" system with dependent according to the LLS rule components of their series subsystems is discussed. Next, a parallel system following the ELS (equal load sharing) dependency rule, an "m out of n" system following the ELS dependency rule, a parallel-series system with dependent according to the ELS rule components of its parallel subsystems and an "m out of l"-series system with dependent according to the ELS rule components of its "m out of l" subsystems are considered. The availability analysis of multistate parallel-series and "m out of l"-series systems, assuming the MLS (mixed load sharing) dependency model, is discussed as well. The obtained theoretical results are applied to availability analysis of an exemplary technical system.

Finally, the maintenance costs for a renewable system with negligible and non-negligible renewal durations are analyzed. The procedure for determining the total cost of maintenance up to the fixed time point is given. Moreover, the procedure comparing the costs of repair and operation of the new system until the fixed moment, for different

Multistate System Reliability with Dependencies
https://doi.org/10.1016/B978-0-12-821260-8.00004-X

configuration of the number of repairs to various reliability states, in both cases for a renewable system with negligible renewal time and for a system with non-negligible renewal time, is proposed.

4.1 Renewal stream and renewal process

Availability analysis is often associated with multistate systems, whose components degrade ranging from the state of perfect reliability to entire failure state (Kołowrocki, 2014; Li and Pham, 2006; Lisnianski and Levitin, 2003; Pourret et al., 1999; Szymkowiak, 2018, 2019; Xue, 1985). Considering maintenance and renewal of multistate systems, the replacement of a system component by a new one is often assumed. For a multistate aging system, it means that the component after repair is in the best "as good as new" reliability state, which in fact is not always possible. Moreover, even the perfect repair, due to conditions resulting from the state of the remaining components, may cause the component not to behave as new. In addition, its lifetimes in the reliability state subsets can be shorter than for a component in a new system (Blokus-Roszkowska, 2019). Liu et al. (2013) analyze redundancy and imperfect maintenance strategy for multistate systems assuming that even if the component is replaced by a new one, it cannot be restored to a virtually new condition. Imperfect maintenance of a deteriorating system is widely discussed in Castro (2009), Høyland and Rausand (2004), Crespo Marquez (2007), Pham and Wang (1996), and Wang and Pham (2011). In this book, it is assumed that a multistate system is repaired after exceeding its critical reliability state. Taking into account this assumption under imperfect repair, the renewal of the system means that the system is restored to the critical reliability state or better, not necessarily to the best state. Combining the results of the reliability analysis of multistate systems with dependent components and the results of the classical renewal theory (Badoux, 1987; Gertsbakh, 2000; Zio, 2007; Zio and Compare, 2013), the availability analysis of aging-dependent systems under imperfect repair (Finkelstein, 2008; Pham et al., 1997; Levitin and Lisnianski, 2000; Muhammad et al., 2010; Nourelfath et al., 2012; Soro et al., 2010) is proposed.

It is assumed that all the components and the system under consideration have the reliability state set $\{0,1, \ldots, z\}$, $z \geq 1$, where state 0 is the worst and state z is the best. The state of the system and components degrades over time.

Similarly as in Blokus (2006), Blokus-Roszkowska and Kołowrocki (2009), Kołowrocki and Soszyńska (2006), Kołowrocki and Soszyńska (2010), and Kołowrocki and Soszyńska-Budny (2011), it is assumed that a system is repaired after exceeding its critical reliability state r, $r \in \{1, 2, \ldots, z-1\}$. $\boldsymbol{T}(r)$ denotes the system's lifetime in the reliability state subset $\{r, r+1, \ldots, z\}$.

A random variable $\boldsymbol{T}^{(N)}(r)$, $r \in \{1,2, \ldots, z-1\}$, $N=2,3, \ldots$, describes the time between the moment of the N-1 system renovation and the moment when the system critical state is exceeded for Nth time. Wherein $\boldsymbol{T}^{(1)}(r)$ denotes the time between the commencement of the system operation and the moment of its first renovation. It is assumed that the random variables $\boldsymbol{T}^{(1)}(r)$, $\boldsymbol{T}^{(2)}(r)$, …, $r \in \{1,2, \ldots, z-1\}$ are independent.

System renovation often means the operation that causes the system to return to its full reliability state z. However, in reality after some time or after a number of renewals, the return of a system to the best "as good as new" state is no longer possible. Thus, we assume that after N_1 system's renovations to the full reliability state z, the next system renovation will only be possible to state $z-1$. Similarly, we assume that the system renewal from the renovation N_1+1 up to the renovation N_2 takes place to state $z-1$, so the number of such repairs is N_2-N_1. Further, by the renewal of the system from the renovation N_2+1 to the renovation N_3, we mean the operation that causes the system to return to the reliability state $z-2$. Generally, we assume that the system renewal from the renovation N_x+1 to the renovation N_{x+1} takes place to state $z-x$, $x=1,2, \ldots, z-r$ and $r \in \{1, 2, \ldots, z-1\}$. From the renovation $N_{z-r}+1$ to renovation N_{z-r+1}, system renewal can only be carried out as a renewal to the critical reliability state r. After N_{z-r+1} renovations, another system renewal is no longer possible and after $N_{z-r+1}+1$ times of exceeding the reliability critical state r, the system remains unable to work.

The system renewal is characterized by a sequence of random variables (Blokus-Roszkowska, 2019; Badoux, 1987; Bobrowski, 1985; Gertsbakh, 2000; Kołowrocki and Soszyńska, 2006; Kopociński, 1973; Zio, 2007), the moment of successive renovations.

$$S^{(N)}(r) = \boldsymbol{T}^{(1)}(r) + \boldsymbol{T}^{(2)}(r) + \ldots + \boldsymbol{T}^{(N)}(r), r \in \{1, 2, \ldots, z-1\}, \quad N = 1, 2, \ldots. \tag{4.1.1}$$

Subsequently, a sequence of random variables $S^{(1)}(r)$, $S^{(2)}(r)$,…, $r \in \{1, 2, \ldots, z-1\}$ is a renewal stream. With this renewal stream, there is associated the random process $\{N(t,r),\ t \geq 0\}$, called a renewal process of a multistate system, where $N(t,r)$ is the number of systems exceeding the critical state r, i.e., the number of renewals of the system, up to the time point t.

4.2 Characteristics of renewal stream with imperfect repair

In this section, the expected values of the times until the successive times that the reliability critical state is exceeded are determined under assumptions presented in Section 4.1. We analyze two cases of repairable systems. First case of a system with negligible renewal time and second case of a system in which the renewal time is significant and cannot be omitted. We start with determining renewal and availability characteristics in first of those cases and assume that time of system renovation can be ignored comparing to its lifetime in the reliability state subset $\{r, r+1, \ldots, z\}$.

It is assumed that first N_1 system's renewals include operations that allow the system to return to the full reliability state z. Then if r, $r \in \{1, 2, \ldots, z-1\}$, is reliability critical state of a system, $\boldsymbol{T}^{(1)}(r)$, $\boldsymbol{T}^{(2)}(r), \ldots, \boldsymbol{T}^{(N_1+1)}(r)$ are independent random variables from the same distribution with expected value $\mu(r)$ and standard deviation $\sigma(r) \neq 0$, the same as for a new system. The variable $S^{(N_1+1)}(r)$ representing the time until the N_1+1 exceeding the critical state r by a system, for sufficiently large N_1, has an approximately normal distribution with the expected value and the variance, respectively given by (Blokus, 2006; Bobrowski, 1985; Kołowrocki and Soszyńska, 2010; Kołowrocki and Soszyńska-Budny, 2011)

$$E\left[S^{(N_1+1)}(r)\right] \cong (N_1+1)\mu(r), D\left[S^{(N_1+1)}(r)\right] \cong (N_1+1)\sigma^2(r), r \in \{1, 2, \ldots, z-1\}. \quad (4.2.1)$$

As first N_1 system's renewals include operations that allow the system to return to the full reliability state z, the expected value of a random variable $S^{(\omega)}(r)$, $\omega = 1, 2, \ldots, N_1, N_1+1$, representing the time until the ωth time the system exceeds the reliability critical state r, is (Blokus, 2006; Bobrowski, 1985; Kołowrocki and Soszyńska, 2010; Kołowrocki and Soszyńska-Budny, 2011).

$$E\left[S^{(\omega)}(r)\right] \cong \omega \cdot \mu(r), \omega = 1, 2, \ldots, N_1, N_1+1, r \in \{1, 2, \ldots, z-1\}. \quad (4.2.2)$$

Subsequently, the system renewal from the renovation N_1+1 to the renovation N_2 takes place to state $z-1$. If a random variable $\boldsymbol{T}^{(\omega)}(r)$, $\omega=N_1+2, \ldots, N_2+1$, represents the time between the moment of the $\omega-1$ system renovation and the ωth time that the system critical state is exceeded, then its expected value is given by (Blokus-Roszkowska, 2019)

$$E\left[\boldsymbol{T}^{(\omega)}(r)\right]=\mu(r)-\mu(z), \omega=N_1+2, \ldots, N_2+1, \quad r \in\{1,2, \ldots, z-1\}. \tag{4.2.3}$$

Therefore, the expected value of a random variable $S^{(\omega)}(r)$, $\omega=N_1+2, \ldots, N_2+1$, is

$$E\left[S^{(\omega)}(r)\right] \cong\left(N_1+1\right) \mu(r)+\left(\omega-N_1-1\right)(\mu(r)-\mu(z)), \quad \omega=N_1+2, \ldots, N_2+1, r \in\{1,2, \ldots, z-1\}. \tag{4.2.4}$$

More generally, random variables $\boldsymbol{T}^{(\omega)}(r)$, $\omega=N_x+2, \ldots, N_{x+1}+1$, $x=1,2, \ldots, z-r$, have the same distribution with expected values (Blokus-Roszkowska, 2019).

$$E\left[\boldsymbol{T}^{(\omega)}(r)\right]=\mu(r)-\mu(z-x+1), \omega=N_x+2, \ldots, N_{x+1}+1, \quad x=1,2, \ldots, z-r, r \in\{1,2, \ldots, z-1\} \tag{4.2.5}$$

and the expected values of random variables $S^{(\omega)}(r)$ are

$$E\left[S^{(\omega)}(r)\right] \cong\left(N_1+1\right) \mu(r)+\sum_{j=1}^{x-1}\left(N_{j+1}-N_j\right)(\mu(r)-\mu(z-j+1)) +\left(\omega-N_x-1\right)(\mu(r)-\mu(z-x+1)), \omega=N_x+2, \ldots, N_{x+1} +1, x=1,2, \ldots, z-r, r \in\{1,2, \ldots, z-1\}. \tag{4.2.6}$$

As assumed before, the system cannot be repaired after $N_{z-r+1}+1$ exceeding the reliability critical state r. Thus, time up to complete system damage is (Blokus-Roszkowska, 2019)

$$E\left[S^{\left(N_{z-r+1}+1\right)}(r)\right] \cong\left(N_1+1\right) \mu(r) +\sum_{j=1}^{z-r}\left(N_{j+1}-N_j\right)(\mu(r)-\mu(z-j+1)), \quad r \in\{1,2, \ldots, z-1\}. \tag{4.2.7}$$

Now, we consider a repairable system with non-negligible renewal time. Assumptions related to number of system's renewals to particular state

$z, z-1, \ldots, r, r\in\{1,2,\ldots,z-1\}$, are the same as before in Section 4.1 and in this section for a repairable system with negligible renewal time. $\boldsymbol{T}^{(1)}(r)$, $\boldsymbol{T}^{(2)}(r),\ldots, \boldsymbol{T}^{(N_1+1)}(r)$ are independent random variables from the same distribution with expected value $\mu(r)$ and standard deviation $\sigma(r)$. The successive times of system's renovations to the reliability state z after exceeding the critical state r, are independent and have an identical distribution function with expected value $\mu_R(z)$ and standard deviation $\sigma_R(z)$. Subsequently, the variable $S^{(N_1+1)}(r)$ representing the time until the N_1+1 exceeding the critical state r, has, for sufficiently large N_1, an approximately normal distribution with expected value (Blokus, 2006; Bobrowski, 1985; Kołowrocki and Soszyńska, 2010; Kołowrocki and Soszyńska-Budny, 2011).

$$E\left[S^{(N_1+1)}(r)\right] \cong (N_1+1)\mu(r) + N_1\mu_R(z), r\in\{1,2,\ldots,z-1\}, \quad (4.2.8)$$

and variance given by (Blokus, 2006; Bobrowski, 1985; Kołowrocki and Soszyńska, 2010; Kołowrocki and Soszyńska-Budny, 2011)

$$D\left[S^{(N_1+1)}(r)\right] \cong (N_1+1)\sigma^2(r) + N_1\sigma_R^2(z), r\in\{1,2,\ldots,z-1\}. \quad (4.2.9)$$

Using (4.2.8) it can be concluded that the expected value of a random variable $S^{(\omega)}(r)$, $\omega=1, 2, \ldots, N_1, N_1+1$, representing the time until the ωth exceeding the reliability critical state r, is

$$E\left[S^{(\omega)}(r)\right] \cong \omega\cdot\mu(r) + (\omega-1)\cdot\mu_R(z),$$
$$\omega=1,2,\ldots,N_1,N_1+1, r\in\{1,2,\ldots,z-1\}. \quad (4.2.10)$$

The expected value and the variance of time $S_R^{(N_1)}$ until the N_1th renovation of this system are, respectively, given by (Blokus, 2006; Bobrowski, 1985; Kołowrocki and Soszyńska, 2010; Kołowrocki and Soszyńska-Budny, 2011

$$E\left[S_R^{(N_1)}\right] \cong N_1(\mu(r)+\mu_R(z)), D\left[S_R^{(N_1)}(r)\right]$$
$$\cong N_1\left(\sigma^2(r)+\sigma_R^2(z)\right), r\in\{1,2,\ldots,z-1\}. \quad (4.2.11)$$

Similarly using (4.2.11), the expected value of a random variable $S_R^{(\omega)}$, $\omega=1$, 2, …, N_1, representing the time until the ωth system's renovation, is

$$E\left[S_R^{(\omega)}\right] \cong \omega\cdot(\mu(r)+\mu_R(z)), \omega=1,2,\ldots,N_1,$$
$$r\in\{1,2,\ldots,z-1\}. \quad (4.2.12)$$

The system renewal from the renovation N_1+1 to the renovation N_2 takes place to state $z-1$, and expected value of system recovery time to the reliability state $z-1$ after exceeding the critical state r is $\mu_R(z-1)$. The expected value of a random variable $\boldsymbol{T}^{(\omega)}(r)$, $\omega=N_1+2, \ldots, N_2+1$, is given by (4.2.3); thus, the expected value of a random variable $S^{(\omega)}(r)$, $\omega=N_1+2, \ldots, N_2+1$, is estimated from the formula

$$\begin{aligned} E\left[S^{(\omega)}(r)\right] &\cong (N_1+1)\mu(r)+N_1\mu_R(z) \\ &\quad+(\omega-N_1-1)(\mu(r)-\mu(z)+\mu_R(z-1)), \\ \omega &= N_1+2,\ldots,N_2+1, r\in\{1,2,\ldots,z-1\}. \end{aligned} \tag{4.2.13}$$

The expected value of time $S_R^{(\omega)}$, $\omega=N_1+1, \ldots, N_2$, until the ωth system's renovation is

$$\begin{aligned} E\left[S_R^{(\omega)}\right] &\cong (N_1+1)\mu(r)+N_1\mu_R(z)+(\omega-N_1-1)(\mu(r)-\mu(z)) \\ &+(\omega-N_1)\mu_R(z-1), \omega=N_1+1,\ldots,N_2, r\in\{1,2,\ldots,z-1\}. \end{aligned} \tag{4.2.14}$$

And more generally, from (4.2.5) it can be concluded that the expected values of random variables $S^{(\omega)}(r)$, $\omega=N_x+2, \ldots, N_{x+1}+1$, $x=1, 2, \ldots, z-r$, are

$$\begin{aligned} E\left[S^{(\omega)}(r)\right] &\cong (N_1+1)\mu(r)+N_1\mu_R(z) \\ &+\sum_{j=1}^{x-1}(N_{j+1}-N_j)(\mu(r)-\mu(z-j+1)+\mu_R(z-j)) \\ &+(\omega-N_x-1)(\mu(r)-\mu(z-x+1)+\mu_R(z-x)), \\ &\omega=N_x+2,\ldots,N_{x+1}+1, x=1,2,\ldots,z-r, \\ &r\in\{1,2,\ldots,z-1\}, \end{aligned} \tag{4.2.15}$$

where $\mu_R(u)$, $u=z, z-1, \ldots, r$, denotes expected value of system recovery time to the reliability state u after exceeding the critical state r.

As assumed before, the system cannot be repaired after $N_{z-r+1}+1$ exceeding the reliability critical state r. Thus, time up to complete system damage is

$$\begin{aligned} &E\left[S^{(N_{z-r+1}+1)}(r)\right] \cong (N_1+1)\mu(r)+N_1\mu_R(z) \\ &+\sum_{j=1}^{z-r}(N_{j+1}-N_j)(\mu(r)-\mu(z-j+1)+\mu_R(z-j)), r\in\{1,2,\ldots,z-1\}. \end{aligned} \tag{4.2.16}$$

Similarly generalizing, the expected value of time $S_R^{(\omega)}$, $\omega = N_x + 1, \ldots, N_{x+1}$, $x = 1, 2, \ldots, z - r$, until the ωth system's renovation is

$$E\left[S_R^{(\omega)}\right] \cong (N_1+1)\mu(r) + N_1\mu_R(z) + \sum_{j=1}^{x-1}\left(N_{j+1} - N_j\right)(\mu(r) - \mu(z-j+1) + \mu_R(z-j))$$
$$+(\omega - N_x - 1)(\mu(r) - \mu(z-x+1)) + (\omega - N_x)\mu_R(z-x), \omega = N_x+2, \ldots, N_{x+1}+1,$$
$$x = 1, 2, \ldots, z-r, r \in \{1, 2, \ldots, z-1\}. \tag{4.2.17}$$

4.3 Renewal function under imperfect repair

The expected value of the renewal process $N(t,r)$ is called the renewal function. As assumed in Sections 4.1 and 4.2, first N_1 system's renewals cause the system to return to the full reliability state z, and $\boldsymbol{T}^{(1)}(r)$, $\boldsymbol{T}^{(2)}(r), \ldots$, $\boldsymbol{T}^{(N_1+1)}(r)$ are independent random variables from the same distribution with expected value $\mu(r)$ and standard deviation $\sigma(r)$. We assume that system recovery time is negligible. In that case, the expected value of the number $N(t,r)$ of exceeding the reliability critical state r up to the time point t, $0 \le t \le S^{(N_1+1)}(r)$, i.e., up to time not larger than the time until the $N_1 + 1$ exceeding the reliability critical state r, is given by (Blokus, 2006; Bobrowski, 1985; Kołowrocki and Soszyńska, 2010; Kołowrocki and Soszyńska-Budny, 2011)

$$E[N(t,r)] \cong \frac{t}{\mu(r)}, 0 \le t \le S^{(N_1+1)}(r), r \in \{1, 2, \ldots, z-1\}. \tag{4.3.1}$$

Similarly, the expected value of the number $N(t,r)$ of exceeding the reliability critical state r up to the time point t, $S^{(N_1+1)}(r) < t \le S^{(N_2+1)}(r)$, is estimated from following formula

$$E[N(t,r)] \cong N_1 + 1 + \frac{t - (N_1+1)\mu(r)}{\mu(r) - \mu(z)}, S^{(N_1+1)}(r) < t$$
$$\le S^{(N_2+1)}(r), r \in \{1, 2, \ldots, z-1\}. \tag{4.3.2}$$

Generalizing, if random variables $\boldsymbol{T}^{(\omega)}(r)$, $\omega = N_x + 2, \ldots, N_{x+1} + 1$, $x = 1, 2, \ldots, z - r$, are independent random variables from the same distribution with expected value (4.2.5), then expected value of the number $N(t,r)$ of exceeding the reliability critical state r by a system, up to the time point t, $S^{(N_x+1)}(r) < t \le S^{(N_{x+1}+1)}(r)$, $x = 1, 2, \ldots, z - r$, is (Blokus-Roszkowska, 2019)

$$E[N(t,r)] \cong N_x + 1 + \frac{t - (N_1+1)\mu(r) - \sum_{j=1}^{x-1}(N_{j+1} - N_j)(\mu(r) - \mu(z-j+1))}{\mu(r) - \mu(z-x+1)},$$

$$S^{(N_x+1)}(r) < t \le S^{(N_{x+1}+1)}(r), x = 1,2,\ldots,z-r, r \in \{1,2,\ldots,z-1\}. \quad (4.3.3)$$

Taking into account obtained results, the expected value of the number $N(t,r)$ of exceeding the reliability critical state r up to the time point t, $t \geq 0$, is determined using the following procedure (Blokus-Roszkowska, 2019):

```
x = 1;
E[S^(N_x+1)(r)] ≅ (N_1 + 1)μ(r);
if (t ≤ E[S^(N_x+1)(r)])
   then E[N(t,r)] ≅ t/μ(r);
else {
   do{
      x := x + 1;
      E[S^(N_x+1)(r)] ≅ (N_1 + 1)μ(r) + Σ_{j=1}^{x-1} (N_{j+1} − N_j)(μ(r) − μ(z−j+1));
   } while (t > E[S^(N_x+1)(r)]) and (x < z−r+1);
   E[N(t,r)] ≅ N_{x−1} + 1 + [t − (N_1 + 1)μ(r) − Σ_{j=1}^{x-2} (N_{j+1} − N_j)(μ(r) − μ(z−j+1))] / [μ(r) − μ(z−x+2)];
   if (t > E[S^(N_{z−r+1}+1)(r)])
      then E[N(t,r)] ≅ N_{z−r+1} + 1; }
```

As a second case, a repairable system with non-negligible renewal time is analyzed. $\boldsymbol{T}^{(1)}(r)$, $\boldsymbol{T}^{(2)}(r), \ldots$, $\boldsymbol{T}^{(N_1+1)}(r)$ are independent random variables from the same distribution with expected value $\mu(r)$ and standard deviation $\sigma(r)$ for critical state of a system r $r \in \{1,2,\ldots,z-1\}$. The successive times of system's renovations to the reliability state z after exceeding the critical state r, are independent and have an identical distribution function with expected value $\mu_R(z)$ and standard deviation $\sigma_R(z)$.

In that case, the expected value of the number $N(t,r)$ of exceeding the reliability critical state r up to the time point t, $0 \leq t \leq S^{(N_1+1)}(r)$, is given by (Blokus, 2006; Bobrowski, 1985; Kołowrocki and Soszyńska, 2010; Kołowrocki and Soszyńska-Budny, 2011)

$$E[N(t,r)] \cong \frac{t+\mu_R(z)}{\mu(r)+\mu_R(z)}, 0 \le t \le S^{(N_1+1)}(r), r \in \{1,2,\ldots,z-1\}. \quad (4.3.4)$$

The expected value of the number $N(t,r)$ of exceeding the reliability critical state r up to the time point t, $S^{(N_1+1)}(r) < t \le S^{(N_2+1)}(r)$, is determined from following formula

$$E[N(t,r)] \cong N_1+1+\frac{t-(N_1+1)\mu(r)-N_1\mu_R(z)}{\mu(r)-\mu(z)+\mu_R(z-1)}, S^{(N_1+1)}(r) < t \le S^{(N_2+1)}(r),$$
$$r \in \{1,2,\ldots,z-1\}. \quad (4.3.5)$$

Generalizing, if random variables $T^{(\omega)}(r)$, $\omega = N_x+2, \ldots, N_{x+1}+1$, $x = 1, 2, \ldots, z-r$, are independent random variables from the same distribution with expected value (4.2.5), then expected value of the number $N(t,r)$ of exceeding the reliability critical state r up to the time point t, $S^{(N_x+1)}(r) < t \le S^{(N_{x+1}+1)}(r)$, $x = 1, 2, \ldots, z-r$, is

$$E[N(tr)] \cong N_x+1$$
$$+\frac{t-(N_1+1)\mu(r)-N_1\mu_R(z)-\sum_{j=1}^{x-1}(N_{j+1}-N_j)(\mu(r)-\mu(z-j+1)+\mu_R(z-j))}{\mu(r)-\mu(z-x+1)+\mu_R(z-x)},$$
$$S^{(N_x+1)}(r) < t \le S^{(N_{x+1}+1)}(r), \quad x = 1,2,\ldots,z-r, \; r \in \{1,2,\ldots,z-1\}. \quad (4.3.6)$$

Taking into account obtained results, the expected value of the number $N(t,r)$ of exceeding the reliability critical state r up to the time point t, $t \ge 0$, is determined using the following procedure:

$x = 1$;

$E[S^{(N_x+1)}(r)] \cong (N_1+1)\mu(r)+N_1\mu_R(z)$;

`if` $(t \le E[S^{(N_x+1)}(r)])$

  `then` $E[N(t,r)] \cong \frac{t+\mu_R(z)}{\mu(r)+\mu_R(z)}$;

`else {`

  `do{`

  $x := x+1$;

  $E[S^{(N_x+1)}(r)] \cong (N_1+1)\mu(r)+N_1\mu_R(z)$

  $+\sum_{j=1}^{x-1}(N_{j+1}-N_j)(\mu(r)-\mu(z-j+1)+\mu_R(z-j))$;

  `} while` $(t > E[S^{(N_x+1)}(r)])$ `and` $(x < z-r+1)$;

$E[N(t,r)] \cong N_{x-1} + 1$

$+\dfrac{t-(N_1+1)\mu(r)-N_1\mu_R(z)-\sum_{j=1}^{x-2}(N_{j+1}-N_j)(\mu(r)-\mu(z-j+1)+\mu_R(z-j))}{\mu(r)-\mu(z-x+2)+\mu_R(z-x+1)};$

if $(t > E[S^{(N_{z-r+1}+1)}(r)])$

then $E[N(t,r)] \cong N_{z-r+1} + 1;\}$

Another availability characteristic of repairable system with non-negligible renewal time is the expected value of the number of system renovations up to the time point t, $t \geq 0$. Next, from results given in (Blokus, 2006; Bobrowski, 1985; Kołowrocki and Soszyńska, 2010; Kołowrocki and Soszyńska-Budny, 2011), the expected value of the number $N_R(t)$ of system renovations up to the time point t, $0 \leq t \leq S^{(N_1+1)}(r)$, is determined from

$$E[N_R(t)] \cong \frac{t}{\mu(r)+\mu_R(z)}, 0 \leq t \leq S^{(N_1+1)}(r), r \in \{1, 2, \ldots, z-1\}. \quad (4.3.7)$$

Similarly as before, we propose procedure for determining expected value of the number $N_R(t)$ of system renovations up to the time point t, $t \geq 0$:

$x = 1;$

$E[S^{(N_x+1)}(r)] \cong (N_1+1)\mu(r) + N_1\mu_R(z);$

if $(t \leq E[S^{(N_x+1)}(r)])$

then $E[N_R(t)] \cong \dfrac{t}{\mu(r)+\mu_R(z)};$

else {

do{

$x := x+1;$

$E[S^{(N_x+1)}(r)] \cong (N_{1+1})\mu(r) + N_1\mu_R(z)$

$+\sum_{j=1}^{x-1}(N_{j+1}-N_j)(\mu(r)-\mu(z-j+1)+\mu_R(z-j));$

} while $(t > E[S^{(N_x+1)}(r)])$ and $(x < z-r+1);$

if $(t < (E[S^{(N_{x-1}+1)}(r)] + \mu_R(z-x+1)))$

then $E[N_R(t))] \cong N_{x-1};$

else {

$E[N_R(t)] \cong N_{x-1} + 1$

$+\dfrac{t-(N_1+1)\mu(r)-N_1\mu_R(z)-\sum_{j=1}^{x-2}(N_{j+1}-N_j)(\mu(r)-\mu(z-j+1)+\mu_R(z-j))-\mu_R(z-x+1)}{\mu(r)-\mu(z-x+2)+\mu_R(z-x+1)};$

```
        if (t>(E[S^(N_{z-r+1})(r)]+μ_R(z-x+1)))
          then E[N_R(t)]≅N_{z-r+1};
    }
  }
```

Finally, the steady availability coefficient of the system at the moment t, $t \geq 0$, is determined from the formula

$$A(t,r) \cong \frac{\mu(r)}{\mu(r)+\mu_R(z)}, 0 \leq t \leq S^{(N_1+1)}(r), r \in \{1,2,\ldots,z-1\}. \tag{4.3.8}$$

As the expected values of time between the moment of the $\omega - 1$ system's renovation and the ω-th exceeding the system critical state, after some time can be shorter, the availability coefficient can also change. Next, assuming as before that after N_1 system's renewals to the full reliability state, i.e., z, subsequent repairs can cause the system to return only up to the reliability state $z-1$ or worse state, we get

$$\begin{aligned} A(t,r) \cong & \frac{\mu(r)-\mu(z)}{\mu(r)-\mu(z)+\mu_R(z-1)}, S^{(N_1+1)}(r) < t \\ & \leq S^{(N_2+1)}(r), r \in \{1,2,\ldots,z-1\}, \end{aligned} \tag{4.3.9}$$

and

$$\begin{aligned} A(tr) \cong & \frac{\mu(r)-\mu(z-x+1)}{\mu(r)-\mu(z-x+1)+\mu_R(z-x)}, \quad S^{(N_x+1)}(r) < t \\ & \leq S^{(N_{x+1}+1)}(r), \quad x = 1,2,\ldots,z-r, \ r \in \{1,2,\ldots,z-1\}. \end{aligned} \tag{4.3.10}$$

4.4 Local load-sharing dependency model

The results of availability analysis of multistate renewal systems, presented in Sections 4.2 and 4.3, apply to systems with independent components. In this section and next two sections, multistate systems assuming the dependence of components' departures from the reliability state subsets are considered. More exactly, in Section 4.4 local load-sharing (LLS) dependency model, widely described in Section 3.1, is analyzed. In Section 4.5 equal load-sharing (ELS) dependency model among components, introduced in Section 3.2, is considered. And the last dependency model, described in Section 3.3, is mixed load sharing (MLS) that is applied in Section 4.6 to availability analysis of multistate parallel-series and "m out of l"-series systems assuming the dependency among subsystems of a system and the dependency among components in subsystems.

4.4.1 Availability of multistate series system with dependent components

We assume that components E_i, $i=1, \ldots, n$, of a series system have piecewise exponential reliability functions given by

$$R_i(t\cdot) = [1, R_i(t1), \ldots, R_i(t,z)], \quad t \geq 0, \ i=1,2,\ldots..n, \tag{4.4.1}$$

with the coordinates

$$R_i(t, u) = \exp[-\lambda_i(u)t], u=1,2,\ldots,z, \tag{4.4.2}$$

where $\lambda_i(u)$, $\lambda_i(u) \geq 0$, $i=1, \ldots, n$, are components' intensities of departure from the reliability state subsets $\{u,u+1, \ldots, z\}$, $u=1,2, \ldots, z$.

From *Corollary 3.1.1* in Chapter 3, and using results given in (Blokus-Roszkowska and Kołowrocki, 2015b), for a multistate series system with components dependent according the LLS rule and having piecewise exponential reliability functions (4.4.1)–(4.4.2), the system's mean lifetime in the reliability state subsets $\{u,u+1, \ldots, z\}$, $u=1,2, \ldots, z$, is given by

$$\mu_{LLS}(u) = \frac{1}{\sum_{i=1}^{n} \lambda_i(u+1)} + \sum_{j=1}^{n} \frac{\lambda_j(u+1) - \lambda_j(u)}{\sum_{i=1}^{n} (\lambda_i(u+1) - \lambda_i(u))}$$
$$\cdot \left[\frac{1}{\sum_{i=1}^{n} \frac{\lambda_i(u)}{q(u, E_j, E_i)}} - \frac{1}{\sum_{i=1}^{n} \left(\lambda_i(u+1) - \lambda_i(u) + \frac{\lambda_i(u)}{q(u, E_j, E_i)} \right)} \right], \tag{4.4.3}$$
$$u=1,2,\ldots,z-1,$$

$$\mu_{LLS}(z) = \frac{1}{\sum_{i=1}^{n} \lambda_i(z)}. \tag{4.4.4}$$

The assumptions about the system renovation are the same as in Section 4.1, and there are used similar notations. Thus, a random variable $S_{LLS}^{(N)}(r)$ represents the time until the Nth exceeding the reliability critical state r by a system following the LLS dependency rule, and $N_{LLS}(t,r)$ is the number of exceeding the critical state r up to the time point t, by the system with components dependent according to the LLS rule. Numbers of system renovations $N_1, N_2, \ldots, N_{z-r+1}$, have the same interpretation as in Section 4.1

and mean, respectively, the operations that cause a system to return to its full reliability state z, to a state $z-1$, and to the critical state r.

Under these assumptions, the expected value of a random variable $S_{LLS}^{(\omega)}(r)$, $\omega=1, 2, \ldots, N_1, N_1+1$, representing the time until the ωth exceeding the reliability critical state r, applying (4.2.2), is (Blokus-Roszkowska, 2019).

$$E\left[S_{LLS}^{(\omega)}(r)\right] \cong \omega \cdot \mu_{LLS}(r),$$
$$\omega=1,2,\ldots,N_1,N_1+1, r\in\{1,2,\ldots,z-1\}, \tag{4.4.5}$$

where $\mu_{LLS}(r)$ is given by (4.4.3) for $u=r$, $r\in\{1,2,\ldots,z-1\}$.

The expected values of random variables $S_{LLS}^{(\omega)}(r)$, $\omega=N_x+2, \ldots, N_{x+1}+1$, $x=1, 2, \ldots, z-r$, applying (4.2.6), are

$$E\left[S_{LLS}^{(\omega)}(r)\right] \cong (N_1+1)\mu_{LLS}(r) + \sum_{j=1}^{x-1}\left(N_{j+1}-N_j\right)\left(\mu_{LLS}(r)-\mu_{LLS}(z-j+1)\right)$$
$$+(\omega-N_x-1)(\mu_{LLS}(r)-\mu_{LLS}(z-x+1)), \omega=N_x+2,\ldots,N_{x+1}$$
$$+1, x=1,2,\ldots,z-r, \tag{4.4.6}$$

and the time up to complete system damage, by (4.2.7), is

$$E\left[S_{LLS}^{(N_{z-r+1}+1)}(r)\right] \cong (N_1+1)\mu_{LLS}(r)$$
$$+\sum_{j=1}^{z-r}\left(N_{j+1}-N_j\right)\left(\mu_{LLS}(r)-\mu_{LLS}(z-j+1)\right), \tag{4.4.7}$$

for $r\in\{1,2,\ldots,z-1\}$, where $\mu_{LLS}(u)$, $u=1, 2, \ldots, z-1$, is given by (4.4.3) and $\mu_{LLS}(z)$ by (4.4.4).

Next, the expected value of the number $N_{LLS}(t,r)$ of times the system following LLS dependency rule exceeds the reliability critical state r up to the time point t, $0\leq t\leq S_{LLS}^{(N_1+1)}(r)$, i.e., up to time not larger than the time until the N_1+1-th exceeding the reliability critical state r, applying (4.3.1), is

$$E[N_{LLS}(t,r)] \cong \frac{t}{\mu_{LLS}(r)}, 0\leq t\leq S_{LLS}^{(N_1+1)}(r), r\in\{1,2,\ldots,z-1\}, \tag{4.4.8}$$

where $\mu_{LLS}(r)$ is given by (4.4.3) for $u=r$, $r\in\{1,2,\ldots,z-1\}$.

Subsequently, the expected value of the number $N_{LLS}(t,r)$ of exceeding the reliability critical state r up to the time point t by a system following the

LLS dependency rule, $S_{LLS}^{(N_1+1)}(r) < t \le S_{LLS}^{(N_2+1)}(r)$, applying (4.3.2) and using (4.4.3)–(4.4.4), is estimated from following formula

$$E[N_{LLS}(tr)] \cong N_1 + 1 + \frac{t-(N_1+1)\mu_{LLS}(r)}{\mu_{LLS}(r)-\mu_{LLS}(z)}, \quad S_{LLS}^{(N_1+1)}(r) < t \le S_{LLS}^{(N_2+1)}(r), \quad r \in \{1, 2, \ldots, z-1\}. \tag{4.4.9}$$

Generalizing, the expected value of the number $N_{LLS}(t,r)$ of exceeding the reliability critical state r by a system following the LLS dependency rule, up to the time point t, $S_{LLS}^{(N_x+1)}(r) < t \le S_{LLS}^{(N_{x+1}+1)}(r)$, $x = 1, 2, \ldots, z-r$, by (4.3.3), is

$$E[N_{LLS}(t,r)] \cong N_x + 1 + \frac{t-(N_1+1)\mu_{LLS}(r) - \sum_{j=1}^{x-1}(N_{j+1}-N_j)(\mu_{LLS}(r)-\mu_{LLS}(z-j+1))}{\mu_{LLS}(r)-\mu_{LLS}(z-x+1)},$$
$$S_{LLS}^{(N_x+1)}(r) < t \le S_{LLS}^{(N_{x+1}+1)}(r), x = 1,2,\ldots,z-r, r \in \{1, 2, \ldots, z-1\}, \tag{4.4.10}$$

where $\mu_{LLS}(u)$, $u = 1, 2, \ldots, z-1$, is given by (4.4.3) and $\mu_{LLS}(z)$ by (4.4.4).

Considering series system as a repairable system with non-negligible renewal time, we estimate its availability coefficient. The steady availability coefficient of a series system following the LLS rule, at the moment t, $t \ge 0$, by (4.3.8)–(4.3.10), is determined as follows

$$A_{LLS}(t,r) \cong \frac{\mu_{LLS}(r)}{\mu_{LLS}(r)+\mu_R(z)}, 0 \le t \le S_{LLS}^{(N_1+1)}(r), \tag{4.4.11}$$

$$A_{LLS}(t,r) \cong \frac{\mu_{LLS}(r)-\mu_{LLS}(z)}{\mu_{LLS}(r)-\mu_{LLS}(z)+\mu_R(z-1)}, S_{LLS}^{(N_1+1)}(r) < t \le S_{LLS}^{(N_2+1)}(r), \tag{4.4.12}$$

$$A_{LLS}(t,r) \cong \frac{\mu_{LLS}(r)-\mu_{LLS}(z-x+1)}{\mu_{LLS}(r)-\mu_{LLS}(z-x+1)+\mu_R(z-x)}, S_{LLS}^{(N_x+1)}(r) < t \le S_{LLS}^{(N_{x+1}+1)}(r),$$
$$x = 1,2,\ldots,z-r, r \in \{1, 2, \ldots, z-1\}, \tag{4.4.13}$$

where $\mu_R(u)$, $u = z, z-1, \ldots, r$, denotes expected value of system recovery time to the reliability state u after exceeding the critical state r, $\mu_{LLS}(u)$, $u = 1, 2, \ldots, z-1$, is given by (4.4.3) and $\mu_{LLS}(z)$ by (4.4.4).

To illustrate the influence of the component dependence on availability, the 4-state series system composed of identical components is taken into account. It is assumed that components have piecewise exponential reliability functions with the intensities of departure from the reliability state subsets {1,2,3}, {2,3}, {3}, by the assumption, given by

$$\lambda(1)=0.125, \lambda(2)=0.167, \lambda(3)=0.5. \tag{4.4.14}$$

The coefficients $q(v, E_j, E_i)$, $v=u, u-1, \ldots, 1$, $u=1,2$, of the component E_j, $j=1,2, \ldots, n$, impact on other components E_i, $i=1,2, \ldots, n$, $i \neq j$, lifetimes in the reliability state subsets are given by

$$q\left(v, E_j, E_i\right)=1-\exp[-|i-j|], i=1,2,\ldots,n, j=1,2,\ldots,n, i\neq j, \tag{4.4.15}$$

and

$$q\left(v, E_j, E_j\right)=1, j=1,2,\ldots,n. \tag{4.4.16}$$

In Table 4.1 the expected values of lifetime in the reliability state subsets {1,2,3}, {2,3}, {3} of the exemplary series system for different numbers of components n, $n=3,4, \ldots, 10$ are given. The mean values are determined in two cases, assuming independence of its components and assuming dependence according to LLS rule, described in Section 3.1.1. Because the last coordinate of a system reliability function in the full reliability state, i.e., $u=3$, for a multistate series system following the LLS dependency rule is the same as for a multistate series system with independent components, we get $\mu(3)=\mu_{LLS}(3)$.

A critical reliability state of the system is assumed $r=1$. For the availability analysis of the exemplary series system, the number of its renewals is fixed. It is assumed that, first four system renovations ($N_1=4$) are to the best state $z=3$, next four renewals of the system ($N_2=8$) take place to the state 2, the last two

Table 4.1 The expected values of lifetime in the reliability state subsets {1,2,3}, {2,3}, {3} of an exemplary series system assuming components' independence and dependence according to LLS rule.

n	$\mu(1)$	$\mu_{LLS}(1)$	$\mu(2)$	$\mu_{LLS}(2)$	$\mu(3)=\mu_{LLS}(3)$
3	2.667	2.424	1.996	1.606	0.667
4	2.000	1.832	1.497	1.228	0.500
5	1.600	1.477	1.198	1.002	0.400
6	1.333	1.239	0.998	0.849	0.333
7	1.143	1.069	0.855	0.739	0.286
8	1.000	0.940	0.749	0.655	0.250
9	0.889	0.840	0.665	0.588	0.222
10	0.800	0.759	0.599	0.534	0.200

Table 4.2 The expected values of the time until *N*th exceeding the reliability critical state by an exemplary series system assuming components' independence and dependence according to LLS rule.

	$n=5$		$n=6$		$n=7$		$n=8$		$n=9$		$n=10$	
N	**indep.**	**LLS**	**indep.**	**LLS**	**indep.**	**LLS**	**indep.**	**LLS**	**indep.**	**LLS**	**indep.**	**LLS**
1	1.60	1.48	1.33	1.24	1.14	1.07	1.00	0.94	0.89	0.84	0.80	0.76
2	3.20	2.95	2.67	2.48	2.29	2.14	2.00	1.88	1.78	1.68	1.60	1.52
3	4.80	4.43	4.00	3.72	3.43	3.21	3.00	2.82	2.67	2.52	2.40	2.28
4	6.40	5.91	5.33	4.96	4.57	4.28	4.00	3.76	3.56	3.36	3.20	3.04
5	8.00	7.39	6.67	6.20	5.72	5.35	5.00	4.70	4.45	4.20	4.00	3.80
6	9.20	8.46	7.67	7.10	6.57	6.13	5.75	5.39	5.11	4.82	4.60	4.35
7	10.40	9.54	8.67	8.01	7.43	6.91	6.50	6.08	5.78	5.44	5.20	4.91
8	11.60	10.62	9.67	8.91	8.29	7.69	7.25	6.77	6.45	6.05	5.80	5.47
9	12.80	11.69	10.67	9.82	9.14	8.48	8.00	7.46	7.11	6.67	6.40	6.03
10	13.20	12.17	11.00	10.21	9.43	8.81	8.25	7.75	7.34	6.92	6.60	6.26
11	13.60	12.64	11.34	10.60	9.72	9.14	8.50	8.03	7.56	7.18	6.80	6.48

renewals ($N_3=10$) include operations that causes the system to return to the critical state 1. After the system has exceeded the critical state eleventh times ($N_3+1=11$), the system remains in the state of complete failure. The results for the exemplary series system with negligible renewal time, assuming components' independence and dependence according to the LLS dependency rule, are compared in Table 4.2 and graphically in Fig. 4.1.

Comparing these results, the expected values of the time until *N*th exceeding the reliability critical state by a series system with dependent components are less by about 5% than for a system with independent components, when the system consists of 10 components. These differences are greater for systems with fewer components. And so in the case of a 5-component series system following the LLS rule of dependency, the expected values of time until successive exceeding the critical state are less by about 8% compared to a system with independent components.

As a second case, we consider this series system as a repairable system with non-negligible renewal time, and assume that the expected value of system recovery time to the reliability state 3, 2, and 1 after exceeding the critical state $r=1$ is 0.04, i.e.,

$$\mu_R(3)=0.04, \mu_R(2)=0.04, \mu_R(1)=0.04.$$

Taking the assumptions regarding reliability parameters and number of renewals, as before for a system with negligible renewal time, we get the

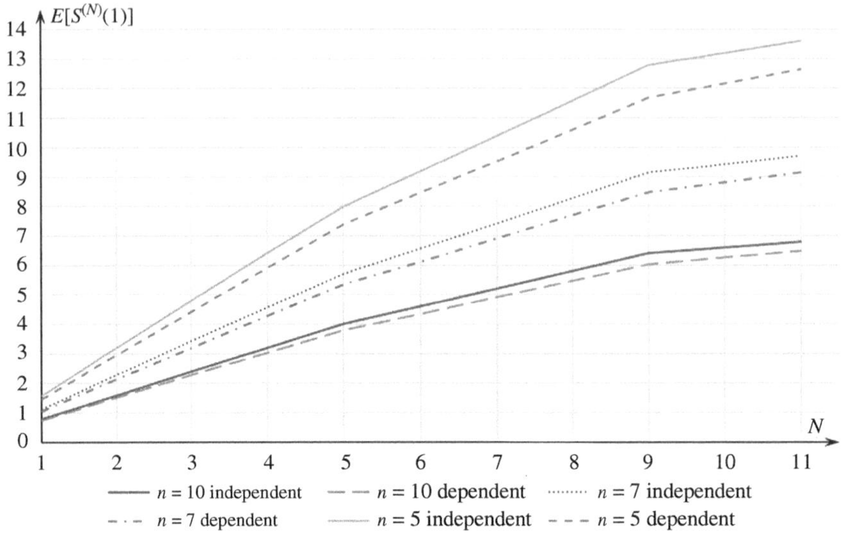

Fig. 4.1 The graph of the expected values of the time until *N*th exceeding the reliability critical state by an exemplary series system assuming components' independence and dependence according to LLS rule.

values of steady availability coefficient of the system with non-negligible renewal time, given in Table 4.3. The availability coefficient of the exemplary series system is determined, assuming components' independence and dependence according to the LLS rule, for different numbers of components, and the results are compared graphically in Figs. 4.2 and 4.3.

4.4.2 Availability of multistate series-parallel system with dependent components of its subsystems

Now we consider a multistate series-parallel system, described in Section 3.1.3, composed of k multistate series subsystems. We assume that these subsystems are working independently and are linked in parallel. In the ith series subsystem S_i, $i=1,2, \ldots, k$, there are l_i components dependent according to the LLS rule. We denote by E_{ij}, $i=1,2, \ldots, k$, $j=1,2, \ldots, l_i$, the system components and assume that they have piecewise exponential reliability functions

$$R_{ij}(t, \cdot) = [1, R_{ij}(t, 1), \ldots, R_{ij}(t, z)], t \geq 0, \tag{4.4.17}$$

Table 4.3 The values of steady availability coefficient of an exemplary series system, assuming components' independence and dependence according to LLS rule (%).

	$n=5$		$n=6$		$n=7$		$n=8$		$n=9$		$n=10$	
t	**indep.**	**LLS**	**indep.**	**LLS**	**indep.**	**LLS**	**indep.**	**LLS**	**indep.**	**LLS**	**indep.**	**LLS**
0.5	97.56	97.36	97.09	96.87	96.62	96.39	96.15	95.92	95.69	95.45	95.24	94.99
1.0	97.56	97.36	97.09	96.87	96.62	96.39	96.15	95.92	95.69	95.45	95.24	94.99
1.5	97.56	97.36	97.09	96.87	96.62	96.39	96.15	95.92	95.69	95.45	95.24	94.99
2.0	97.56	97.36	97.09	96.87	96.62	96.39	96.15	95.92	95.69	95.45	95.24	94.99
2.5	97.56	97.36	97.09	96.87	96.62	96.39	96.15	95.92	95.69	95.45	95.24	94.99
3.0	97.56	97.36	97.09	96.87	96.62	96.39	96.15	95.92	95.69	95.45	95.24	94.99
3.5	97.56	97.36	97.09	96.87	96.62	96.39	96.15	95.92	95.69	95.45	95.24	94.99
4.0	97.56	97.36	97.09	96.87	96.62	96.39	96.15	95.92	95.69	95.45	95.24	93.32
4.5	97.56	97.36	97.09	96.87	96.62	96.39	96.15	95.92	95.69	93.92	93.75	93.32
5.0	97.56	97.36	97.09	96.87	96.62	96.39	96.15	94.52	94.34	93.92	93.75	93.32
5.5	97.56	97.36	97.09	96.87	96.62	96.39	94.94	94.52	94.34	93.92	93.75	93.32
6.0	97.56	97.36	97.09	96.87	95.54	95.14	94.94	94.52	94.34	93.92	93.75	93.32
6.5	97.56	97.36	97.09	95.77	95.54	95.14	94.94	94.52	94.34	93.92	93.75	84.91
7.0	97.56	97.36	96.15	95.77	95.54	95.14	94.94	94.52	94.34	86.30	83.40	0.00
7.5	97.56	97.36	96.15	95.77	95.54	95.14	94.94	94.52	84.85	86.30	0.00	0.00
8.0	97.56	96.42	96.15	95.77	95.54	95.14	94.94	87.69	0.00	0.00	0.00	0.00
8.5	96.77	96.42	96.15	95.77	95.54	95.14	86.25	0.00	0.00	0.00	0.00	0.00
9.0	96.77	96.42	96.15	95.77	95.54	89.19	0.00	0.00	0.00	0.00	0.00	0.00
9.5	96.77	96.42	96.15	95.77	87.80	89.19	0.00	0.00	0.00	0.00	0.00	0.00
10.0	96.77	96.42	96.15	95.77	87.80	0.00	0.00	0.00	0.00	0.00	0.00	0.00
10.5	96.77	96.42	96.15	90.70	0.00	0.00	0.00	0.00	0.00	0.00	0.00	0.00
11.0	96.77	96.42	89.33	0.00	0.00	0.00	0.00	0.00	0.00	0.00	0.00	0.00

Continued

Table 4.3 The values of steady availability coefficient of an exemplary series system, assuming components' independence and dependence according to LLS rule (%)—cont'd

	$n=5$		$n=6$		$n=7$		$n=8$		$n=9$		$n=10$	
t	indep.	LLS	indep.	LLS	indep.	LLS	indep.	LLS	indep.	LLS	indep.	LLS
11.5	96.77	96.42	89.33	0.00	0.00	0.00	0.00	0.00	0.00	0.00	0.00	0.00
12.0	96.77	96.42	0.00	0.00	0.00	0.00	0.00	0.00	0.00	0.00	0.00	0.00
12.5	96.77	92.23	0.00	0.00	0.00	0.00	0.00	0.00	0.00	0.00	0.00	0.00
13.0	96.77	92.23	0.00	0.00	0.00	0.00	0.00	0.00	0.00	0.00	0.00	0.00
13.5	90.95	0.00	0.00	0.00	0.00	0.00	0.00	0.00	0.00	0.00	0.00	0.00
14.0	90.95	0.00	0.00	0.00	0.00	0.00	0.00	0.00	0.00	0.00	0.00	0.00
14.5	0.00	0.00	0.00	0.00	0.00	0.00	0.00	0.00	0.00	0.00	0.00	0.00
15.0	0.00	0.00	0.00	0.00	0.00	0.00	0.00	0.00	0.00	0.00	0.00	0.00

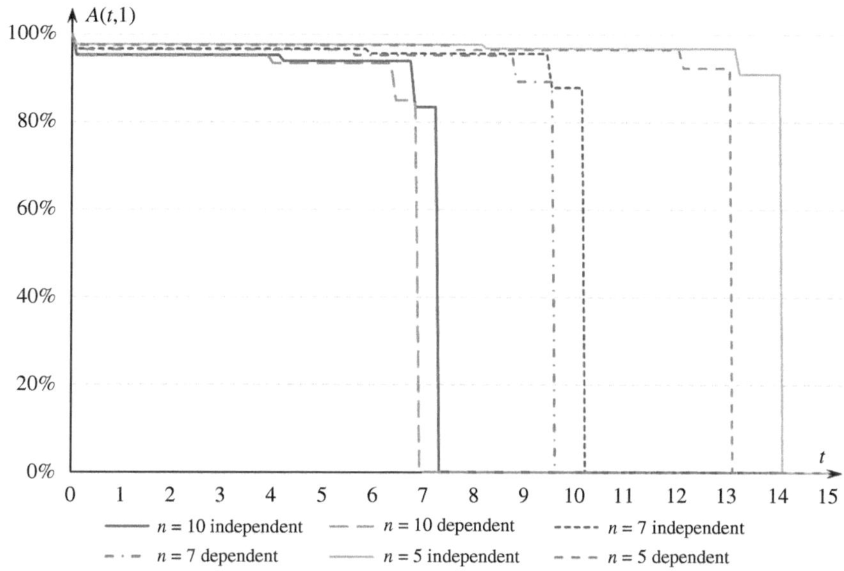

Fig. 4.2 The graphs of steady availability coefficient of an exemplary series system, assuming components' independence and dependence according to LLS rule.

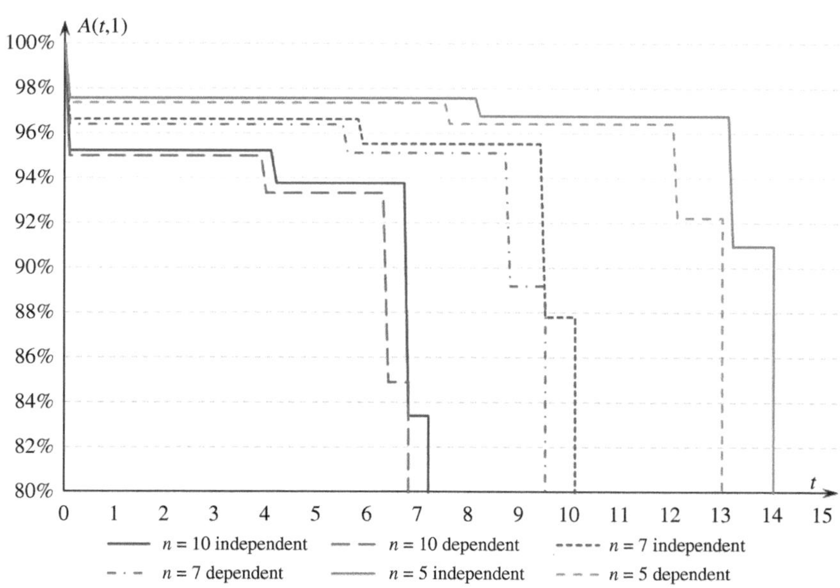

Fig. 4.3 The graphs of steady availability coefficient of an exemplary series system, assuming components' independence and dependence according to LLS rule (for a scale of values from 80% to 100%).

with the coordinates

$$R_{ij}(t, u) = \exp\left[-\lambda_{ij}(u)t\right], \quad u = 1, 2, \ldots, z, \tag{4.4.18}$$

where $\lambda_{ij}(u)$, $i = 1,2, \ldots, k$, $j = 1,2, \ldots, l_i$, are E_{ij} components' intensities of departure from the reliability state subsets $\{u, u+1, \ldots, z\}$, $u = 1,2, \ldots, z$.

Using the results of *Proposition 3.1.9*, i.e., the reliability function of an aging series-parallel system with dependent according to the LLS rule components of its subsystems, the mean lifetime of the system in the reliability state subsets $\{u, u+1, \ldots, z\}$, $u = 1,2, \ldots, z$, is given by the formula

$$\mu_{LLS}(u) = \int_0^{+\infty} \overline{\boldsymbol{R}}_{LLS}(t, u)\,dt, \quad u = 1, 2, \ldots, z, \tag{4.4.19}$$

where $\overline{\boldsymbol{R}}_{LLS}(t, u)$ for $u = 1, 2, \ldots, z-1$, is given by (3.1.100) and $\overline{\boldsymbol{R}}_{LLS}(t, z)$ is given by (3.1.101).

In that case, the expected values of the time until *N*th exceeding the reliability critical state *r* by a multistate series-parallel system following the LLS dependency rule, are given by (4.4.5)–(4.4.7), where $\mu_{LLS}(u)$, $u = 1, 2, \ldots, z$, is given by (4.4.19) and in particular case $\mu_{LLS}(r)$ is determined from (4.4.19) for $u = r$, where the critical reliability state *r* is fixed and $r \in \{1, 2, \ldots, z-1\}$.

Similarly, the expected value of the number $N_{LLS}(t,r)$ of exceeding the reliability critical state *r* by a multistate series-parallel system following the LLS dependency rule, up to the time point *t*, $t \geq 0$, is determined from (4.4.8) and (4.4.10), substituting mean value obtained from (4.4.19).

4.4.3 Availability of multistate series-"*m* out of *k*" system with dependent components of its subsystems

In this section, we analyze a multistate series-"*m* out of *k*" system, described in Section 3.1.4, as an "*m* out of *k*" system composed of *k* series subsystems with dependent according to the LLS rule components. We assume its components E_{ij}, $i = 1,2, \ldots, k$, $j = 1,2, \ldots, l_i$, have piecewise exponential reliability functions given by (4.4.17)–(4.4.18).

Under these assumptions, using the reliability function of an aging series-"*m* out of *k*" system with the coordinates given by (3.1.111)–(3.1.112) or by (3.1.113)–(3.1.114) in *Proposition 3.1.13*, the system's mean lifetime in the state subsets $\{u, u+1, \ldots, z\}$, $u = 1,2, \ldots, z$, is given (4.4.19), similarly as for a series-parallel system in Section 4.4.2. Next, the expected values of the time until *N*th exceeding the reliability critical state *r* by a multistate series-"m out of k" system following the LLS dependency rule, and the expected value of the number $N_{LLS}(t,r)$ of exceeding the reliability critical state r up to the time point t, $t \geq 0$, are determined substituting obtained from (4.4.19) value of system's mean lifetime into (4.4.5)–(4.4.7) and (4.4.8)–(4.4.10), respectively.

4.5 Equal load-sharing dependency model

4.5.1 Availability of multistate parallel system with dependent components

In this section, we consider a homogeneous and aging parallel system with components having piecewise exponential reliability functions

$$R(t, \cdot) = [1, R(t, 1), \ldots, R(t, z)], t \geq 0, \tag{4.5.1}$$

with the coordinates

$$R(t, u) = \exp[-\lambda(u)t], t \geq 0, \lambda(u) \geq 0, u = 1, 2, \ldots, z. \tag{4.5.2}$$

From Corollary 3.2.2 in Chapter 3, and using results given in (Blokus-Roszkowska, 2007), we obtain the mean value of lifetime in the reliability state subsets $\{u, u+1, \ldots, z\}$, $u = 1, 2, \ldots, z$, for a parallel system following the ELS dependency rule

$$\mu_{ELS}(u) = \frac{c(u)}{\lambda(u)}, u = 1, 2, \ldots, z, \tag{4.5.3}$$

where $\lambda(u)$, $\lambda(u) \geq 0$, $u = 1, 2, \ldots, z$, is components' intensity of departure from the reliability state subset $\{u, u+1, \ldots, z\}$, and $c(u)$ is the component stress proportionality correction coefficient for each u, $u = 1, 2, \ldots, z$.

The assumptions and notations related to the system renovation are the same as in Section 4.1. Thereby, a random variable $S_{ELS}^{(N)}(r)$ represents the time until the Nth exceeding the reliability critical state r by a system following the ELS rule, and $N_{ELS}(t,r)$ is the number of exceeding the critical state r up to the time point t, by the system with components dependent according to the ELS dependency rule. Numbers of system renovations N_1, N_2, $\ldots$, N_{z-r+1}, have the same interpretation as in Section 4.1 and mean, respectively, the operations that cause a system to return to its full reliability state z, to a state $z-1$, and to the critical state r.

Using the result (4.5.3) for $u=r$ and applying (4.2.2), we get the expected value of a random variable $S_{ELS}^{(\omega)}(r)$, $\omega=1, 2, \ldots, N_1, N_1+1$, representing the time until the ωth exceeding the reliability critical state r.

$$E\left[S_{ELS}^{(\omega)}(r)\right] \cong \omega \cdot \frac{c(r)}{\lambda(r)}, \quad \omega=1,2,\ldots,N_1,N_1+1, r\in\{1, 2, \ldots, z-1\}. \tag{4.5.4}$$

Random variables $\boldsymbol{T}_{ELS}^{(\omega)}(r)$, $\omega=N_x+2, \ldots, N_{x+1}+1$, $x=1, 2, \ldots, z-r$, by (4.2.5) and (4.5.3), have the same distribution with expected value

$$E\left[\boldsymbol{T}_{ELS}^{(\omega)}(r)\right] = \frac{c(r)}{\lambda(r)} - \frac{c(z-x+1)}{\lambda(z-x+1)}, \quad \omega=N_x+2,\ldots,N_{x+1}+1, \quad x=1,2,\ldots,z-r, \tag{4.5.5}$$

for $r\in\{1,2,\ldots,z-1\}$.

Subsequently, the expected values of random variables $S_{ELS}^{(\omega)}(r)$, $\omega=N_x+2, \ldots, N_{x+1}+1$, $x=1, 2, \ldots, z-r$, applying (4.2.6) and using (4.5.3), are

$$E\left[S_{ELS}^{(\omega)}(r)\right] \cong (N_1+1)\frac{c(r)}{\lambda(r)} + \sum_{j=1}^{x-1}(N_{j+1}-N_j)\left(\frac{c(r)}{\lambda(r)} - \frac{c(z-j+1)}{\lambda(z-j+1)}\right) + (\omega-N_x-1)\left(\frac{c(r)}{\lambda(r)} - \frac{c(z-x+1)}{\lambda(z-x+1)}\right), \quad \omega=N_x+2,\ldots,N_{x+1}+1, x=1,2,\ldots,z-r, \tag{4.5.6}$$

and the time up to complete system damage, by (4.2.7) and (4.5.3), is

$$E\left[S_{ELS}^{(N_{z-r+1}+1)}(r)\right] \cong (N_1+1)\frac{c(r)}{\lambda(r)} + \sum_{j=1}^{z-r}(N_{j+1}-N_j)\left(\frac{c(r)}{\lambda(r)} - \frac{c(z-j+1)}{\lambda(z-j+1)}\right), \tag{4.5.7}$$

for $r\in\{1,2,\ldots,z-1\}$.

Next, the expected value of the number $N_{ELS}(t,r)$ of exceeding the reliability critical state r up to the time point t, $0\le t\le S_{ELS}^{(N_1+1)}(r)$, by a parallel system following ELS dependency rule, using the result (4.5.3) for $u=r$ and applying (4.3.1), is given by

$$E[N_{ELS}(t,r)] \cong \frac{\lambda(r)t}{c(r)}, 0\le t\le S_{ELS}^{(N_1+1)}(r), r\in\{1, 2, \ldots, z-1\}. \tag{4.5.8}$$

Similarly by (4.3.2) and using (4.5.3), the expected value of the number $N_{ELS}(t,r)$ of exceeding the reliability critical state r up to the time point t by a parallel system following ELS dependency rule, where $S_{ELS}^{(N_1+1)}(r) < t \leq S_{ELS}^{(N_2+1)}(r)$, is

$$E[N_{ELS}(tr)] \cong N_1 + 1 + \frac{\lambda(r)\lambda(z)t - (N_1+1)c(r)\lambda(z)}{c(r)\lambda(z) - c(z)\lambda(r)}, \quad (4.5.9)$$
$$S_{ELS}^{(N_1+1)}(r) < t \leq S_{ELS}^{(N_2+1)}(r), \; r \in \{1, 2, \ldots, z-1\}.$$

And generally from (4.3.3) and using (4.5.3), the expected value of the number $N_{ELS}(t,r)$ of exceeding the reliability critical state r by a system following the ELS dependency rule, up to the time point t, $S_{ELS}^{(N_x+1)}(r) < t \leq S_{ELS}^{(N_{x+1}+1)}(r)$, $x = 1, 2, \ldots, z-r$, is

$$E[N_{ELS}(t,r)] \cong N_x + 1 + \frac{t - (N_1+1)\frac{c(r)}{\lambda(r)} - \sum_{j=1}^{x-1}(N_{j+1} - N_j)\left(\frac{c(r)}{\lambda(r)} - \frac{c(z-j+1)}{\lambda(z-j+1)}\right)}{\frac{c(r)}{\lambda(r)} - \frac{c(z-x+1)}{\lambda(z-x+1)}}, \quad (4.5.10)$$
$$S_{ELS}^{(N_x+1)}(r) < t \leq S_{ELS}^{(N_{x+1}+1)}(r), x = 1, 2, \ldots, z-r, r \in \{1, 2, \ldots, z-1\}.$$

Considering parallel system as a repairable system with non-negligible renewal time, we estimate its availability coefficient. The steady availability coefficient of a parallel system following the ELS dependency rule, at the moment t, $t \geq 0$, by (4.3.8)–(4.3.10) and substituting (4.5.3), is determined as follows.

$$A_{ELS}(t,r) \cong \frac{\mu_{ELS}(r)}{\mu_{ELS}(r) + \mu_R(z)} = \frac{c(r)}{c(r) + \lambda(r)\mu_R(z)}, \quad (4.5.11)$$
$$0 \leq t \leq S_{ELS}^{(N_1+1)}(r),$$

$$A_{ELS}(t,r) \cong \frac{\mu_{ELS}(r) - \mu_{ELS}(z)}{\mu_{ELS}(r) - \mu_{ELS}(z) + \mu_R(z-1)}$$
$$= \frac{c(r)\lambda(z) - c(z)\lambda(r)}{c(r)\lambda(z) - c(z)\lambda(r) + \lambda(r)\lambda(z)\mu_R(z-1)}, \; S_{ELS}^{(N_1+1)}(r) < t \leq S_{ELS}^{(N_2+1)}(r), \quad (4.5.12)$$

$$A_{ELS}(t,r) \cong \frac{\mu_{ELS}(r) - \mu_{ELS}(z-x+1)}{\mu_{ELS}(r) - \mu_{ELS}(z-x+1) + \mu_R(z-x)}$$
$$= \frac{c(r)\lambda(z-x+1) - c(z-x+1)\lambda(r)}{c(r)\lambda(z-x+1) - c(z-x+1)\lambda(r) + \lambda(r)\lambda(z-x+1)\mu_R(z-x)},$$
$$S_{ELS}^{(N_x+1)}(r) < t \leq S_{ELS}^{(N_{x+1}+1)}(r), x = 1, 2, \ldots, z-r, r \in \{1, 2, \ldots, z-1\}, \quad (4.5.13)$$

where $\mu_R(u)$, $u=z, z-1, \ldots, r$, denotes expected value of system recovery time to the reliability state u after exceeding the critical reliability state r.

As an example, a parallel system with components having identical piecewise exponential reliability functions is analyzed. It is assumed that the system and its components have four reliability states and their critical state is $r=1$. Component intensities of departure from the reliability state subsets $\{1,2,3\}$, $\{2,3\}$, $\{3\}$, by the assumption are given by (4.4.14). The component stress proportionality correction coefficient $c(u)$ for each u, $u=1,2,3$, is equal to 0.9. In Table 4.4 the expected values of lifetime in the reliability state subsets $\{1,2,3\}$, $\{2,3\}$, $\{3\}$ of the exemplary parallel system for different numbers of components n, $n=3,4, \ldots, 10$ are given. The mean values are determined in two cases, assuming independence of its components and assuming dependence according to ELS rule, described in Section 3.2.1.

To compare the results of availability analysis of a multistate parallel system with independent components and in a second case with components dependent according to ELS rule, the assumed number of system renewals is $N_1=4$, $N_2=8$, $N_3=10$. The expected values of time until successive exceeding the reliability critical state $r=1$ by the system in both cases are given in Table 4.5 and presented in Fig. 4.4.

As the mean values of lifetime in subsets, by (4.5.3), do not depend on the number of components in case of a parallel system following the ELS dependency rule, it can be noticed that the expected values of the time until Nth exceeding the reliability critical state by the system are also the same.

For a parallel system with dependent components following the ELS rule, assuming the component stress proportionality correction coefficient equals 0.9, the expected values of the time until successive exceeding the reliability critical state, are much worse than in case of the system with

Table 4.4 The expected values of lifetime in the reliability state subsets {1,2,3}, {2,3}, {3} of an exemplary parallel system assuming components' independence and dependence according to ELS rule.

N	$\mu(1)$	$\mu_{ELS}(1)$	$\mu(2)$	$\mu_{ELS}(2)$	$\mu(3)$	$\mu_{ELS}(3)$
3	14.667	7.200	10.978	5.389	3.667	1.800
4	16.667	7.200	12.475	5.389	4.167	1.800
5	18.267	7.200	13.673	5.389	4.567	1.800
6	19.600	7.200	14.671	5.389	4.900	1.800
7	20.743	7.200	15.526	5.389	5.186	1.800
8	21.743	7.200	16.275	5.389	5.436	1.800
9	22.632	7.200	16.940	5.389	5.658	1.800
10	23.432	7.200	17.539	5.389	5.858	1.800

Table 4.5 The expected values of the time until *N*th exceeding the reliability critical state by an exemplary parallel system assuming components' independence and dependence according to ELS rule.

	$n=5$		$n=6$		$n=7$		$n=8$		$n=9$		$n=10$	
N	**indep.**	**ELS**	**indep.**	**ELS**	**indep.**	**ELS**	**indep.**	**ELS**	**indep.**	**ELS**	**indep.**	**ELS**
1	18.27	7.20	19.60	7.20	20.74	7.20	21.74	7.20	22.63	7.20	23.43	7.20
2	36.53	14.40	39.20	14.40	41.49	14.40	43.49	14.40	45.26	14.40	46.86	14.40
3	54.80	21.60	58.80	21.60	62.23	21.60	65.23	21.60	67.90	21.60	70.30	21.60
4	73.07	28.80	78.40	28.80	82.97	28.80	86.97	28.80	90.53	28.80	93.73	28.80
5	91.34	36.00	98.00	36.00	103.72	36.00	108.72	36.00	113.16	36.00	117.16	36.00
6	105.04	41.40	112.70	41.40	119.27	41.40	125.02	41.40	130.13	41.40	134.73	41.40
7	118.74	46.80	127.40	46.80	134.83	46.80	141.33	46.80	147.11	46.80	152.31	46.80
8	132.44	52.20	142.10	52.20	150.39	52.20	157.64	52.20	164.08	52.20	169.88	52.20
9	146.14	57.60	156.80	57.60	165.94	57.60	173.94	57.60	181.06	57.60	187.46	57.60
10	150.73	59.41	161.73	59.41	171.16	59.41	179.41	59.41	186.75	59.41	193.35	59.41
11	155.32	61.22	166.66	61.22	176.38	61.22	184.88	61.22	192.44	61.22	199.24	61.22

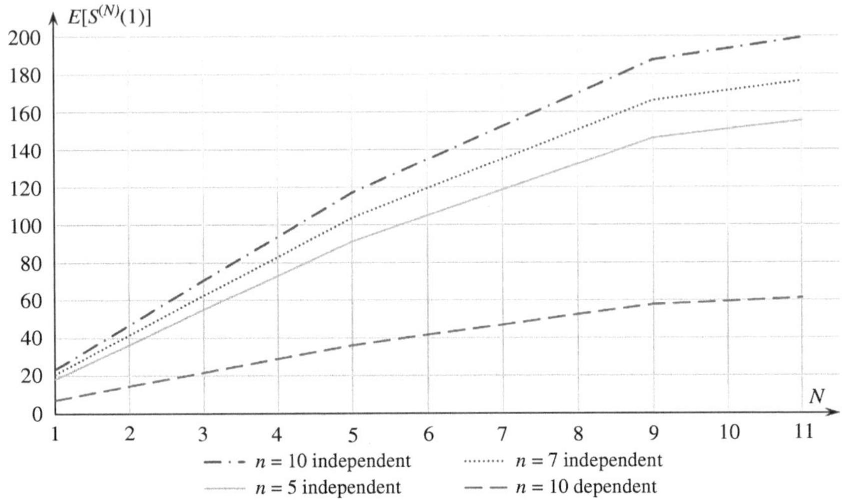

Fig. 4.4 The graph of the expected values of the time until *N*th exceeding the reliability critical state by an exemplary parallel system assuming components' independence and dependence according to ELS rule.

independent components. Namely, for a system composed of 5 components these values are less by 61%. This difference is still increasing as the number of system components increases, and for a parallel system consisting of 10 components is 69%.

As a second case, we consider this parallel system as a repairable system with non-negligible renewal time, and assume that the expected value of system recovery time to the reliability state 3, 2, and 1 after exceeding the critical state $r=1$ is 0.04.

Taking the assumptions regarding reliability parameters and number of renewals, as before for a system with negligible renewal time, we get the values of steady availability coefficient of the system with non-negligible renewal time, given in Table 4.6. The availability coefficient of the exemplary parallel system is determined, assuming components' independence and dependence according to the ELS rule, for different numbers of components, and the results are compared graphically in Fig. 4.5.

4.5.2 Availability of multistate "*m* out of *n*" system with dependent components

We consider an aging "*m* out of *n*" system with components dependent according to the ELS rule. Similarly as in Section 4.5.1, it is assumed that

Table 4.6 The values of steady availability coefficient of an exemplary parallel system, assuming components' independence and dependence according to ELS rule (%).

	$n=5$		$n=6$		$n=7$		$n=8$		$n=9$		$n=10$	
t	indep.	ELS	indep.	ELS	indep.	ELS	indep.	ELS	indep.	ELS	indep.	ELS
10	99.78	99.45	99.80	99.45	99.81	99.45	99.82	99.45	99.82	99.45	99.83	99.45
20	99.78	99.45	99.80	99.45	99.81	99.45	99.82	99.45	99.82	99.45	99.83	99.45
30	99.78	99.45	99.80	99.45	99.81	99.45	99.82	99.45	99.82	99.45	99.83	99.45
40	99.78	99.26	99.80	99.26	99.81	99.26	99.82	99.26	99.82	99.26	99.83	99.26
50	99.78	99.26	99.80	99.26	99.81	99.26	99.82	99.26	99.82	99.26	99.83	99.26
60	99.78	97.84	99.80	97.84	99.81	97.84	99.82	97.84	99.82	97.84	99.83	97.84
70	99.78	0.00	99.80	0.00	99.81	0.00	99.82	0.00	99.82	0.00	99.83	0.00
80	99.78	0.00	99.80	0.00	99.81	0.00	99.82	0.00	99.82	0.00	99.83	0.00
90	99.78	0.00	99.80	0.00	99.81	0.00	99.82	0.00	99.82	0.00	99.83	0.00
100	99.71	0.00	99.73	0.00	99.81	0.00	99.82	0.00	99.82	0.00	99.83	0.00
110	99.71	0.00	99.73	0.00	99.74	0.00	99.76	0.00	99.82	0.00	99.83	0.00
120	99.71	0.00	99.73	0.00	99.74	0.00	99.76	0.00	99.76	0.00	99.77	0.00
130	99.71	0.00	99.73	0.00	99.74	0.00	99.76	0.00	99.76	0.00	99.77	0.00
140	99.71	0.00	99.73	0.00	99.74	0.00	99.76	0.00	99.76	0.00	99.77	0.00
150	99.14	0.00	99.73	0.00	99.74	0.00	99.76	0.00	99.76	0.00	99.77	0.00
160	0.00	0.00	99.20	0.00	99.74	0.00	99.76	0.00	99.76	0.00	99.77	0.00
170	0.00	0.00	0.00	0.00	99.24	0.00	99.76	0.00	99.76	0.00	99.77	0.00
180	0.00	0.00	0.00	0.00	0.00	0.00	99.27	0.00	99.76	0.00	99.77	0.00
190	0.00	0.00	0.00	0.00	0.00	0.00	0.00	0.00	99.30	0.00	99.33	0.00
200	0.00	0.00	0.00	0.00	0.00	0.00	0.00	0.00	0.00	0.00	0.00	0.00

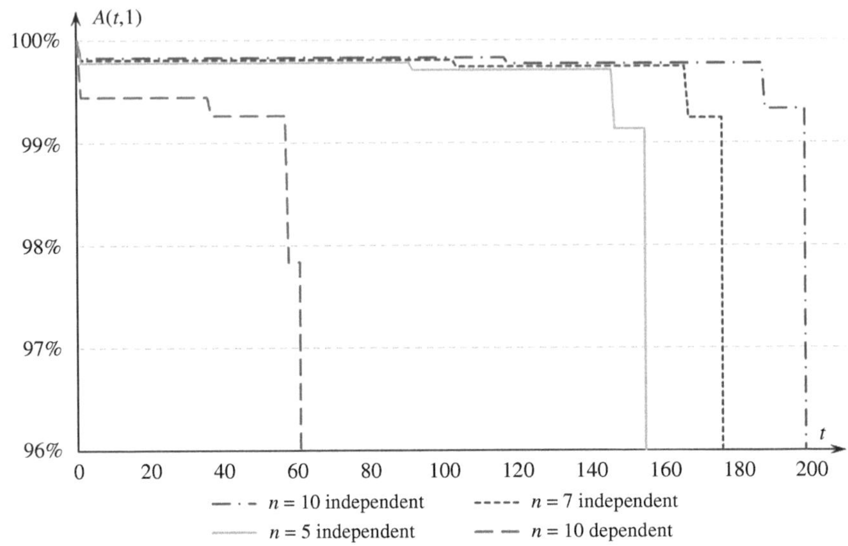

Fig. 4.5 The graphs of steady availability coefficient of an exemplary parallel system, assuming components' independence and dependence according to ELS rule (for a scale of values from 96% to 100%).

its components have identical piecewise exponential reliability functions (4.5.1)–(4.5.2).

From Corollary 3.2.4 in Chapter 3, we obtain the mean value of the system's lifetime in the state subsets $\{u, u+1, \ldots, z\}$, $u=1,2, \ldots, z$, given by

$$\mu_{ELS}(u) = \frac{(n-m+1)\cdot c(u)}{n\lambda(u)}, u=1,2,\ldots,z, \tag{4.5.14}$$

where $\lambda(u)$, $\lambda(u) \geq 0$, $u=1,2, \ldots, z$, is components' intensity of departure from the reliability state subset $\{u, u+1, \ldots, z\}$, and $c(u)$ is the component stress proportionality correction coefficient for each u, $u=1,2, \ldots, z$.

Using the result (4.5.14) for $u=r$ and applying (4.2.2), the expected value of a random variable $S_{ELS}^{(\omega)}(r)$, $\omega=1, 2, \ldots, N_1, N_1+1$, representing the time until the ωth exceeding the reliability critical state r, is given by

$$E\left[S_{ELS}^{(\omega)}(r)\right] \cong \omega \cdot \frac{(n-m+1)\cdot c(r)}{n\lambda(r)}, \quad \omega = 1,2,\ldots,N_1,N_1+1, r\in\{1, 2, \ldots, z-1\}. \tag{4.5.15}$$

Random variables $\boldsymbol{T}_{ELS}^{(\omega)}(r)$, $\omega=N_x+2, \ldots, N_{x+1}+1$, $x=1, 2, \ldots, z-r$, by (4.2.5) and (4.5.14), have the same distribution with expected value

$$E\left[\boldsymbol{T}_{ELS}^{(\omega)}(r)\right] = \frac{(n-m+1)\cdot c(r)}{n\lambda(r)} - \frac{(n-m+1)\cdot c(z-x+1)}{n\lambda(z-x+1)},$$
$$\omega = N_x + 2, \ldots, N_{x+1} + 1,$$
$$x = 1, 2, \ldots, z-r, r \in \{1, 2, \ldots, z-1\}. \tag{4.5.16}$$

Subsequently, the expected values of random variables $S_{ELS}^{(\omega)}(r)$, $\omega = N_x + 2, \ldots, N_{x+1} + 1$, $x = 1, 2, \ldots, z-r$, applying (4.2.6) and using (4.5.14), are

$$E\left[S_{ELS}^{(\omega)}(r)\right] \cong (N_1+1)\frac{(n-m+1)\cdot c(r)}{n\lambda(r)}$$
$$+\sum_{j=1}^{x-1}(N_{j+1}-N_j)\left(\frac{(n-m+1)\cdot c(r)}{n\lambda(r)} - \frac{(n-m+1)\cdot c(z-j+1)}{n\lambda(z-j+1)}\right)$$
$$+(\omega - N_x - 1)\left(\frac{(n-m+1)\cdot c(r)}{n\lambda(r)} - \frac{(n-m+1)\cdot c(z-x+1)}{n\lambda(z-x+1)}\right),$$
$$\omega = N_x + 2, \ldots, N_{x+1} + 1, x = 1, 2, \ldots, z-r, \tag{4.5.17}$$

and the time up to complete system damage, by (4.2.7) and (4.5.14), is

$$E\left[S_{ELS}^{(N_{z-r+1}+1)}(r)\right] \cong (N_1+1)\frac{(n-m+1)\cdot c(r)}{n\lambda(r)}$$
$$+\sum_{j=1}^{z-r}(N_{j+1}-N_j)\left(\frac{(n-m+1)\cdot c(r)}{n\lambda(r)} - \frac{(n-m+1)\cdot c(z-j+1)}{n\lambda(z-j+1)}\right), \tag{4.5.18}$$

for $r \in \{1, 2, \ldots, z-1\}$.

Similarly as in Section 4.5.1, using the result (4.5.14) for $u = r$ and applying (4.3.1), we obtain the expected value of the number $N_{ELS}(t,r)$ of exceeding the reliability critical state r by an "m out of n" system following ELS rule, up to the time point t, $0 \le t \le S_{ELS}^{(N_1+1)}(r)$.

$$E[N_{ELS}(t,r)] \cong \frac{n\lambda(r)t}{(n-m+1)\cdot c(r)}, 0 \le t \le S_{ELS}^{(N_1+1)}(r),$$
$$r \in \{1, 2, \ldots, z-1\}. \tag{4.5.19}$$

From (4.3.3) and using (4.5.14), the expected value of number $N_{ELS}(t,r)$ of exceeding the reliability critical state r by an "m out of n" system following ELS dependency rule, up to the time point t for $S_{ELS}^{(N_x+1)}(r) < t \le S_{ELS}^{(N_{x+1}+1)}(r)$, $x = 1, 2, \ldots, z-r$, is

$$E[N_{ELS}(t,r)] \cong N_x + 1$$
$$+\frac{t-(N_1+1)\frac{(n-m+1)\cdot c(r)}{n\lambda(r)}-\sum_{j=1}^{x-1}(N_{j+1}-N_j)\left(\frac{(n-m+1)\cdot c(r)}{n\lambda(r)}-\frac{(n-m+1)\cdot c(z-j+1)}{n\lambda(z-j+1)}\right)}{\frac{(n-m+1)\cdot c(r)}{n\lambda(r)}-\frac{(n-m+1)\cdot c(z-x+1)}{n\lambda(z-x+1)}},$$

$$S_{ELS}^{(N_x+1)}(r) < t \le S_{ELS}^{(N_{x+1}+1)}(r), \quad x=1,2,\ldots,z-r, r\in\{1,2,\ldots,z-1\}. \tag{4.5.20}$$

Considering "m out of n" system as a repairable system with non-negligible renewal time, we estimate its availability coefficient at the moment t, $t\ge 0$, by (4.3.8)–(4.3.10) and (4.5.14), using following formulas

$$A_{ELS}(t,r) \cong \frac{\mu_{ELS}(r)}{\mu_{ELS}(r)+\mu_R(z)} = \frac{(n-m+1)c(r)}{(n-m+1)c(r)+n\lambda(r)\mu_R(z)}, \quad 0 \le t \le S_{ELS}^{(N_1+1)}(r), \tag{4.5.21}$$

$$A_{ELS}(tr) \cong \frac{\mu_{ELS}(r)-\mu_{ELS}(z)}{\mu_{ELS}(r)-\mu_{ELS}(z)+\mu_R(z-1)}$$
$$= \frac{(n-m+1)[c(r)\lambda(z)-c(z)\lambda(r)]}{(n-m+1)[c(r)\lambda(z)-c(z)\lambda(r)]+n\lambda(r)\lambda(z)\mu_R(z-1)},$$
$$S_{ELS}^{(N_1+1)}(r) < t \le S_{ELS}^{(N_2+1)}(r), \tag{4.5.22}$$

$$A_{ELS}(t,r) \cong \frac{\mu_{ELS}(r)-\mu_{ELS}(z-x+1)}{\mu_{ELS}(r)-\mu_{ELS}(z-x+1)+\mu_R(z-x)}$$
$$= \frac{(n-m+1)[c(r)\lambda(z-x+1)-c(z-x+1)\lambda(r)]}{(n-m+1)[c(r)\lambda(z-x+1)-c(z-x+1)\lambda(r)]+n\lambda(r)\lambda(z-x+1)\mu_R(z-x)},$$
$$S_{ELS}^{(N_x+1)}(r) < t \le S_{ELS}^{(N_{x+1}+1)}(r), x=1,2,\ldots,z-r, r\in\{1,2,\ldots,z-1\}, \tag{4.5.23}$$

where $\mu_R(u)$, $u=z, z-1, \ldots, r$, denotes expected value of system recovery time to the reliability state u after exceeding the critical reliability state r, $r\in\{1,2,\ldots,z-1\}$.

Similarly as for a parallel system in Section 4.5.1, availability analysis of an exemplary "m out of n" system is performed. As before, it is assumed that components have identical piecewise exponential reliability functions with the intensities of departure from the reliability state subsets $\{1,2,3\}$, $\{2,3\}$,

Table 4.7 The expected values of lifetime in the reliability state subsets {1,2,3}, {2,3}, {3} of an exemplary "*m* out of *n*" system assuming components' independence and dependence according to ELS rule.

n	$\mu(1)$	$\mu_{ELS}(1)$	$\mu(2)$	$\mu_{ELS}(2)$	$\mu(3)$	$\mu_{ELS}(3)$
5	3.600	2.880	2.695	2.156	0.900	0.720
6	4.933	3.600	3.693	2.695	1.233	0.900
7	6.076	4.114	4.548	3.080	1.519	1.029
8	7.076	4.500	5.297	3.368	1.769	1.125
9	7.965	4.800	5.962	3.593	1.991	1.200
10	8.765	5.040	6.561	3.772	2.191	1.260

{3}, given by (4.4.14), and the component stress proportionality correction coefficient $c(u)$ for each u, $u=1,2,3$, are equal to 0.9. It is assumed that the critical state is $r=1$. In Table 4.7 the expected values of lifetime in the reliability state subsets {1,2,3}, {2,3}, {3} of the exemplary "*m* out of *n*" system for different numbers of components n, $n=5,6, \ldots, 10$, and for threshold number of components $m=4$ are given. The mean values are determined assuming independence of its components and assuming dependence according to ELS rule.

The results for the exemplary "*m* out of *n*" system assuming components' independence and dependence according to ELS rule are compared in Table 4.8 for the number of system renewals $N_1=4$, $N_2=8$, $N_3=10$.

The expected values of time until successive exceeding the reliability critical state $r=1$ by the system in both cases are also compared graphically for 5, 7, and 10 components in Fig. 4.6.

For an "*m* out of *n*" system, the impact of the dependence among components on system's availability characteristics, depends mainly on the structure of the system, including the number of components and the threshold number of components. For fixed number $m=4$ and for assumed value 0.9 of the component stress proportionality correction coefficient, the difference between expected values of the time until the Nth exceeding the reliability critical state, for a series system with dependent components and for a system with independent components, is 20% when the system consists of 5 components, and this difference grows to 42% when the system consists of 10 components.

As a second case, we consider the "*m* out of *n*" system as a repairable system with non-negligible renewal time, and assume that the expected value of system recovery time to the reliability state 3, 2, and 1 after exceeding the critical state $r=1$ is 0.04.

Table 4.8 The expected values of the time until *N*th exceeding the reliability critical state by an exemplary parallel system assuming components' independence and dependence according to ELS rule.

	$n=5$		$n=6$		$n=7$		$n=8$		$n=9$		$n=10$	
N	indep.	ELS	indep.	ELS	indep.	ELS	indep.	ELS	indep.	ELS	indep.	ELS
1	3.60	2.88	4.93	3.60	6.08	4.11	7.08	4.50	7.97	4.80	8.77	5.04
2	7.20	5.76	9.87	7.20	12.15	8.23	14.15	9.00	15.93	9.60	17.53	10.08
3	10.80	8.64	14.80	10.80	18.23	12.34	21.23	13.50	23.90	14.40	26.30	15.12
4	14.40	11.52	19.73	14.40	24.30	16.46	28.30	18.00	31.86	19.20	35.06	20.16
5	18.00	14.40	24.67	18.00	30.38	20.57	35.38	22.50	39.83	24.00	43.83	25.20
6	20.70	16.56	28.37	20.70	34.94	23.66	40.69	25.88	45.80	27.60	50.40	28.98
7	23.40	18.72	32.07	23.40	39.49	26.74	45.99	29.25	51.77	31.20	56.97	32.76
8	26.10	20.88	35.77	26.10	44.05	29.83	51.30	32.63	57.75	34.80	63.55	36.54
9	28.80	23.04	39.47	28.80	48.61	32.91	56.61	36.00	63.72	38.40	70.12	40.32
10	29.71	23.76	40.71	29.71	50.14	33.94	58.39	37.13	65.72	39.61	72.33	41.59
11	30.61	24.49	41.95	30.61	51.66	34.98	60.17	38.26	67.73	40.81	74.53	42.86

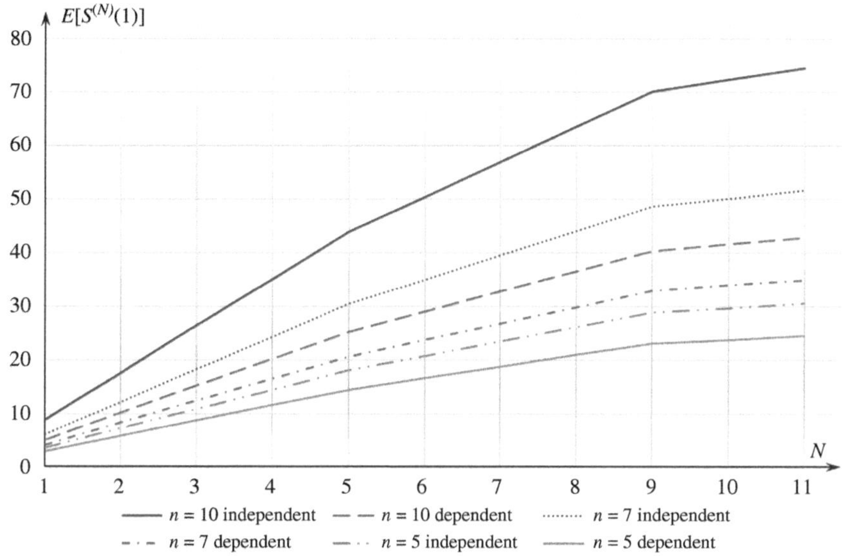

Fig. 4.6 The graph of the expected values of the time until *N*th exceeding the reliability critical state by an exemplary "*m* out of *n*" system assuming components' independence and dependence according to ELS rule.

Taking the assumptions regarding reliability parameters and number of renewals, as before for a system with negligible renewal time, we get the values of steady availability coefficient of the system with non-negligible renewal time, given in Table 4.9. The availability coefficient of the exemplary "*m* out of *n*" system is determined, assuming components' independence and dependence according to the ELS rule, for different numbers of components, and the results are compared graphically in Fig. 4.7.

4.5.3 Availability of multistate parallel-series system with dependent components of its subsystems

Here, we consider a multistate parallel-series system, described in Section 3.2.4, composed of k parallel subsystems that are working independently and linked in series. In the ith subsystem S_i, $i=1,2, \ldots, k$, there are l_i components E_{ij}, $j=1,2, \ldots, l_i$, dependent according to the ELS rule. Additionally, we assume that these components in ith subsystem S_i, $i=1,2, \ldots, k$, have identical piecewise exponential reliability functions

$$R_{ij}(t, \cdot)=\left[1, R_{ij}(t, 1), \ldots, R_{ij}(t, z)\right], t \geq 0, i=1,2, \ldots, k,$$
$$j=1,2, \ldots, l_i, \tag{4.5.24}$$

Table 4.9 The values of steady availability coefficient of an exemplary "m out of n" system, assuming components' independence and dependence according to ELS rule (%).

	$n=5$		$n=6$		$n=7$		$n=8$		$n=9$		$n=10$	
t	**indep.**	**ELS**	**indep.**	**ELS**	**indep.**	**ELS**	**indep.**	**ELS**	**indep.**	**ELS**	**indep.**	**ELS**
5	98.90	98.63	99.20	98.90	99.35	99.04	99.44	99.12	99.50	99.17	99.55	99.21
10	98.90	98.63	99.20	98.90	99.35	99.04	99.44	99.12	99.50	99.17	99.55	99.21
15	98.90	98.18	99.20	98.90	99.35	99.04	99.44	99.12	99.50	99.17	99.55	99.21
20	98.54	98.18	99.20	98.54	99.35	99.04	99.44	99.12	99.50	99.17	99.55	99.21
25	98.54	0.00	98.93	98.54	99.35	98.72	99.44	98.83	99.50	98.90	99.55	99.21
30	95.77	0.00	98.93	95.77	99.35	98.72	99.44	98.83	99.50	98.90	99.55	98.95
35	0.00	0.00	98.93	0.00	99.13	96.28	99.44	98.83	99.50	98.90	99.55	98.95
40	0.00	0.00	96.88	0.00	99.13	0.00	99.25	0.00	99.33	96.79	99.55	98.95
45	0.00	0.00	0.00	0.00	99.13	0.00	99.25	0.00	99.33	0.00	99.40	0.00
50	0.00	0.00	0.00	0.00	97.45	0.00	99.25	0.00	99.33	0.00	99.40	0.00
55	0.00	0.00	0.00	0.00	0.00	0.00	99.25	0.00	99.33	0.00	99.40	0.00
60	0.00	0.00	0.00	0.00	0.00	0.00	97.80	0.00	99.33	0.00	99.40	0.00
65	0.00	0.00	0.00	0.00	0.00	0.00	0.00	0.00	98.04	0.00	99.40	0.00
70	0.00	0.00	0.00	0.00	0.00	0.00	0.00	0.00	0.00	0.00	99.40	0.00
75	0.00	0.00	0.00	0.00	0.00	0.00	0.00	0.00	0.00	0.00	0.00	0.00
80	0.00	0.00	0.00	0.00	0.00	0.00	0.00	0.00	0.00	0.00	0.00	0.00

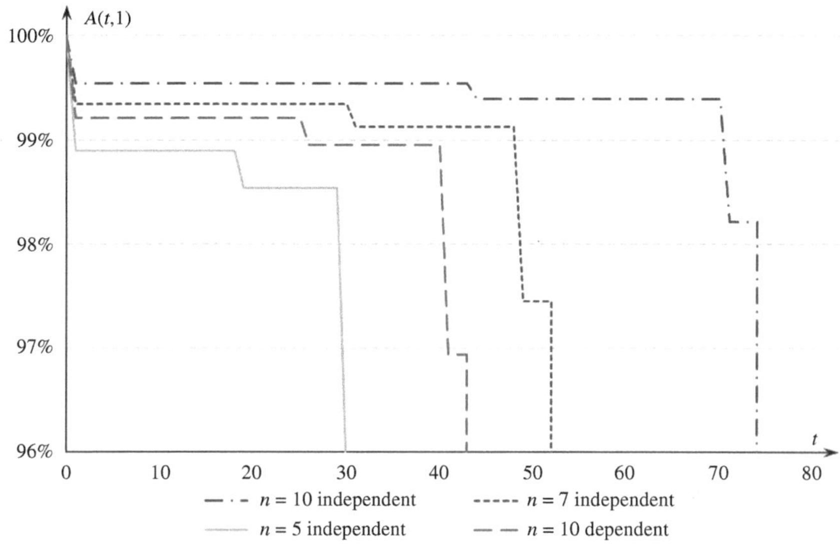

Fig. 4.7 The graphs of steady availability coefficient of an exemplary "*m* out of *n*" system, assuming components' independence and dependence according to ELS rule (for a scale of values from 96% to 100%).

with the coordinates

$$R_{ij}(t,u) = \exp[-\lambda_i(u)t], t \geq 0, u = 1,2,\ldots,z, \tag{4.5.25}$$

where $\lambda_i(u)$, $i=1,2, \ldots, k$, $\lambda_i(u) \geq 0$, are intensities of departure from the reliability state subsets $\{u,u+1, \ldots,z\}$, $u=1,2, \ldots, z$, of components in the subsystem S_i.

Under these assumptions, the reliability function of a multistate parallel-series system following the ELS dependency rule is given by (3.2.22)–(3.2.23) in *Proposition 3.2.3*. Using this result, the system's mean lifetime in the reliability state subsets $\{u,u+1, \ldots, z\}$, $u=1,2, \ldots, z$, is given by the formula

$$\mu_{ELS}(u) = \int_0^{+\infty} \boldsymbol{R}_{ELS}(t,u)dt, u = 1,2,\ldots,z, \tag{4.5.26}$$

where $\boldsymbol{R}_{ELS}(t,u)$, $u=1, 2, \ldots, z$, is given by (3.2.23).

Next, the expected value of a random variable $S_{ELS}^{(\omega)}(r)$, $\omega = 1, 2, \ldots, N_1$, $N_1 + 1$, representing the time until successive exceeding the reliability critical state r, by a multistate parallel-series system with dependent components of its subsystems, applying (4.2.2), is

$$E\left[S_{ELS}^{(\omega)}(r)\right] \cong \omega \cdot \mu_{ELS}(r), \omega = 1,2,\ldots,N_1, N_1+1, r \in \{1, 2, \ldots, z-1\}, \tag{4.5.27}$$

where $\mu_{ELS}(r)$ is given by (4.5.26) for $u = r$, $r \in \{1,2,\ldots,z-1\}$.

The expected values of random variables $S_{ELS}^{(\omega)}(r)$, $\omega = N_x + 2, \ldots, N_{x+1} + 1$, $x = 1, 2, \ldots, z - r$, applying (4.2.6), are

$$E\left[S_{ELS}^{(\omega)}(r)\right] \cong (N_1+1)\mu_{ELS}(r) + \sum_{j=1}^{x-1}\left(N_{j+1} - N_j\right)(\mu_{ELS}(r) - \mu_{ELS}(z-j+1)) + (\omega - N_x - 1)(\mu_{ELS}(r) - \mu_{ELS}(z-x+1)), \omega = N_x+2,\ldots,N_{x+1}+1, x = 1,2,\ldots,z-r, \tag{4.5.28}$$

and the time up to complete system damage, by (4.2.7), is

$$E\left[S_{ELS}^{(N_{z-r+1}+1)}(r)\right] \cong (N_1+1)\mu_{ELS}(r) + \sum_{j=1}^{z-r}\left(N_{j+1} - N_j\right)(\mu_{ELS}(r) - \mu_{ELS}(z-j+1)), \tag{4.5.29}$$

for $r \in \{1,2,\ldots,z-1\}$, where $\mu_{ELS}(u)$, $u = 1, 2, \ldots, z$, is given by (4.5.26).

Substituting the obtained from (4.5.26) mean value $\mu_{ELS}(r)$ for $u = r$, $r \in \{1,2,\ldots,z-1\}$, into the following formula

$$E[N_{ELS}(t,r)] \cong \frac{t}{\mu_{ELS}(r)}, 0 \le t \le S_{ELS}^{(N_1+1)}(r), r \in \{1, 2, \ldots, z-1\}, \tag{4.5.30}$$

we get the expected value of the number $N_{ELS}(t,r)$ of exceeding the reliability critical state r up to the time point t, $0 \le t \le S_{ELS}^{(N_1+1)}(r)$.

Subsequently, the expected value of the number $N_{ELS}(t,r)$ of exceeding the critical state r by a system with components following the ELS dependency rule, up to the time point t, $S_{ELS}^{(N_x+1)}(r) < t \le S_{ELS}^{(N_{x+1}+1)}(r)$, $x = 1, 2, \ldots, z - r$, by (4.3.3), is

$$E[N_{ELS}(t,r)] \cong N_x + 1 + \frac{t - (N_1+1)\mu_{ELS}(r) - \sum_{j=1}^{x-1}\left(N_{j+1} - N_j\right)\left(\mu_{ELS}(r) - \mu_{ELS}(z-j+1)\right)}{\mu_{ELS}(r) - \mu_{ELS}(z-x+1)},$$

$$S_{ELS}^{(N_x+1)}(r) < t \le S_{ELS}^{(N_{x+1}+1)}(r), x = 1, 2, \ldots, z-r, r \in \{1, 2, \ldots, z-1\}, \quad (4.5.31)$$

where $\mu_{ELS}(u)$, $u = 1, 2, \ldots, z$, is given by (4.5.26).

4.5.4 Availability of multistate "*m* out of *l*"-series system with dependent components of its subsystems

In this section, we consider a multistate "m_i out of l_i"-series system, described in Section 3.2.5, composed of k linked in series "m_i out of l_i", $i = 1, 2, \ldots, k$, subsystems with components dependent according to the ELS rule. We assume that components in i-th subsystem S_i, $i = 1, 2, \ldots, k$, have identical piecewise exponential reliability functions given by (4.5.24)–(4.5.25).

Under these assumptions, using the reliability function of an aging "m_i out of l_i"-series system with the coordinates given by (3.2.30) in *Proposition 3.2.5*, the system's mean lifetime in the reliability state subsets $\{u, u+1, \ldots, z\}$, $u = 1, 2, \ldots, z$, is given by (4.5.26), similarly as for a parallel-series system in Section 4.5.3. Therefore, the expected values of the time until successive exceeding the reliability critical state r, by a multistate "m_i out of l_i"-series system with dependent according to the ELS rule components of its subsystems, are determined applying (4.5.27)–(4.5.29) and using obtained from (4.5.26) value of system's mean lifetime. Similarly, the expected value of the number $N_{ELS}(t,r)$ of exceeding the critical state r by a multistate "m_i out of l_i"-series system with components following the ELS dependency rule, up to the time point t, $t \ge 0$, is estimated substituting the obtained from (4.5.26) mean value $\mu_{ELS}(u)$, $u = 1, 2, \ldots, z$, into (4.5.30) and (4.5.31).

4.6 Mixed load-sharing dependency model

4.6.1 Availability of multistate parallel-series system with dependent components and subsystems

In this section, we present the availability analysis of a multistate parallel-series system assuming the MLS dependency model, described in Section 3.3. Namely, we assume dependence among its parallel subsystems according to the local load-sharing rule and the dependence among components in these subsystems according to the equal load sharing rule.

We consider an aging parallel-series system following the MLS dependency rule, composed of k parallel subsystems S_i, $i=1,2, \ldots, k$, and assume that components in i-th subsystem have identical piecewise exponential reliability functions given by (4.5.24)–(4.5.25).

Under these assumptions, the reliability function of a multistate parallel-series system following the MLS dependency rule is given by (3.3.3)–(3.3.6) in *Proposition 3.3.1*. In that case, the system's mean lifetime in the reliability state subsets $\{u,u+1, \ldots,z\}$, $u=1,2, \ldots,z$, is given by the formula

$$\mu_{MLS}(u) = \int_0^{+\infty} \boldsymbol{R}_{MLS}(t, u)dt, u=1,2,\ldots,z, \tag{4.6.1}$$

where $\boldsymbol{R}_{MLS}(t, u)$ for $u=1, 2, \ldots, z-1$, is given by (3.3.4)–(3.3.5) and $\boldsymbol{R}_{MLS}(t, z)$ is given by (3.3.6).

Next, the expected value of a random variable $S_{MLS}^{(\omega)}(r)$, $\omega=1, 2, \ldots, N_1$, N_1+1, representing the time until successive exceeding the reliability critical state r, by a multistate parallel-series system following the MLS dependency rule, applying (4.2.2), is

$$E\left[S_{MLS}^{(\omega)}(r)\right] \cong \omega \cdot \mu_{MLS}(r), \quad \omega=1,2,\ldots,N_1,N_1+1, r\in\{1, 2, \ldots, z-1\}, \tag{4.6.2}$$

where $\mu_{MLS}(r)$ is given by (4.6.1) for $u=r$, $r\in\{1,2,\ldots,z-1\}$.

The expected values of random variables $S_{MLS}^{(\omega)}(r)$, $\omega=N_x+2, \ldots, N_{x+1}+1$, $x=1, 2, \ldots, z-r$, applying (4.2.6), are

$$\begin{aligned} E\left[S_{MLS}^{(\omega)}(r)\right] &\cong (N_1+1)\mu_{MLS}(r) + \sum_{j=1}^{x-1}(N_{j+1}-N_j)(\mu_{MLS}(r)-\mu_{MLS}(z-j+1)) \\ &+(\omega-N_x-1)(\mu_{MLS}(r)-\mu_{MLS}(z-x+1)), \omega=N_x+2,\ldots,N_{x+1}+1, \\ &x=1,2,\ldots,z-r, \end{aligned} \tag{4.6.3}$$

and the time up to complete system damage, by (4.2.7), is

$$\begin{aligned} E\left[S_{MLS}^{(N_{z-r+1}+1)}(r)\right] &\cong (N_1+1)\mu_{MLS}(r) \\ &+ \sum_{j=1}^{z-r}(N_{j+1}-N_j)(\mu_{MLS}(r)-\mu_{MLS}(z-j+1)), \end{aligned} \tag{4.6.4}$$

for $r\in\{1,2,\ldots,z-1\}$, where $\mu_{MLS}(u)$, $u=1, 2, \ldots, z$, is given by (4.6.1).

The expected value of the number $N_{MLS}(t,r)$ of exceeding the reliability critical state r by a system following the MLS dependency rule, up to the time point t, $0 \le t \le S_{MLS}^{(N_1+1)}(r)$, applying (4.3.1), is

$$E[N_{MLS}(t,r)] \cong \frac{t}{\mu_{MLS}(r)}, 0 \le t \le S_{MLS}^{(N_1+1)}(r), r \in \{1,2,\ldots,z-1\}, \quad (4.6.5)$$

where $\mu_{MLS}(r)$ is given by (4.6.1) for $u = r$, $r \in \{1,2,\ldots,z-1\}$.

Subsequently, the expected value of the number $N_{MLS}(t,r)$ of exceeding the reliability critical state r by a system following the MLS dependency rule, up to the time point t, $S_{MLS}^{(N_x+1)}(r) < t \le S_{MLS}^{(N_{x+1}+1)}(r)$, $x = 1, 2, \ldots, z - r$, by (4.3.3), is estimated from the formula

$$E[N_{MLS}(t,r)] \cong N_x + 1$$
$$+ \frac{t - (N_1+1)\mu_{MLS}(r) - \sum_{j=1}^{x-1}(N_{j+1} - N_j)(\mu_{MLS}(r) - \mu_{MLS}(z-j+1))}{\mu_{MLS}(r) - \mu_{MLS}(z-x+1)},$$
$$S_{MLS}^{(N_x+1)}(r) < t \le S_{MLS}^{(N_{x+1}+1)}(r), x = 1,2,\ldots,z-r, r \in \{1,2,\ldots,z-1\}, \quad (4.6.6)$$

where $\mu_{MLS}(u)$, $u = 1, 2, \ldots, z$, is given by (4.6.1).

4.6.2 Availability of multistate "*m* out of *l*"-series system with dependent components and subsystems

Here, we apply the MLS model of components and subsystems dependency, to the availability analysis of a multistate "m_i out of l_i"-series system. We assume that components in ith subsystem S_i, $i = 1,2, \ldots, k$, have identical piecewise exponential reliability functions given by (4.5.24)–(4.5.25).

Under these assumptions, using the reliability function of an aging "m_i out of l_i"-series system with the coordinates given by (3.3.12)–(3.3.14) in *Proposition 3.3.3*, the system's mean lifetime in the reliability state subsets $\{u,u+1, \ldots,z\}$, $u = 1,2,\ldots,z$, is given by (4.6.1), similarly as for a parallel-series system in Section 4.6.1. Therefore, the expected values of the time until successive exceeding the reliability critical state r, by a multistate "m_i out of l_i"-series system with the MLS dependency rule, are determined applying (4.6.2)–(4.6.4) and using obtained from (4.6.1) value of the system's mean lifetime. Similarly, the expected value of the number $N_{MLS}(t,r)$ of exceeding the critical state r by a multistate "m_i out of l_i"-series system following the MLS dependency rule, up to the time point t, $t \geq 0$, is estimated substituting the obtained from (4.6.1) mean value $\mu_{MLS}(u)$, $u=1, 2, \ldots, z$, into (4.6.5) and (4.6.6).

4.7 Application

Theoretical results are applied to the availability evaluation of an exemplary transport navigation system. A navigation system with a back-up system is analyzed under assumption of imperfect repair, for various configurations of system's renewals. Moreover, the results of availability analysis for a shipyard rope elevator in case when its components are independent and in case of the MLS dependency model are compared.

4.7.1 Availability of transport navigation system under imperfect repair

As an application, a navigation system (Weintrit et al., 2015) consisting of four independently operating and connected in series systems: data collection system S_1, data processing system S_2, data presentation system S_3, and user interface S_4, is considered. It is assumed that a navigation system and those four systems are multistate systems and following four reliability states are distinguished (Blokus-Roszkowska, 2019; Dziula et al., 2015):

- state 3—full reliability of a system,
- state 2—partially degraded reliability of a system and impendency over safety,
- state 1—significantly worse reliability of a system,
- state 0—a system entire failure.

Further, in reliability analysis of a navigation system, it is assumed that the four systems S_1, S_2, S_3, and S_4 forming the navigation system have piecewise exponential distribution. Their reliability functions are given by a vector (Blokus-Roszkowska, 2019; Kołowrocki, 2014).

$$R_i(t,\ \cdot) = [1, R_i(t,1), R_i(t,2), R_i(t,3)],\ t \geq 0, i=1,2,3,4, \qquad (4.7.1)$$

with the coordinates

$$R_i(t,u) = \exp[-\lambda_i(u)t],\ i=1,2,3,4, u=1,2,3. \qquad (4.7.2)$$

The lifetimes in the safety states of systems S_1, S_2, S_3, S_4 are expressed in years and they have the piecewise exponential reliability functions (4.7.1) and (4.7.2) with the intensities of departure from the safety subsets, by the assumption, given by (Blokus-Roszkowska, 2019)

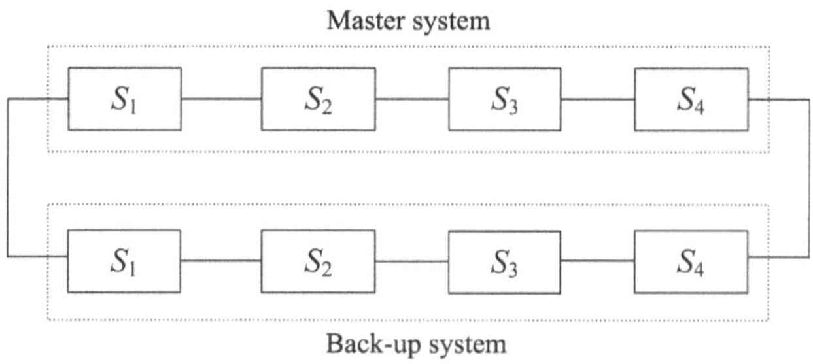

Fig. 4.8 The reliability structure scheme of a navigation system with a back-up system (Blokus-Roszkowska, 2019).

$$\lambda_i(1) = 0.125, \lambda_i(2) = 0.167, \lambda_i(3) = 0.5, i = 1,4, \tag{4.7.3}$$

$$\lambda_i(1) = 0.1, \lambda_i(2) = 0.125, \lambda_i(3) = 0.25, i = 2,3. \tag{4.7.4}$$

Assuming the navigation system is a multistate series system, in case S_1, S_2, S_3, S_4 have piecewise exponential reliability functions (4.7.1) and (4.7.2) with the intensities of departure (4.7.3) and (4.7.4), the reliability function of a single navigation system is given by the vector

$$\boldsymbol{R}_M(t, \cdot) = [1, \boldsymbol{R}_M(t, 1), \boldsymbol{R}_M(t, 2), \boldsymbol{R}_M(t, 3)], t \geq 0, \tag{4.7.5}$$

where

$$\boldsymbol{R}_M(t, 1) = \exp[-0.45t], \boldsymbol{R}_M(t, 2) = \exp[-0.58t], \boldsymbol{R}_M(t, 3) = \exp[-1.5t]. \tag{4.7.6}$$

The mean lifetimes of a single navigation system in the subsets {1,2,3}, {2,3},{3}, determined according to (2.1.16) and using (4.7.6), in years are:

$$\mu_M(1) \cong 2.22, \mu_M(2) \cong 1.71, \mu_M(3) \cong 0.67. \tag{4.7.7}$$

Considering a navigation system (Weintrit et al., 2015) consisting of a master system and a back-up system (Fig. 4.8), we assume that both systems have identical reliability functions, given by (4.7.5) and (4.7.6), and are linked in parallel.

Taking into account the reliability structure of a navigation system (Fig. 4.8) its reliability function is as follows (Blokus-Roszkowska, 2019; Kołowrocki, 2014).

$$\boldsymbol{R}_{NS}(t, \cdot) = [1, \boldsymbol{R}_{NS}(t, 1), \boldsymbol{R}_{NS}(t, 2), \boldsymbol{R}_{NS}(t, 3)], t \geq 0, \tag{4.7.8}$$

with the coordinates

$$\begin{aligned}\boldsymbol{R}_{NS}(t,1) &= 1-[1-\exp[-0.45t]]^2,\\ \boldsymbol{R}_{NS}(t,2) &= 1-[1-\exp[-0.58t]]^2,\\ \boldsymbol{R}_{NS}(t,3) &= 1-[1-\exp[-1.5t]]^2.\end{aligned} \tag{4.7.9}$$

The mean lifetimes of the navigation system in the subsets $\{1,2,3\},\{2,3\}$, $\{3\}$, determined according to (2.1.16) and using (4.7.9), in years are:

$$\mu_{NS}(1) \cong 3.33, \mu_{NS}(2) \cong 2.57, \mu_{NS}(3) \cong 1.00. \tag{4.7.10}$$

and the mean values of navigation system lifetimes in individual states 1, 2, 3, determined by (2.1.20) and using (4.7.10), respectively, in years are:

$$\bar{\mu}_{NS}(1) \cong 0.76, \bar{\mu}_{NS}(2) \cong 1.57, \bar{\mu}_{NS}(3) \cong 1.00. \tag{4.7.11}$$

It is assumed that a critical reliability state is the state $r=1$. We assume that navigation system is repaired after exceeding the critical state, i.e., reliability state 1. It is assumed that recovery time is negligible comparing to system's lifetime in the reliability state subset $\{1,2,3\}$. The system is repaired 10 times ($N_3=10$) and after the eleventh ($N_3+1=11$) exceeding the critical state, the system remains in a state of complete damage and is unfit for operation. Using the notations and assumptions from Section 4.1, N_1 denotes the number of system repairs to the state of full reliability, i.e., state 3. Consequently, N_2 is the number of the last repair to the state of partially degraded reliability, state 2. Thus, the renovations from N_1+1 to N_2 allow the system to return to the reliability state 2. As a result of renovations from N_2+1 to N_3 ($N_3=10$), the system returns to critical state 1.

Applying the formulas (4.2.2) and (4.2.6), the expected values of variable $S^{(N)}(1)$, representing time until Nth exceeding the reliability critical state $r=1$, are determined. The expected values $E[S^{(N)}(1)]$ are given in years in Table 4.10 for fixed numbers $N_2=8$ and $N_3=10$, and for N_1 ranging from zero to 6. In Table 4.11 the expected values for N_2 ranging from 3 to 9, and fixed $N_1=2$, $N_3=10$, are presented.

Assuming that the system can be repaired 10 times ($N_3=10$), and once the critical state of the system is exceeded 11 times, it remains unusable, based on the results presented in Table 4.10, it can be seen that the system's lifetime, i.e., the time after which the system remains failed in state 0 for $N_1=6$ is 25.5% longer than when $N_1=0$, with fixed values $N_2=8$, $N_3=10$. For fixed $N_1=2$ and $N_3=10$ with N_2 ranging from 3 to 9 the differences are much larger. The time until the eleventh exceeding the critical state is 53.4% longer for $N_2=9$ than when $N_2=3$.

Table 4.10 The expected values of the time until Nth exceeding the reliability critical state by a navigation system for N_1 ranging from zero to 6, $N_2=8$, $N_3=10$.

N	$N_1=0$	$N_1=1$	$N_1=2$	$N_1=3$	$N_1=4$	$N_1=5$	$N_1=6$
1	3.33	3.33	3.33	3.33	3.33	3.33	3.33
2	5.67	6.67	6.67	6.67	6.67	6.67	6.67
3	8.00	9.00	10.00	10.00	10.00	10.00	10.00
4	10.33	11.33	12.33	13.33	13.33	13.33	13.33
5	12.67	13.67	14.67	15.67	16.67	16.67	16.67
6	15.00	16.00	17.00	18.00	19.00	20.00	20.00
7	17.33	18.33	19.33	20.33	21.33	22.33	23.33
8	19.67	20.67	21.67	22.67	23.67	24.67	25.67
9	22.00	23.00	24.00	25.00	26.00	27.00	28.00
10	22.76	23.76	24.76	25.76	26.76	27.76	28.76
11	23.52	24.52	25.52	26.52	27.52	28.52	29.52

Table 4.11 The expected values of the time until Nth exceeding the reliability critical state by a navigation system for N_2 ranging from 3 to 9, $N_1=2$, $N_3=10$.

N	$N_2=3$	$N_2=4$	$N_2=5$	$N_2=6$	$N_2=7$	$N_2=8$	$N_2=9$
1	3.33	3.33	3.33	3.33	3.33	3.33	3.33
2	6.67	6.67	6.67	6.67	6.67	6.67	6.67
3	10.00	10.00	10.00	10.00	10.00	10.00	10.00
4	12.33	12.33	12.33	12.33	12.33	12.33	12.33
5	13.10	14.67	14.67	14.67	14.67	14.67	14.67
6	13.86	15.43	17.00	17.00	17.00	17.00	17.00
7	14.62	16.19	17.76	19.33	19.33	19.33	19.33
8	15.38	16.95	18.52	20.10	21.67	21.67	21.67
9	16.14	17.71	19.29	20.86	22.43	24.00	24.00
10	16.90	18.48	20.05	21.62	23.19	24.76	26.33
11	17.67	19.24	20.81	22.38	23.95	25.52	27.10

The expected values of time until successive exceeding the critical state are also compared graphically in Fig. 4.9 for four different configurations of navigation system's renewals.

Furthermore, the expected values of the number $N(t,1)$ of exceeding the critical state $r=1$ up to time point t, are determined applying formulas (4.3.1) and (4.3.3). Their values are given in years for different values of N_1 and fixed $N_2=8$, $N_3=10$ in Table 4.12, and respectively, for different values of N_2 and fixed $N_1=2$, $N_3=10$ in Table 4.13.

In case of negligible time of system's renovation, the number of exceeding the critical state is equivalent to the number of system renewals. Then, the expected values of the number $N(t,1)$ of system's renewals after

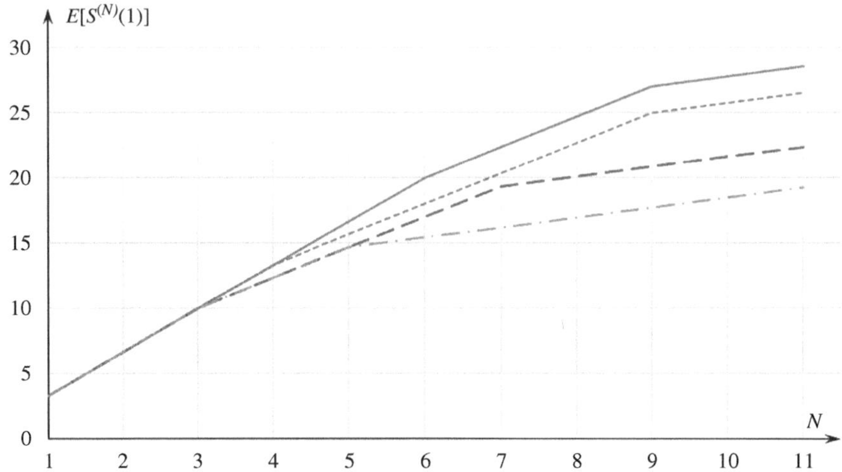

Fig. 4.9 The graph of expected values of the time until *N*th exceeding the reliability critical state for different configurations of navigation system's renewals.

Table 4.12 The expected values of the number $N(t,1)$ of exceeding the critical state by a navigation system for N_1 ranging from zero to 6, $N_2=8$, $N_3=10$.

t (years)	$N_1=0$	$N_1=1$	$N_1=2$	$N_1=3$	$N_1=4$	$N_1=5$	$N_1=6$
0	0.00	0.00	0.00	0.00	0.00	0.00	0.00
2	0.60	0.60	0.60	0.60	0.60	0.60	0.60
4	1.29	1.20	1.20	1.20	1.20	1.20	1.20
6	2.14	1.80	1.80	1.80	1.80	1.80	1.80
8	3.00	2.57	2.40	2.40	2.40	2.40	2.40
10	3.86	3.43	3.00	3.00	3.00	3.00	3.00
12	4.71	4.29	3.86	3.60	3.60	3.60	3.60
14	5.57	5.14	4.71	4.29	4.20	4.20	4.20
16	6.43	6.00	5.57	5.14	4.80	4.80	4.80
18	7.29	6.86	6.43	6.00	5.57	5.40	5.40
20	8.14	7.71	7.29	6.86	6.43	6.00	6.00
22	9.00	8.57	8.14	7.71	7.29	6.86	6.60
24	11.00	10.31	9.00	8.57	8.14	7.71	7.29
26	11.00	11.00	11.00	10.31	9.00	8.57	8.14
28	11.00	11.00	11.00	11.00	11.00	10.31	9.00
30	11.00	11.00	11.00	11.00	11.00	11.00	11.00

exceeding the critical state up to time point *t*, for different configurations of navigation system's renewals, are presented in Fig. 4.10.

A multistate approach to the assessment of system availability allows for a more accurate analysis of the operation of systems and the diagnostics of their

Table 4.13 The expected values of the number $N(t,1)$ of exceeding the critical state by a navigation system for N_2 ranging from 3 to 9, $N_1=2$, $N_3=10$.

t (years)	$N_2=3$	$N_2=4$	$N_2=5$	$N_2=6$	$N_2=7$	$N_2=8$	$N_2=9$
0	0.00	0.00	0.00	0.00	0.00	0.00	0.00
2	0.60	0.60	0.60	0.60	0.60	0.60	0.60
4	1.20	1.20	1.20	1.20	1.20	1.20	1.20
6	1.80	1.80	1.80	1.80	1.80	1.80	1.80
8	2.40	2.40	2.40	2.40	2.40	2.40	2.40
10	3.00	3.00	3.00	3.00	3.00	3.00	3.00
12	3.86	3.86	3.86	3.86	3.86	3.86	3.86
14	6.19	4.71	4.71	4.71	4.71	4.71	4.71
16	8.81	6.75	5.57	5.57	5.57	5.57	5.57
18	11.00	9.38	7.31	6.43	6.43	6.43	6.43
20	11.00	11.00	9.94	7.88	7.29	7.29	7.29
22	11.00	11.00	11.00	10.50	8.44	8.14	8.14
24	11.00	11.00	11.00	11.00	11.00	9.00	9.00
26	11.00	11.00	11.00	11.00	11.00	11.00	9.86
28	11.00	11.00	11.00	11.00	11.00	11.00	11.00
30	11.00	11.00	11.00	11.00	11.00	11.00	11.00

work safety. Many technical systems are multistate systems due to the deteriorating technical condition of their components. Thus, it is possible to consider intermediate states of a system and its components, in addition to the state of system full reliability and the state of entire failure. The number of such states should be determined after consulting with the experts operating the system. Similarly, establishing a critical state of the system, the exceeding of which, does not ensure an adequate level of operational efficiency requires consultation with an expert (Blokus-Roszkowska, 2019).

For further availability analysis of navigation system, we assume that the first N_1 renewals of system to state 3 take place every period of time corresponding to the mean lifetime $\mu(1)$ in the subset of states $\{1,2,3\}$. Because the renovations from N_1+1 to N_2 result in a system renewal to state 2, the time between these renovations can be estimated as expected value of time $\mu(1)-\mu(3)$ between the moment of system renovation and the time of next exceeding critical state. Finally, renewals numbered from N_2+1 to N_3 cause the system to return to state 1 and we assume that the time between these renovations is equal to the expected value of time $\mu(1)-\mu(2)$ between the moment of system renovation and the time of next exceeding critical state.

With these assumptions, we can propose a function determining the probability of the system staying in the subset of states $\{1,2,3\}$, equivalent

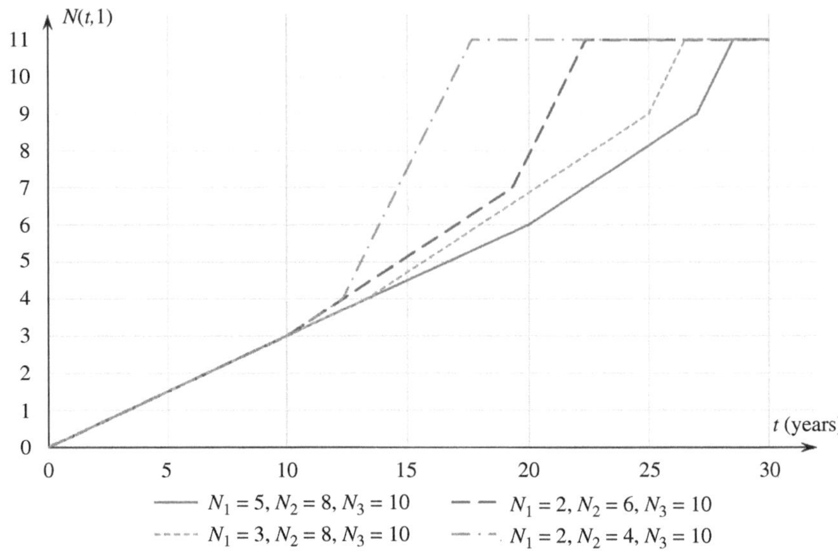

Fig. 4.10 The graph of expected values of the number of exceeding the critical state for different configurations of navigation system's renewals.

to the probability that the system is not failed (i.e., it is not in state 0). This function takes the following form:

$$\boldsymbol{AF}_{NS}(t,1) = \boldsymbol{R}_{NS}(t-\omega\cdot\mu(1),1), \omega\cdot\mu(1) \le t < (\omega+1)\cdot\mu(1), \omega = 0,1,\ldots,N_1, \tag{4.7.12}$$

$$\boldsymbol{AF}_{NS}(t,1) = \widetilde{\boldsymbol{R}}_{NS}(t-a-\omega\cdot(\mu(1)-\mu(3)),1), \quad a+\omega\cdot(\mu(1)-\mu(3)) \le t < a+(\omega+1)\cdot(\mu(1)-\mu(3)), \quad a = (N_1+1)\cdot\mu(1), \omega = 0,1,\ldots,N_2-N_1-1, \tag{4.7.13}$$

$$\boldsymbol{AF}_{NS}(t,1) = \hat{\boldsymbol{R}}_{NS}(t-b-\omega\cdot(\mu(1)-\mu(2)),1), \quad b+\omega\cdot(\mu(1)-\mu(2)) \le t < b+(\omega+1)\cdot(\mu(1)-\mu(2)), \quad b = (N_1+1)\cdot\mu(1)+(N_2-N_1)\cdot(\mu(1)-\mu(3)), \quad \omega = 0,1,\ldots,N_3-N_2-2, \tag{4.7.14}$$

$$\boldsymbol{AF}_{NS}(t,1) = \hat{\boldsymbol{R}}_{NS}(t-c,1), t \ge c,$$
$$c = (N_1+1)\cdot\mu(1)+(N_2-N_1)\cdot(\mu(1)-\mu(3))+(N_3-N_2-1)\cdot(\mu(1)-\mu(2)). \tag{4.7.15}$$

The coordinate of system reliability function $\boldsymbol{R}_{NS}(t,1)$, given in (4.7.9), according to Definition 2.1.2 in Chapter 2, expresses the probability that the system is in state subset {1,2,3}, under condition that at the beginning the system was in state 3. Thus, in the first period of system recovery, when system is repaired to state 3, this coordinate $\boldsymbol{R}_{NS}(t,1)$ can be used in (4.7.12). In second period of system's recovery to reliability state 2, the coordinate $\widetilde{\boldsymbol{R}}_{NS}(t, 1)$ in (4.7.13) expresses the probability of system being in state subset {1,2,3}, assuming that the system after recovery is in state 2. In that case, the intensity of departure from the state subset {1,2,3} should be corrected and it is further denoted in (4.7.17) by $\widetilde{\lambda}(1)$. Similarly, in the last period of system's recovery, when the system is repaired only to state 1, $\hat{\boldsymbol{R}}_{NS}(t, 1)$ in (4.7.14) and (4.7.15) expresses the probability of system being in state subset {1,2,3}, while system after recovery is in state 1, and corrected intensity in (4.7.18) and (4.7.19) is denoted by $\hat{\lambda}(1)$..

After substituting into (4.7.12)–(4.7.15) the mean lifetimes values given in (4.7.10), we get

$$\begin{aligned}\boldsymbol{AF}_{NS}(t, 1) &= \boldsymbol{R}_{NS}(t-3.33\omega, 1)\\ &= 1-[1-\exp[-0.45(t-3.33\omega)]]^2, 3.33\omega \le t\\ &\quad < 3.33(\omega+1), \omega = 0, 1, \ldots, N_1,\end{aligned} \tag{4.7.16}$$

$$\begin{aligned}\boldsymbol{AF}_{NS}(t, 1) &= \widetilde{\boldsymbol{R}}_{NS}(t-a-2.33\omega, 1)\\ &= 1-\left[1-\exp\left[-\widetilde{\lambda}(1)\cdot(t-a-2.33\omega)\right]\right]^2,\\ &\quad a+2.33\omega \le t < a+2.33(\omega+1),\\ &\quad a = 3.33(N_1+1), \omega = 0, 1, \ldots, N_2-N_1-1,\end{aligned} \tag{4.7.17}$$

$$\begin{aligned}\boldsymbol{AF}_{NS}(t, 1) &= \hat{\boldsymbol{R}}_{NS}(t-b-0.76\omega, 1)\\ &= 1-\left[1-\exp\left[-\hat{\lambda}(1)\cdot(t-b-0.76\omega)\right]\right]^2,\\ &\quad b+0.76\omega \le t < b+0.76(\omega+1),\\ &\quad b = 3.33(N_1+1)+2.33(N_2-N_1),\\ &\quad \omega = 0, 1, \ldots, N_3-N_2-2,\end{aligned} \tag{4.7.18}$$

$$\begin{aligned}\boldsymbol{AF}_{NS}(t, 1) = \hat{\boldsymbol{R}}_{NS}(t-c, 1) = 1-\left[1-\exp\left[-\hat{\lambda}(1)\cdot(t-c)\right]\right]^2, t \ge c, c\\ = 3.33(N_1+1)+2.33(N_2-N_1)\\ +0.76(N_3-N_2-1).\end{aligned} \tag{4.7.19}$$

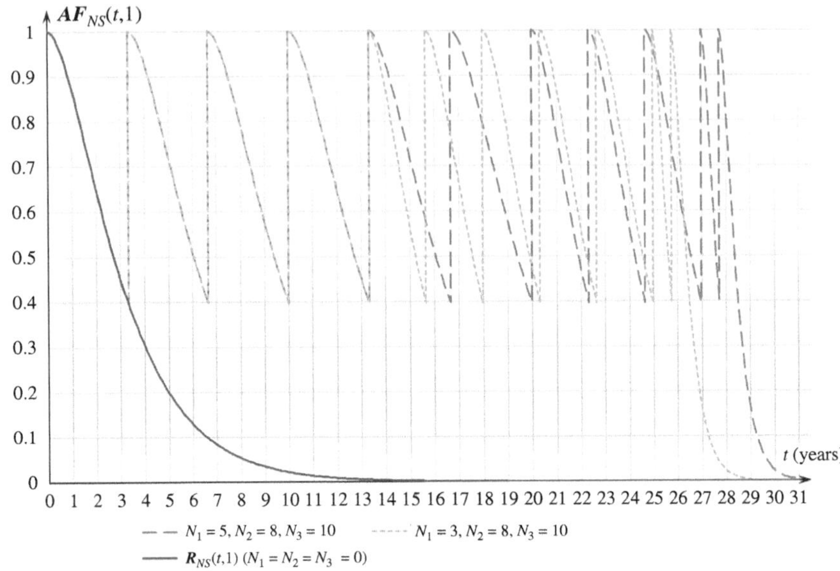

Fig. 4.11 The graph of the probability of navigation system being in the subset of reliability states {1,2,3} for different configurations of system's renewal.

Based on the mean lifetimes of a single navigation system in the reliability state subsets, given by (4.7.7), the intensities described earlier were estimated as follows

$$\tilde{\lambda}(1) \cong 0.645, \ \hat{\lambda}(1) \cong 1.96. \tag{4.7.20}$$

Fig. 4.11, using (4.7.16)–(4.7.19) for values given in (4.7.20), shows a function expressing the probability that navigation system is in state subset of reliability states $\{1,2,3\}$ over a period of 31 years, for two configurations of system's renewal ($N_1 = 5$, $N_2 = 8$, $N_3 = 10$ and $N_1 = 3$, $N_2 = 8$, $N_3 = 10$). In addition, Fig. 4.11 presents the coordinate of system reliability function $\boldsymbol{R}_{NS}(t,1)$, determining this probability for navigation system without repairs.

Comparing the system availability for selected scenarios of various configurations of system's renewal, a maintenance and renewal strategy of systems can be developed. In practical terms, such an analysis would gain, taking into account the costs of constructing the system, repair costs, and other associated costs. Detailed analysis, based on the history of system failures and renewals, could be used to plan the operation of the system and to avoid exceeding the critical state of the system.

4.7.2 Availability of shipyard rope elevator under imperfect repair

As a second application of this chapter results, we consider the shipyard rope elevator. The reliability of shipyard rope elevator under the assumption that its components are independent is analyzed in Chapter 2 and in (Blokus-Roszkowska and Kołowrocki, 2015a).

Considering strands as the basic components, four reliability states are distinguished of these system components. It is assumed, as in Chapter 2, that the components E_{ij}, $i=1,2, \ldots, 10, j=1,2, \ldots, 22$, have reliability functions with the following piecewise exponential reliability function coordinates:

$$R(t,1)=\exp[-0.1613t], R(t,2)=\exp[-0.2041t],$$
$$R(t,3)=\exp[-0.2326t], t\geq 0. \tag{4.7.21}$$

Taking into account the reliability structure of the shipyard rope elevator, described in Chapter 2, and the results of reliability analysis of this system under assumption that its components are independent, we use in availability analysis of this system values of reliability characteristics estimated in Chapter 2. Namely, the mean lifetimes of the shipyard rope elevator in the reliability state subsets $\{1,2,3\}$, $\{2,3\}$, $\{3\}$, in years, respectively, are:

$$\mu(1)\cong 6.441, \mu(2)\cong 5.091, \mu(3)\cong 4.467. \tag{4.7.22}$$

It is assumed, as in previous chapters, that a critical reliability state is $r=2$.

Next, the number of renewals for the shipyard rope elevator is fixed. We consider the shipyard rope elevator as a repairable system with negligible renewal time. It is assumed that, first $N_1=10$ system renovations are to the best state $z=3$. Subsequent repairs of the shipyard rope elevator, up to the repair number $N_2=60$, allow to restore the system to the reliability state $u=2$. After the system has exceeded the critical state for the sixty-first time, the shipyard rope elevator remains in the state of entire failure and cannot be repaired.

Subsequently, applying the formulas (4.2.2) and (4.2.4), and using (4.7.22), the expected values of variables, representing time until successive exceeding the reliability critical state $r=2$ by the shipyard rope elevator, are determined:

$$E\left[S^{(\omega)}(2)\right]\cong \omega\cdot\mu(2)\cong 5.091\omega, \omega=1,2,\ldots,11, \tag{4.7.23}$$

$$\begin{aligned} E\left[S^{(\omega)}(2)\right] &\cong (N_1+1)\mu(2)+(\omega-N_1-1)(\mu(2)-\mu(3)) \\ &\cong 56.001+0.624(\omega-11), \omega=12,\ldots,61. \end{aligned} \tag{4.7.24}$$

Next, the expected value of the number $N(t,r)$ of exceeding the reliability critical state $r=2$ by the shipyard rope elevator, up to the time point t, applying (4.3.1), (4.3.2), and using (4.7.22), is

$$E[N(t,2)] \cong \frac{t}{\mu(2)} \cong 0.196t, \; 0 \le t \le 56.0, \tag{4.7.25}$$

$$E[N(t,2)] \cong N_1 + 1 + \frac{t-(N_1+1)\mu(2)}{\mu(2)-\mu(3)} \cong 11 + \frac{t-56.001}{0.624}, \quad 56.0 < t \le 87.2. \tag{4.7.26}$$

According to the previously adopted assumption, after exceeding 61 times the reliability critical state $r=2$ by the shipyard rope elevator, the system cannot be repaired. Thus, for $t > 87.2$, the number $N(t,2)$ is constant and equals 61.

As a second case, the shipyard rope elevator following the MLS dependency rule is analyzed. Considering the system reliability structure, similarly as in Chapter 3, it is assumed that subsystems, i.e., ropes, are dependent according to the LLS rule, and components in these subsystems, i.e., strands in the ropes, are dependent according to the ELS rule.

Using the results of the reliability analysis of the shipyard rope elevator, performed in Section 3.4.2, the expected values of the system lifetimes in the reliability state subsets {1,2,3}, {2,3}, {3}, assuming the MLS dependency model, estimated in years, respectively, are:

$$\mu_{MLS}(1) \cong 2.514, \mu_{MLS}(2) \cong 2.149, \mu_{MLS}(3) \cong 2.124. \tag{4.7.27}$$

The numbers of renewals is assumed the same as for a system with independent components, i.e., $N_1 = 10$ and $N_2 = 60$.

Next, applying the formulas (4.2.2) and (4.2.4), respectively, and using (4.7.27), the expected values of time until successive exceeding the reliability critical state $r=2$ by the shipyard rope elevator with the MLS dependency model, are:

$$E\left[S_{MLS}^{(\omega)}(2)\right] \cong \omega \cdot \mu_{MLS}(2) \cong 2.149\omega, \omega = 1,2,\ldots,11, \quad \omega = 1,2,\ldots,11, \tag{4.7.28}$$

$$E\left[S_{MLS}^{(\omega)}(2)\right] \cong (N_1+1)\mu_{MLS}(2) + (\omega - N_1 - 1)(\mu_{MLS}(2) - \mu_{MLS}(3)) \cong 23.639 + 0.025(\omega - 11), \; \omega = 12,\ldots,61. \tag{4.7.29}$$

Next, the expected value of the number $N_{MLS}(t,r)$ of exceeding the reliability critical state $r=2$ by the shipyard rope elevator with the MLS dependency model, up to the time point t, applying (4.3.1), (4.3.2) and using (4.7.27), is

$$E[N_{MLS}(t,2)] \cong \frac{t}{\mu_{MLS}(2)} \cong 0.465t,\ 0 \le t \le 23.6, \tag{4.7.30}$$

$$E[N_{MLS}(t,2)] \cong N_1 + 1 + \frac{t - (N_1 + 1)\mu_{MLS}(2)}{\mu_{MLS}(2) - \mu_{MLS}(3)}$$

$$\cong 11 + \frac{t - 23.639}{0.025}, 23.6 < t \le 24.9. \tag{4.7.31}$$

After exceeding 61 times the reliability critical state $r=2$, the shipyard rope elevator cannot be repaired and the number $N_{MLS}(t,2)$ is constant and equals 61.

The expected values of time until the *Nth* exceeding the critical reliability state by the shipyard rope elevator, estimated in years, respectively, by (4.7.23) and (4.7.24) in case when the system components are independent, and by (4.7.28) and (4.7.29) assuming the MLS dependency model, are compared graphically in Fig. 4.12.

The results for the shipyard rope elevator in these two cases, assuming components independence and the MLS dependency model, are also

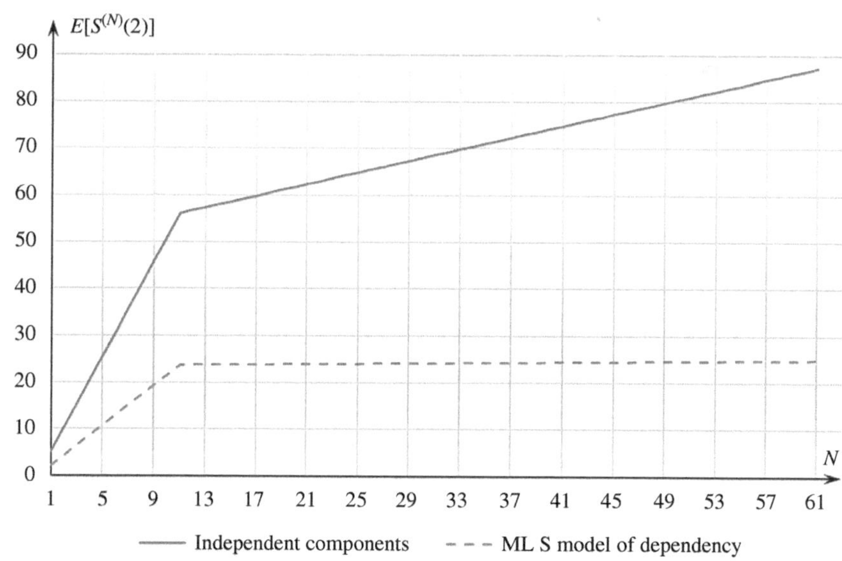

Fig. 4.12 The graph of the expected values of time until the *N*th exceeding the reliability critical state by the shipyard rope elevator assuming components independence and the MLS dependency model.

Table 4.14 The expected values of time until the *N*th exceeding the reliability critical state by the shipyard rope elevator assuming components independence and the MLS dependency model.

	$N_1=10, N_2=60$		$N_1=20, N_2=60$		$N_1=30, N_2=60$	
N	**indep.**	**MLS**	**indep.**	**MLS**	**indep.**	**MLS**
1	5.09	2.15	5.09	2.15	5.09	2.15
6	30.55	12.89	30.55	12.89	30.55	12.89
11	56.00	23.64	56.00	23.64	56.00	23.64
16	59.12	23.76	81.46	34.38	81.46	34.38
21	62.24	23.89	106.91	45.13	106.91	45.13
26	65.36	24.01	110.03	45.25	132.37	55.87
31	68.48	24.14	113.15	45.38	157.82	66.62
36	71.60	24.26	116.27	45.50	160.94	66.74
41	74.72	24.39	119.39	45.63	164.06	66.87
46	77.84	24.51	122.51	45.75	167.18	66.99
51	80.96	24.64	125.63	45.88	170.30	67.12
56	84.08	24.76	128.75	46.00	173.42	67.24
61	87.20	24.89	131.87	46.13	176.54	67.37

compared in Table 4.14 for other values of the number of system renewals $N_1=20$, $N_2=60$ and $N_1=30$, $N_2=60$.

The expected values of time until the moment of the *N*th exceeding the reliability critical state for the shipyard rope elevator, assuming the MLS dependency model, are 58% lower than for the system with independent components, in case of first 10 renewals. Next, this difference increases for subsequent renewals up to the level of 71%. If the number of renewals to the best state $z=3$ would be higher $N_1=20$, the difference between the results in both cases slightly decreases and ranges from 58% to 65%. And in case $N_1=30$, the difference of these expected values of time ranges from 58% to 62%.

The expected value of the number of exceeding the reliability critical state $r=2$ by the shipyard rope elevator, up to the time point *t*, calculated from (4.7.25) to (4.7.26) for the system with independent components, and from (4.7.30) to (4.7.31) for the system with the MLS dependency model, are given in Table 4.15 and illustrated in Fig. 4.13.

The results for the shipyard rope elevator in these two cases, assuming components independence and the MLS dependency model, are also compared in Table 4.15 for other numbers of system renewals $N_1=20$, $N_2=60$ and $N_1=30$, $N_2=60$.

Table 4.15 The expected values of the number $N(t,2)$ of exceeding the reliability critical state up to time point t, by the shipyard rope elevator assuming components independence and the MLS dependency model.

	$N_1=10, N_2=60$		$N_1=20, N_2=60$		$N_1=30, N_2=60$	
***t* (years)**	**indep.**	**MLS**	**indep.**	**MLS**	**indep.**	**MLS**
0	0	0	0	0	0	0
5	0.98	2.33	0.98	2.33	0.98	2.33
10	1.96	4.65	1.96	4.65	1.96	4.65
15	2.95	6.98	2.95	6.98	2.95	6.98
20	3.93	9.31	3.93	9.31	3.93	9.31
25	4.91	61.00	4.91	11.63	4.91	11.63
30	5.89	61.00	5.89	13.96	5.89	13.96
35	6.87	61.00	6.87	16.29	6.87	16.29
40	7.86	61.00	7.86	18.61	7.86	18.61
45	8.84	61.00	8.84	20.94	8.84	20.94
50	9.82	61.00	9.82	61.00	9.82	23.27
55	10.80	61.00	10.80	61.00	10.80	25.59
60	17.41	61.00	11.79	61.00	11.79	27.92
65	25.42	61.00	12.77	61.00	12.77	30.25
70	33.43	61.00	13.75	61.00	13.75	61.00
75	41.45	61.00	14.73	61.00	14.73	61.00
80	49.46	61.00	15.71	61.00	15.71	61.00
85	57.47	61.00	16.70	61.00	16.70	61.00
90	61.00	61.00	17.68	61.00	17.68	61.00
100	61.00	61.00	19.64	61.00	19.64	61.00
110	61.00	61.00	25.95	61.00	21.61	61.00
120	61.00	61.00	41.98	61.00	23.57	61.00
130	61.00	61.00	58.00	61.00	25.54	61.00
140	61.00	61.00	61.00	61.00	27.50	61.00
150	61.00	61.00	61.00	61.00	29.46	61.00
160	61.00	61.00	61.00	61.00	34.49	61.00
170	61.00	61.00	61.00	61.00	50.52	61.00
180	61.00	61.00	61.00	61.00	61.00	61.00

The expected values of the number of times the shipyard rope elevator with independent components and subsystems exceed the reliability critical state $r=2$ up to the time point t, is lower by 58% in case when the system is restored to the best state 3, relative to that values for the shipyard rope elevator assuming the MLS dependency model. This difference increases even up to about 90%, in case when the shipyard rope elevator can be repaired only to reliability state 2.

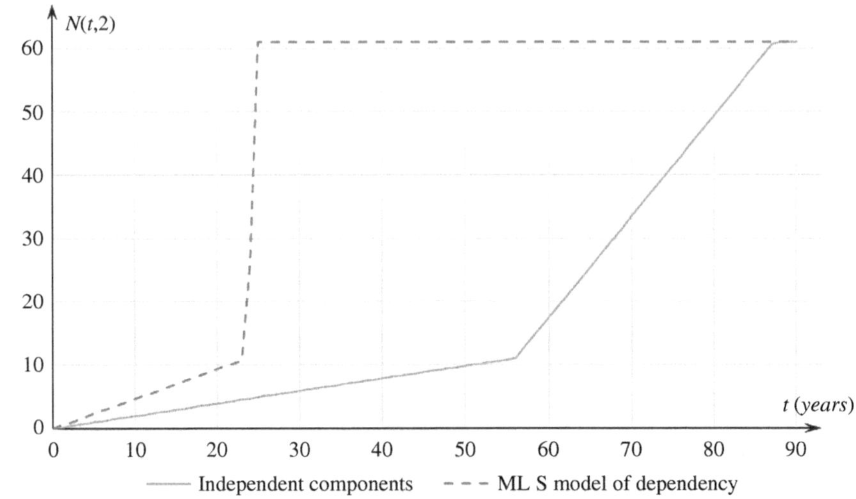

Fig. 4.13 The graph of the number $N(t,2)$ of exceeding the reliability critical state up to time point t, by the shipyard rope elevator assuming components independence and the MLS dependency model.

4.8 Related problems—System cost optimization under imperfect repair

In the reliability and availability analysis of technical systems, cost analysis and optimization are an indispensable issue. Moreover, not only the initial costs of acquiring the system itself and its initial operation are important, but also analysis of the costs of operating and repairing a system over a longer period of use and optimization of total life cycle costs. When analyzing the costs for repairable systems, it is necessary to take into account a number of additional costs associated with the use and renewal of the system, such as the costs of stopping the system or failing to complete the order. The importance of these associated costs is highlighted by the authors in (Lisnianski et al., 2010), calling them reliability-associated costs. In Chapter 6 of that book (Lisnianski et al., 2010), the authors present basic life cycle cost concepts. The basic cost types include design and development costs, production, operation and support costs, and disposal costs. When considering the life cycle cost as a function of system reliability, an optimal system reliability level can be determined. The authors (Lisnianski et al., 2010) also present a comparative analysis based on cost type criteria, such as economic losses and investment cost.

Optimization covering all costs incurred throughout the system's lifetime is also analyzed in Barone and Frangopol (2014), Frangopol and Liu (2007), Kong and Frangopol (2004), and Okasha and Frangopol (2009). Creating an optimal maintenance plan, that takes into account the minimization of total costs and maximization of system performance, one of the crucial factors is accurate modeling system reliability and availability. As the authors Barone and Frangopol (2014) emphasize, this modeling should also include loads acting on the system during its life-cycle. Time-varying loads are taken into account by Wen and Kang (2001) while minimizing the total life-cycle cost. Barone and Frangopol (2014) carry out biobjective optimizations based on total maintenance cost and annual performance indicators (annual reliability index, annual risk) or lifetime distribution (availability and hazard functions).

The topic of complex systems maintenance and its management assessment is widely discussed in de Almeida et al. (2015) and Crespo Marquez (2007). Crespo Marquez (2007) examines, among others, the impact of failures and system reliability on life cycle cost. de Almeida et al. (2015) present multicriteria models for effective maintenance planning. The authors de Almeida et al. (2015) highlight the importance of maintenance priorities during its planning, and even more, they indicate setting priorities and choosing a maintenance strategy as an important point before establishing a maintenance plan.

In this section, we use assumptions and notations the same as in Sections 4.1–4.3. All the components and the system under consideration have the reliability state set $\{0,1, \ldots, z\}$, $z \geq 1$. It is assumed that a system is repaired after exceeding its critical reliability state r, $r \in \{1,2,\ldots,z-1\}$ and in first case the time of its renovation is negligible. $\boldsymbol{T}(r)$ denotes the system's lifetime in the reliability state subset $\{r, r+1, \ldots, z\}$.

As assumed in Sections 4.1 and 4.2, first N_1 system's renewals cause the system to return to the full reliability state z, and $\boldsymbol{T}^{(1)}(r)$, $\boldsymbol{T}^{(2)}(r), \ldots, \boldsymbol{T}^{(N_1+1)}(r)$ are independent random variables from the same distribution with expected value $\mu(r)$ and standard deviation $\sigma(r)$. Next, the system renewal from the renovation N_1+1 up to the renovation N_2 takes place to state $z-1$. And more generally, it is assumed that the system renewal from the renovation N_x+1 to the renovation N_{x+1} takes place to state $z-x$, $x=1, 2, \ldots, z-r$ and $r \in \{1,2,\ldots,z-1\}$. From the renovation $N_{z-r}+1$ up to renovation N_{z-r+1}, system renewal can only be carried out as a renewal to the critical reliability state r. After N_{z-r+1} renovations, another system renewal is no longer possible and after $N_{z-r+1}+1$ times of exceeding the reliability critical state r, the system remains unable to work.

4.8.1 System with negligible renewal time

Taking into account cost of maintenance for a system with negligible renewal time, we consider cost of repair and costs related to maintaining the continuity of the system and performing tasks. Further, there are used following notations:

$C_R(z)$—cost of repair, including tradesmen and materials that causes the system to return to its "as good as new" reliability state z, after exceeding a reliability critical state r,

$C_R(u)$—cost of repair, including tradesmen and materials that causes the system to return to its reliability state u, $u=z-1, z-2, \ldots, r$, after exceeding a reliability critical state r,

$C_{order\_Ex}(n_{order})$—costs of interrupting the execution of orders or delays in their implementation depend on the average number of orders executed at a given moment.

It is assumed that following condition holds

$$C_R(z) \geq C_R(z-1) \geq C_R(z-2) \geq \cdots \geq C_R(r). \tag{4.8.1}$$

It means that cost of the system's repair to the full reliability state is no less than the cost of repairing to a worse reliability state and the system's repair to the critical state r is the cheapest.

For a renewable system with negligible renewal period total cost of repair after exceeding a reliability critical state by a system concerned with operations that cause the system to return to its full reliability state z, is.

$$C_T(z) = C_R(z) + C_{order\_Ex}(n_{order}). \tag{4.8.2}$$

And similarly, total cost of system's repair to other reliability states u, $u=z-1, z-2, \ldots, r$, is

$$C_T(u) = C_R(u) + C_{order\_Ex}(n_{order}), u = z-1, z-2, \ldots, r. \tag{4.8.3}$$

Total cost of maintenance up to the time point t, $0 \leq t \leq S^{(N_1+1)}(r)$, using formula (4.3.1) for expected value of the number $N(t,r)$ of exceeding the reliability critical state r, $r \in \{1, 2, \ldots, z-1\}$, is estimated as

$$C(t,r) \cong C_T(z) \cdot \left\lfloor \frac{t}{\mu(r)} \right\rfloor, \quad 0 \leq t \leq S^{(N_1+1)}(r), \tag{4.8.4}$$

where total cost of operations $C_T(z)$ that cause the system to return to the state z is given by (4.8.2).

Using notations and assumptions from Section 4.2, random variables $\boldsymbol{T}^{(\omega)}(r)$, $\omega = N_1+2, \ldots, N_2+1$, are independent and have the same distribution with expected value (4.2.3). In that case, total cost of maintenance up to the time point t, $S^{(N_1+1)}(r) < t \leq S^{(N_2+1)}(r)$, for fixed critical state r, $r \in \{1,2, \ldots, z-1\}$, using formula (4.3.2), is estimated from following formula

$$C(t,r) \cong C_T(z) \cdot N_1 + C_T(z-1) \cdot \left\lfloor \frac{t-(N_1+1)\mu(r)}{\mu(r)-\mu(z)} \right\rfloor, \quad S^{(N_1+1)}(r) < t \leq S^{(N_2+1)}(r), \tag{4.8.5}$$

where total cost of operations $C_T(z-1)$ that cause the system to return to the state $z-1$ is given by (4.8.3) for $u = z-1$, and $C_T(z)$ is given by (4.8.2).

And more generally, if variables $\boldsymbol{T}^{(\omega)}(r)$, $\omega = N_x+2, \ldots, N_{x+1}+1$, $x=1, 2, \ldots, z-r$, are independent random variables from the same distribution with expected value (4.2.5), then total cost of maintenance up to the time point t, $S^{(N_x+1)}(r) < t \leq S^{(N_{x+1}+1)}(r)$, $x = 1, 2, \ldots, z-r$, for fixed critical state r, $r \in \{1,2,\ldots,z-1\}$, using (4.3.3), is estimated from the formula

$$C(tr) \cong C_T(z) \cdot N_1 + \sum_{j=1}^{x-1} C_T(z-j) \cdot \left(N_{j+1} - N_j\right) + C_T(z-x) \cdot \left\lfloor \frac{t-(N_1+1)\mu(r) - \sum_{j=1}^{x-1} \left(N_{j+1}-N_j\right)\left(\mu(r)-\mu(z-j+1)\right)}{\mu(r)-\mu(z-x+1)} \right\rfloor, \tag{4.8.6}$$

$$S^{(N_x+1)}(r) < t \leq S^{(N_{x+1}+1)}(r), \; x = 1,2,\ldots,z-r,$$

where total cost of operations $C_T(u)$, $u = z-1, z-2, \ldots, r$, are given by (4.8.3), and $C_T(z)$ is given by (4.8.2).

Taking into account obtained results, total cost of maintenance up to the time point t, $t \geq 0$, for fixed critical state r, $r \in \{1,2,\ldots,z-1\}$, is determined using the following procedure:

```
x=1;
E[S^(N_x+1)(r)] ≅ (N_1 + 1)μ(r);
if (t ≤ E[S^(N_x+1)(r)])
  then {
    C_T(z) = C_R(z) + C_order_Ex(n_order);
    C(t,r) ≅ C_T(z) · ⌊t/μ(r)⌋;
  }
```

```
else {
  do{
    x := x + 1;
    E[S^(N_x+1)(r)] ≅ (N_1 + 1)μ(r) + Σ_{j=1}^{x-1} (N_{j+1} − N_j)(μ(r) − μ(z−j+1));
  } while (t > E[S^(N_x+1)(r)]) and (x < z−r+1);
  for (j=0;j<x;j++)
    C_T(z−j) = C_R(z−j) + C_order_Ex(n_order);
  C(t,r) ≅ C_T(z)·N_1 + Σ_{j=1}^{x-2} C_T(z−j)·(N_{j+1} − N_j)
    + C_T(z−x+1)·⌊(t − (N_1+1)μ(r) − Σ_{j=1}^{x-2}(N_{j+1} − N_j)(μ(r) − μ(z−j+1))) / (μ(r) − μ(z−x+2))⌋;
  if (t > E[S^(N_{z−r+1}+1)(r)])
    then{
      for (j=0;j<z-r+1;j++)
        C_T(z−j) = C_R(z−j) + C_order_Ex(n_order);
      C(t,r) ≅ C_T(z)·N_1 + Σ_{j=1}^{z-r} C_T(z−j)·(N_{j+1} − N_j); }
}
```

In case of availability analysis of systems with dependent components, in which due to fatigue or partial failure of some components remaining components operate with increased failure rates, the procedure can be easily modified. Namely, in order to estimate the performance costs of a system with dependent components, it is necessary to replace mean lifetimes in reliability state subsets by the values of mean lifetimes determined taking into account the interactions among system components. Depending on the system structure and dependency model, as discussed in Sections 4.4–4.6, we can analogously estimate the total costs of maintenance up to the time point t, $t \geq 0$, for a system following the LLS, ELS, or MLS model of dependency.

Analysis of maintenance cost for system taking into account load sharing among components is performed in (Olde Keizer et al., 2018). The authors raise the problem of component failure dependency through load sharing. They emphasize that component failure increases the load on the remaining components and can cause them to fail faster. However, they analyze so-called redundant dependency for a parallel system with additional components sharing the total system load to a certain extent. Additionally, Olde Keizer et al. (2018) conducted sensitivity analysis examining, among others, the impact of the degree of load sharing on minimal long-run average cost per time unit.

To compare the costs of repair and operation of the system from the new system to the given moment t, the following procedure is proposed. Choosing the most beneficial repair procedure means determining such numbers of repairs N_1, N_2, ..., N_{z-r+1}, to various states z, $z-1$, ..., r, which minimize these costs.

1. Start with all system repairs only to the full reliability state, i.e., state z. Determine the number of such repairs N_1, and calculate the costs of such system's renewal until time t. Remember/record the cost of system's renewal $C(t, r)$.
2. If $N_1 < 1$, then no system repair until time t was needed and costs of the system's renewal are equal to 0.
3. If critical state $r < z$ and $N_1 > 0$, then set $x = 1$ and reduce the number of repairs N_x (i.e., for $x = 1$, N_1 repairs to the state z) by one and perform the last repair (repairs) to the state $z-x$ (i.e., for $x = 1$, to the state $z-1$). Determine the number of all repairs N_{x+1} (i.e., for $x = 1$ the number N_2) and calculate the costs of such system's renewal until time t.
4. If the cost of repairs from point 3 is lower than the cost of system's renewal from point 1, record it and remember the number of repairs to particular states.
5. Increase the index number x by one.
6. If critical state $r < z - x + 1$ and $N_x > N_{x-1}$, then repeat the procedure from step 3 for the new index x. It means that, reduce the number of repairs to the state $z - x + 1$ $(N_x - N_{x-1})$ by one and perform the last repair (repairs) to the state $z - x$. Determine the number of all repairs N_{x+1} and calculate the costs of such system's renewal until time t.
7. If the repair cost from point 6 is lower than the previously saved cost of system's renewal, update and record this cost. Remember/record the number of repairs to particular states $(N_1, N_2, \ldots, N_{z-r+1})$.
8. Increase the index number x by one.
9. Repeat the procedure from points 6, 7 and 8, when the following conditions are met: $r < z - x + 1$ and $N_x > N_{x-1}$. If $x > z - r + 1$ it means that the last repair took place to the critical state r.
10. If $x > z - r$, then decrease index x one by one until one of the conditions is met $N_x > N_{x-1}$ or $x = 1$.
11. Repeat the procedure given in points from 6 to 10, until $x = 1$ and $N_1 = 0$.

12. Provide the last recorded costs of system's renewal and the number of repairs as the optimal result.

```
N_1 = \lfloor t/\mu(r) \rfloor;
for (i=2;i<z-r+2;i++)
   N_i = N_1;
C(t,r) \cong C_T(z) \cdot N_1;
if (r<z) and (N_1>0) then{
   x=1;
   do{
      do{
         N_x := N_x - 1;
         N_{x+1} := N_x + \left\lfloor \frac{t-(N_1+1)\mu(r)-\sum_{j=1}^{x-1}(N_{j+1}-N_j)(\mu(r)-\mu(z-j+1))}{\mu(r)-\mu(z-x+1)} \right\rfloor;
         C_{new}(t,r) \cong C_T(z) \cdot N_1 + \sum_{j=1}^{x} C_T(z-j) \cdot (N_{j+1}-N_j);
         if (C_{new}(t,r) < C(t,r)) then{
            C(t,r) = C_{new}(t,r);
            for (i=x+2;i<z-r+2;i++)
               N_i = N_{x+1};
            for (i=1;i<z-r+2;i++)
               N[i]:=N_i;
         }
         x := x+1;
      } while (r<z-x+1) and (N_x>N_{x-1});
      if (r>z-x) then{
         N_0 = 0;
         do{
            x:= x-1;
         } while (N_x=N_{x-1}) and (x>1)
      }
   } while (N_1>0)
}
else if (N_1<1) then
   C(t,r) = 0;
System.out.println("Optimal cost of system's renewal"+C(t,r));
for (i=1;i<z-r+2;i++)
   System.out.println("Optimal number of renewals N_i ="+N[i]);
```

4.8.2 System with non-negligible renewal time

In second case, maintenance costs for a renewable system with non-negligible renewal durations are analyzed. As before a system is repaired after exceeding its critical reliability state r, $r \in \{1, 2, \ldots, z-1\}$.

Taking into account cost of system's maintenance we consider cost of repair and downtime cost concerned with system unavailability. Further, thee following notations are used:

$C_R(z)$—cost of repair, including tradesmen and materials that causes the system to return to its "as good as new" reliability state z, after exceeding a reliability critical state r,

$\boldsymbol{T}_R(z)$—time of system's repair to its "as good as new" reliability state z, after exceeding a reliability critical state r,

$C_R(u)$—cost of repair, including tradesmen and materials that causes the system to return to its reliability state u, $u = z-1, z-2, \ldots, r$, after exceeding a reliability critical state r,

$\boldsymbol{T}_R(u)$—time of system's repair to its reliability state u, $u = z-1, z-2, \ldots, r$, after exceeding a reliability critical state r,

C_{Dt}—downtime cost per hour,

$C_{order\_Ex}(n_{order}, \boldsymbol{T}_R(u))$—costs of interrupting the execution of orders or delays in their implementation depend on the average number of orders executed at a given moment and time of system's repair to its reliability state u, $u = z, z-1, z-2, \ldots, r$.

It is assumed that (4.8.1) and following similar condition for time of repair

$$\boldsymbol{T}_R(z) \geq \boldsymbol{T}_R(z-1) \geq \boldsymbol{T}_R(z-2) \geq \cdots \geq \boldsymbol{T}_R(r) \tag{4.8.7}$$

hold.

In that case, total cost of system renewal after exceeding a reliability critical state concerned with operations that cause the system to return to its full reliability state z, is

$$C_T(z) = C_R(z) + C_{Dt} \cdot \boldsymbol{T}_R(z) + C_{order\_Ex}(n_{order}, \boldsymbol{T}_R(z)). \tag{4.8.8}$$

And generally, total cost of system's repair to other reliability states u, $u = z$, $z-1, \ldots, r$, is

$$\begin{aligned} C_T(u) &= C_R(u) + C_{Dt} \cdot \boldsymbol{T}_R(u) + C_{order\_Ex}(n_{order}, \boldsymbol{T}_R(u)), \\ u &= z, z-1, \ldots, r. \end{aligned} \tag{4.8.9}$$

Total cost of maintenance of a renewable system with non-negligible renewal durations up to the time point t, $0 \le t \le S^{(N_1+1)}(r)$, for fixed critical state r, $r \in \{1, 2, \ldots, z-1\}$, using (4.3.4), is estimated as

$$C(t, r) \cong C_T(z) \cdot \left\lfloor \frac{t + \mu_R(z)}{\mu(r) + \mu_R(z)} \right\rfloor, 0 \le t \le S^{(N_1+1)}(r), \tag{4.8.10}$$

where total cost of operations $C_T(z)$ that cause the system to return to the state z is given by (4.8.8).

Next, total cost of maintenance of a renewable system with non-negligible renewal durations up to the time point t, $S^{(N_1+1)}(r) < t \le S^{(N_2+1)}(r)$, for fixed critical state r, $r \in \{1, 2, \ldots, z-1\}$, using (4.3.5), is estimated from following formula

$$C(tr) \cong C_T(z) \cdot N_1 + C_T(z-1) \cdot \left\lfloor \frac{t - (N_1+1)\,\mu(r) - N_1 \mu_R(z)}{\mu(r) - \mu(z) + \mu_R(z-1)} \right\rfloor,$$

$$S^{(N_1+1)}(r) < t \le S^{(N_2+1)}(r), \tag{4.8.11}$$

where total costs $C_T(z)$, $C_T(z-1)$ are given by (4.8.9), respectively, for $u = z$ and $u = z-1$.

More generally, total cost of maintenance of a renewable system with non-negligible renewal durations up to the time point t, $S^{(N_x+1)}(r) < t \le S^{(N_{x+1}+1)}(r)$, $x = 1, 2, \ldots, z-r$, for fixed critical state r, $r \in \{1, 2, \ldots, z-1\}$, using (4.3.6), is estimated from the formula

$$C(t, r) \cong C_T(z) \cdot N_1 + \sum_{j=1}^{x-1} C_T(z-j) \cdot \left(N_{j+1} - N_j\right) + C_T(z-x)$$

$$\cdot \left\lfloor \frac{t - (N_1+1)\mu(r) - N_1 \mu_R(z) - \sum_{j=1}^{x-1} \left(N_{j+1} - N_j\right)\left(\mu(r) - \mu(z-j+1) + \mu_R(z-j)\right)}{\mu(r) - \mu(z-x+1) + \mu_R(z-x)} \right\rfloor,$$

$$S^{(N_x+1)}(r) < t \le S^{(N_{x+1}+1)}(r), x = 1, 2, \ldots, z-r, \tag{4.8.12}$$

where total cost $C_T(u)$, $u = z, z-1, \ldots, r$, are given by (4.8.9).

Taking into account obtained results, total cost of maintenance of a renewable system with non-negligible renewal durations up to the time point t, $t \geq 0$, for fixed critical state r, $r \in \{1, 2, \ldots, z-1\}$, is determined using the following procedure:

$x = 1;$

$E[S^{(N_x+1)}(r)] \cong (N_1+1)\mu(r) + N_1\mu_R(z);$

`if (`$t \leq E[S^{(N_x+1)}(r)]$`)`

`then {`

$$C_T(z) = C_R(z) + C_{Dt} \cdot \boldsymbol{T}_R(z) + C_{order\_Ex}(n_{order}, \boldsymbol{T}_R(z));$$

$$C(t,r) \cong C_T(z) \cdot \left\lfloor \frac{t+\mu_R(z)}{\mu(r)+\mu_R(z)} \right\rfloor;$$

`}`

`else {`

`do{`

$x := x + 1;$

$$E[S^{(N_x+1)}(r)] \cong (N_1+1)\mu(r) + N_1\mu_R(z) + \sum_{j=1}^{x-1} (N_{j+1} - N_j)(\mu(r) - \mu(z-j+1) + \mu_R(z-j));$$

`} while (`$t > E[S^{(N_x+1)}(r)]$`) and (`$x < z - r + 1$`);`

`for (j=0;j<x;j++)`

$$C_T(z-j) = C_R(z-j) + C_{Dt} \cdot \boldsymbol{T}_R(z-j) + C_{order\_Ex}(n_{order}, \boldsymbol{T}_R(z-j));$$

$$C(t,r) \cong C_T(z) \cdot N_1 + \sum_{j=1}^{x-2} C_T(z-j) \cdot (N_{j+1} - N_j)$$

$$+ C_T(z-x+1) \cdot \left\lfloor \frac{t - (N_1+1)\mu(r) - N_1\mu_R(z) - \sum_{j=1}^{x-2} (N_{j+1} - N_j)(\mu(r) - \mu(z-j+1) + \mu_R(z-j))}{\mu(r) - \mu(z-x+2) + \mu_R(z-x+1)} \right\rfloor;$$

`if (`$t > E[S^{(N_{z-r+1}+1)}(r)]$`)`

`then{`

`for (j=0;j<z-r+1;j++)`

$$C_T(z-j) = C_R(z-j) + C_{Dt} \cdot \boldsymbol{T}_R(z-j) + C_{order\_Ex}(n_{order}, \boldsymbol{T}_R(z-j));$$

$$C(t,r) \cong C_T(z) \cdot N_1 + \sum_{j=1}^{z-r} C_T(z-j) \cdot (N_{j+1} - N_j); \}$$

`}`

To minimize the maintenance costs for a renewable system with non-negligible renewal time, a similar procedure as in Section 4.8.1, for a renewable system with negligible renewal time, is proposed. Using this procedure we compare the costs of repair and operation of the system from the new system to the given moment t, for different configuration of the number of repairs N_1, N_2, …, N_{z-r+1} (to various states z, $z-1$, …, r) and select the most beneficial repair procedure.

$N_1 = \left\lfloor \frac{t+\mu_R(z)}{\mu(r)+\mu_R(z)} \right\rfloor$;
for (i=2;i<z-r+2;i++)
    $N_i = N_1$;
$C_T(z) = C_R(z) + C_{Dt} \cdot \boldsymbol{T}_R(z) + C_{order\_Ex}(n_{order}, \boldsymbol{T}_R(z))$;
$C(t,r) \cong C_T(z) \cdot N_1$;
if ($r<z$) and ($N_1>0$) then{
  $x=1$;
  do{
    do{
      $N_x := N_x - 1$;

$$N_{x+1} := N_x + \left\lfloor \frac{t-(N_1+1)\mu(r) - N_1\mu_R(z) - \sum_{j=1}^{x-1}\left(N_{j+1}-N_j\right)\left(\mu(r)-\mu(z-j+1)+\mu_R(z-j)\right)}{\mu(r)-\mu(z-x+1)+\mu_R(z-x)} \right\rfloor;$$

      for (j=0;j<x;j++)
        $C_T(z-j) = C_R(z-j) + C_{Dt} \cdot \boldsymbol{T}_R(z-j) + C_{order\_Ex}(n_{order}, \boldsymbol{T}_R(z-j))$;
      $C_{new}(t,r) \cong C_T(z) \cdot N_1 + \sum_{j=1}^{x} C_T(z-j) \cdot \left(N_{j+1} - N_j\right)$;
      if ($C_{new}(t,r) < C(t,r)$) then{
        $C(t,r) = C_{new}(t,r)$;
        for (i=x+2;i<z-r+2;i++)
          $N_i = N_{x+1}$;
        for (i=1;i<z-r+2;i++)
          N[i]:=N_i;
      }
      x := x+1;
    } while ($r<z-x+1$) and ($N_x > N_{x-1}$);
    if ($r>z-x$) then{
      $N_0 = 0$;
      do{
        x:= x-1;
      } while ($N_x = N_{x-1}$) and (x>1)
    }
  } while ($N_1>0$)
}
else if ($N_1<1$) then
    $C(t,r) = 0$;
System.out.println("Optimal cost of system's renewal"+$C(t,r)$);
for (i=1;i<z-r+2;i++)
    System.out.println("Optimal number of renewals N_i ="+N[i]);

References

Badoux, R.A.J., 1987. Availability and maintainability. In: Colombo, A.G., Keller, A.Z. (Eds.), Reliability Modeling and Applications. Kluwer Academic Publishers Group, Dordrecht, pp. 99–124.

Barone, G., Frangopol, D.M., 2014. Life-cycle maintenance of deteriorating structures by multi-objective optimization involving reliability, risk, availability, hazard and cost. Struct. Saf. 48, 40–50.

Blokus, A., 2006. Reliability and availability evaluation of large renewal systems. Int. J. Mater. Struct. Reliab. 4 (1), 39–52.

Blokus-Roszkowska, A., 2007. Reliability Analysis of Homogenous Large Systems With Component Dependent Failures. PhD Thesis, Gdynia Maritime University – System Research Institute Warsaw (in Polish).

Blokus-Roszkowska, A., 2019. Availability analysis of transport navigation system under imperfect repair. In: Zamojski, W., Mazurkiewicz, J., Sugier, J., Walkowiak, T., Kacprzyk, J. (Eds.), Contemporary Complex Systems and Their Dependability, Proceedings of the Thirteenth International Conference on Dependability and Complex Systems DepCoS-RELCOMEX 2018. Advances in Intelligent Systems and Computing. In: Vol. 761. Springer, Cham, pp. 35–45.

Blokus-Roszkowska, A., Kołowrocki, K., 2009. Reliability and availability of a ground ship-rope transporter in variable operation conditions. Electron. J. Reliab. Risk Anal. Theory Appl. 2 (4), 82–96.

Blokus-Roszkowska, A., Kołowrocki, K., 2015a. Reliability analysis of ship-rope transporter with dependent components. In: Nowakowski, et al., (Eds.), Safety and Reliability: Methodology and Applications – Proceedings of the European Safety and Reliability Conference, ESREL 2014. Taylor & Francis Group, London, pp. 255–263.

Blokus-Roszkowska, A., Kołowrocki, K., 2015b. Reliability analysis of multistate series systems with dependent components. J. Pol. Saf. Reliab. Assoc. Summer Saf. Reliab. Semin. 6 (1), 31–36.

Bobrowski, D., 1985. Mathematical Models and Methods of Reliability Theory in Examples and Exercises. WNT, Warsaw (in Polish).

Castro, I., 2009. A model of imperfect preventive maintenance with dependent failure modes. Eur. J. Oper. Res. 196 (1), 217–224.

Crespo Marquez, A., 2007. The Maintenance Management Framework. Models and Methods for Complex Systems Maintenance. Springer, Londres, Reino Unido.

de Almeida, A.T., Cavalcante, C.A.V., Alencar, M.H., Ferreira, R.J.P., de Almeida-Filho, A.T., Garcez, T.V., 2015. Multicriteria and Multiobjective Models for Risk, Reliability and Maintenance Decision Analysis. International Series in Operations Research & Management Science. 231 Springer, Cham.

Dziula, P., Kołowrocki, K., Rosiński, A., 2015. Issues concerning identification of critical infrastructure systems within the Baltic Sea area. In: Podofillini, L., Sudret, B., Stojadinovic, B., Zio, E., Kröger, W. (Eds.), Safety and Reliability of Complex Engineered Systems ESREL 2015. CRC Press/Balkema, Leiden, pp. 119–126.

Finkelstein, M., 2008. Failure Rate Modelling for Reliability and Risk. Springer, London.

Frangopol, D.M., Liu, M., 2007. Maintenance and management of civil infrastructure based on condition, safety, optimization, and life-cycle cost*. Struct. Infrastruct. Eng. 3 (1), 29–41.

Gertsbakh, I.B., 2000. Reliability Theory With Applications to Preventive Maintenance. Springer, Berlin.

Høyland, A., Rausand, M., 2004. System Reliability Theory: Models, Statistical Methods and Applications. John Wiley & Sons, Inc., Hoboken, NJ.

Kołowrocki, K., 2014. Reliability of Large and Complex Systems, second ed. Elsevier, London.

Kołowrocki, K., Soszyńska, J., 2006. Reliability and availability of complex systems. Qual. Reliab. Eng. Int. 22 (1), 79–99. J. Wiley & Sons Ltd.
Kołowrocki, K., Soszyńska, J., 2010. Reliability, availability and safety of complex technical systems: modeling – identification – prediction – optimization. J. Pol. Saf. Reliab. Assoc. Summer Saf. Reliab. Semin. 4 (1), 133–158.
Kołowrocki, K., Soszyńska-Budny, J., 2011. Reliability and Safety of Complex Technical Systems and Processes: Modeling – Identification – Prediction – Optimization, first ed. Springer-Verlag, London.
Kong, J.S., Frangopol, D.M., 2004. Cost–reliability interaction in life-cycle cost optimization of deteriorating structures. J. Struct. Eng. 130 (11), 1704–1712.
Kopociński, B., 1973. An Outline of Renewal and Reliability Theory. PWN, Warsaw (in Polish).
Levitin, G., Lisnianski, A., 2000. Optimal replacement scheduling in multi-state series-parallel systems. Qual. Reliab. Eng. Int. 16, 157–162.
Li, W., Pham, H., 2006. An inspection-maintenance model for degraded repairable systems. In: Pham, H. (Ed.), Reliability Modeling, Analysis and Optimization. Series on Quality, Reliability and Engineering Statistics. In: Vol. 9. World Scientific Publishing, Singapore, pp. 439–458.
Lisnianski, A., Levitin, G., 2003. Multi-State System Reliability. Assessment, Optimisation and Applications. World Scientific Publishing Co. Pte. Ltd., New Jersey, London, Singapore, Hong Kong.
Lisnianski, A., Frenkel, I., Ding, Y., 2010. Multi-State Systems Reliability Analysis and Optimization for Engineers and Industrial Managers. Springer, London.
Liu, Y., Huang, H., Wang, Z., Li, Y., Yang, Y., 2013. A joint redundancy and imperfect maintenance strategy optimization for multi-state systems. IEEE Trans. Reliab. 62, 368–378.
Muhammad, M., Majid, M.A., Mokhtar, A.A., 2010. Reliability evaluation for a multi-state system subject to imperfect repair and maintenance. Int. J. Eng. Technol. 10 (01), 59–63.
Nourelfath, M., Chatelet, E., Nahas, N., 2012. Joint redundancy and imperfect preventive maintenance optimization for series–parallel multi-state degraded systems. Reliab. Eng. Syst. Saf. 103, 51–60.
Okasha, N.M., Frangopol, D.M., 2009. Lifetime-oriented multi-objective optimization of structural maintenance considering system reliability, redundancy and life-cycle cost using GA. Struct. Saf. 31 (6), 460–474.
Olde Keizer, M.C.A., Teunter, R.H., Veldman, J., Babai, M.Z., 2018. Condition-based maintenance for systems with economic dependence and load sharing. Int. J. Prod. Econ. 195, 319–327.
Pham, H., Wang, H.Z., 1996. Imperfect maintenance. Eur. J. Oper. Res. 94 (3), 425–438.
Pham, H., Suprasad, A., Misra, R.B., 1997. Availability and mean life time prediction of multistage degraded system with partial repairs. Reliab. Eng. Syst. Saf. 56, 169–173.
Pourret, O., Collet, J., Bon, J.L., 1999. Evaluation of the unavailability of a multistate-component system using a binary model. IEEE Trans. Reliab. 64, 13–17.
Soro, I.W., Nourelfath, M., Ait-Kadi, D., 2010. Performance evaluation of multi-state degraded systems with minimal repairs and imperfect preventive maintenance. Reliab. Eng. Syst. Saf. 95 (2), 65–69.
Szymkowiak, M., 2018. Characterizations of distributions through aging intensity. IEEE Trans. Reliab. 67 (2), 446–458.
Szymkowiak, M., 2019. Lifetime Analysis by Aging Intensity Functions. Monograph in Series: Studies in Systems, Decision and Control, Springer.
Wang, Y., Pham, H., 2011. A multi-objective optimization of imperfect preventive maintenance policy for dependent competing risk systems with hidden failure. IEEE Trans. Reliab. 60 (4), 770–781.

Weintrit, A., Dziula, P., Siergiejczyk, M., Rosiński, A., 2015. Reliability and exploitation analysis of navigational system consisting of ECDIS and ECDIS back-up systems. In: Weintrit, A. (Ed.), Activities in Navigation. Marine Navigation and Safety of Sea Transportation. CRC Press/Balkema, Leiden, pp. 109–115.
Wen, Y.K., Kang, Y.J., 2001. Minimum building life-cycle cost design criteria. I. Methodology, and II. Applications. J. Struct. Eng. 127, 330–346.
Xue, J., 1985. On multi-state system analysis. IEEE Trans. Reliab. 34, 329–337.
Zio, E., 2007. An introduction to the basics of reliability and risk analysis. In: Series in Quality, Reliability and Engineering Statistics 13. World Scientific Publishing Co. Pte. Ltd., New Jersey – London – Singapore – Beijing – Shanghai – Hong Kong – Taipei – Chennai.
Zio, E., Compare, M., 2013. Evaluating maintenance policies by quantitative modeling and analysis. Reliab. Eng. Syst. Saf. 109, 53–65.

CHAPTER FIVE

Safety of multistate networks with cascading effects

Basic assumptions related to safety analysis of multistate and aging critical infrastructure (CI) networks, composed of assets, are introduced. The safety functions of a multistate CI network and its assets are defined. The safety functions of a multistate series CI network, composed of independent subnetworks and in case when its subnetworks are dependent according to the local load-sharing (LLS) rule, are determined. Moreover, the safety functions of multistate "m out of k" and parallel CI networks, composed of independent subnetworks and in case when their subnetworks are dependent according to the equal load-sharing (ELS) rule, are determined. Special attention is paid to a multistate series CI network composed of series subnetworks. The safety of this CI network is analyzed in three cases. Namely, in case when subnetworks and their assets are independent, when subnetworks are dependent according to the LLS rule and consist of independent assets, and the last case when subnetworks are independent and assets in the subnetworks follow the LLS dependency rule. Finally, exemplary academic and practical applications of obtained results are presented.

5.1 Assumptions and approach description

In practice, complex systems are often not isolated or self-contained, but form groups of systems or infrastructures that interact with each other. This way, they form a network of connected and interdependent infrastructures. If the systems or infrastructures, forming a network of interdependent infrastructures consist of dependent items, we met a cascade effect in the network. This fact allows us to create new contributions of this chapter in the form of an introduction to the broad topic of safety analysis of critical infrastructure networks with cascading effects. In this context, the concept of cascading failures or cascading effect, as in Baldick et al. (2008) and Billinton and Allan (1996), is understood as a sequence of dependent failures of individual components or subnetworks, consequently affecting safety system or infrastructure networks.

https://doi.org/10.1016/B978-0-12-821260-8.00005-1

In Allen (1999) and Saidi et al. (2018) infrastructures are described as systems that organize and manage complex systems of flows, movement, and exchange. According to Zio (2016), infrastructures are large-scale systems that function interdependently to produce and distribute essential goods and services. If these essential goods and services are necessary for the proper functioning of society, and their damage or interruption may affect the health, safety, security, or economic well-being of citizens, these infrastructures are referred to as critical infrastructures (Council Directive 2008/114/EC, 2008). Some approaches to identification of critical infrastructure systems are presented in Blokus-Roszkowska and Dziula (2016). Johansson and Hassel (2010) describe critical infrastructures as large-scale, spatially distributed systems with high degrees of complexity, resulting mainly from interdependencies among infrastructure systems. Almoghathawi et al. (2019), adopting the definition of President's Commission on Critical Infrastructure Protection (1997), define a critical infrastructure network as a network of independent systems and processes that function collaboratively and synergistically to produce and distribute a continuous flow of essential goods and services.

Torbicki (2018) defines a critical network infrastructure as a critical infrastructure composed of a network of components or assets connected with each other. They can be connected in a physical way, they can interact with each other to perform functions assigned to the critical infrastructure, or share information. This definition has also been adopted. In his dissertation (Torbicki, 2018), the author mainly draws attention to the impact of outside processes on safety of critical infrastructures and critical infrastructure networks. When analyzing the impact of these outside processes, Torbicki (2018) takes simultaneously into account both the critical infrastructure operation process and the weather change process at the critical infrastructure operating area. The joint general safety model of critical infrastructure impacted by these two external processes is also presented in Kołowrocki et al. (2017d). Critical infrastructures in variable operating conditions have been widely discussed in Kołowrocki and Soszyńska-Budny (2011, 2017), while the impact by the weather change process on critical infrastructure is described in Kołowrocki et al. (2017a,c). These models have broad applications in safety analysis of CI and CI networks, such as Baltic oil terminal critical infrastructure (Kołowrocki et al., 2017b), port oil piping transportation system (Kołowrocki et al., 2017e), port transportation systems and port critical infrastructures (Blokus-Roszkowska et al., 2016, 2018a, b; Guze et al., 2008; Guze and Kołowrocki, 2016).

Critical infrastructures are usually interconnected and mutually dependent in various and complex ways, creating critical infrastructure network. They are interacting directly and indirectly at various levels of their complexity and operating activity (Kjølle et al., 2012; Kotzanikolaou et al., 2013; Nieuwenhuijs et al., 2008; Ouyang, 2014; Rinaldi et al., 2001; Zio and Sansavini, 2011). Hernandez-Fajardo and Dueñas-Osorio (2013) propose a methodology that takes into account simultaneously cascading failures in lifeline systems and their internal complexity. They analyze the joint impact of cascading failures, component fragilities, and interdependent effects. Identifying and modeling dependencies depend on the level of analysis. The selected level of analysis can vary from micro- to macrolevel (de Porcellinis et al., 2009; Han and Zio, 2019; Holden et al., 2013). A holistic approach as in Lauge et al. (2015) can be considered or a reductionist approach in which elementary components are identified and their behavior is described. Eusgeld et al. (2011) propose approach to analysis of interdependencies in critical infrastructures at the single system level and the system-of-systems level. Svedsen and Wolthunsen (2007) focus on the components of critical infrastructure networks and they demonstrate several types of multidependency structures.

The key problem is not only to identify relationships between infrastructures or components, but also to determine the type and strength of this relationship. Models of dependency and behavior of components can differ depending on the structural and material properties of the network, operational conditions and many other factors (Huang et al., 2014; Rehak et al., 2018; Utne et al., 2011; Theoharidou et al., 2011; Zhang and Peeta, 2011). The impact of topological properties on the dynamics of cascading in networks and different models of failure propagation can be found in the literature (Ouyang, 2014; Saidi et al., 2018; Zio and Sansavini, 2011). Rehak et al. (2018) assess cascading effect in critical infrastructure system through the quantification of spread of cascading impact. They analyze topology of impacts, including the aspects related to their nature and the way in which this impact continues to spread in a CI structure. Disruptions or failures in CI may have direct impact on a society, may occur simultaneously as synergistic impacts, or further spread across the CI and escalate as cascading impacts (Rehak et al., 2018).

In this context, we propose a multistate approach to safety analysis of systems and infrastructure networks, where cascading leaving safety-state subsets is considered. By extending a cascading effect to the entire system or network, it consequently reduces system lifetime in safety-state subsets and deteriorates its safety. Describing cascading effects in infrastructure networks both the dependencies among subnetworks of this network and among their assets are considered. In that case, after changing the safety-state subset by some of assets in the subnetwork to the worse safety-state subset, the lifetimes of remaining assets in this subnetwork in the safety-state subsets decrease. We assume that the model of cascade effect impact on the functioning and safety of a system or network depends largely on the structure and topological properties. Thus, we consider different models of dependency among assets and subnetworks in networks depending on the network safety structure.

In the multistate safety analysis of critical infrastructure (CI) networks, we assume similarly as in Chapter 2 during reliability analysis of multistate systems, that (Kołowrocki, 2001, 2008, 2014; Kołowrocki and Soszyńska, 2006; Kołowrocki and Soszyńska-Budny, 2011; Blokus-Roszkowska and Kolowrocki, 2017d; Xue, 1985; Xue and Yang, 1995a,b; Yu et al., 1994):
- n is the number of network assets,
- E_i, $i = 1, 2, \ldots, n$, denote the network assets,
- all assets and a CI network under consideration have the safety-state set $\{0, 1, \ldots, z\}$, $z \geq 1$,
- the safety states are ordered, the safety-state 0 is the worst and the safety-state z is the best,
- $T_i(u)$, $i = 1, 2, \ldots, n$, are independent random variables representing the lifetimes of assets E_i in the safety-state subset $\{u, u + 1, \ldots, z\}$, $u = 0, 1, \ldots, z$, while they were in the safety-state z at the moment $t = 0$,
- $\boldsymbol{T}(u)$ is a random variable representing the lifetime of a CI network in the safety-state subset $\{u, u + 1, \ldots, z\}$, $u = 0, 1, \ldots, z$, while it was in the safety-state z at the moment $t = 0$,
- the states of a CI network degrade with time t,
- $s_i(t)$ is an asset E_i safety state at the moment t, $t \in (-\infty, \infty)$, given that it was in the safety-state z at the moment $t = 0$,
- $\boldsymbol{s}(t)$ is a safety state of CI network $\boldsymbol{N}$ at the moment t, $t \in (-\infty, \infty)$, given that it was in the safety-state z at the moment $t = 0$.

The safety function of a multistate asset and of a multistate critical infrastructure (Blokus-Roszkowska and Kolowrocki, 2017b,c; Kołowrocki and Soszyńska-Budny, 2011, 2012, 2013) we define similarly as the reliability function of a multistate component and, respectively, as the system reliability function in Chapter 2.

Definition 5.1.1. The safety function of a multistate asset E_i is a vector (Blokus-Roszkowska and Kolowrocki, 2017b,c; Kołowrocki, 2013; Kołowrocki and Soszyńska-Budny, 2011, 2012)

$$S_i(t,\cdot)=[S_i(t,0),S_i(t,1),\ldots,S_i(t,z)],\quad t\in(-\infty,\infty),\\ i=1,2,\ldots,n, \tag{5.1.1}$$

where

$$S_i(t,u)=P(s_i(t)\geq u\mid s_i(0)=z)=P(T_i(u)>t),\quad u=0,1,\ldots,z, \tag{5.1.2}$$

is the probability that an asset E_i is in the safety-state subset $\{u,u+1,\ldots,z\}$, $u=0,1,\ldots,z$, at the moment t, $t\in(-\infty,\infty)$, while it was in the state z at the moment $t=0$.

The safety functions $S_i(t,u)$, $u=0,1,\ldots,z$, defined by (5.1.2), are called coordinates of the safety function $S_i(t,)$ given by (5.1.1), of the component, that is, asset E_i, $i=1,2,\ldots,n$.

Definition 5.1.2. The safety function of a multistate CI network is a vector (Blokus-Roszkowska and Kolowrocki, 2017b,c; Kołowrocki, 2013; Kołowrocki and Soszyńska-Budny, 2011, 2012)

$$S(t,\cdot)=[S(t,0),S(t,1),\ldots,S(t,z)],\quad t\in(-\infty,\infty), \tag{5.1.3}$$

where

$$\boldsymbol{S}(t,u)=P(\boldsymbol{s}(t)\geq u\mid \boldsymbol{s}(0)=z)=P(\boldsymbol{T}(u)>t),\quad u=0,1,\ldots,z, \tag{5.1.4}$$

is the probability that a network is in the safety-state subset $\{u,u+1,\ldots,z\}$, $u=0,1,\ldots,z$, at the moment t, $t\in(-\infty,\infty)$, while it was in the state z at the moment $t=0$.

Further, we replace $S_i(t,0)$, existing in (5.1.1), and $\boldsymbol{S}(t,0)$, existing in (5.1.3), by 1.

The other safety characteristics of the CI network are similarly defined as reliability characteristics in Chapter 2, and their definitions are presented in Blokus-Roszkowska and Kolowrocki (2017c), Kołowrocki (2013), and Kołowrocki and Soszyńska-Budny (2011, 2012). The basic safety characteristics that we consider in this book include the mean lifetimes and standard deviations of an asset in the safety-state subsets, the mean lifetimes of an asset in particular safety states, the mean lifetimes and standard deviations of a network in the safety-state subsets, the mean lifetimes of a network in particular safety states, and a risk function of a multistate network.

5.2 Basic safety structures of multistate networks and their safety functions

Now, after introducing the notion of multistate safety analysis, we determine the safety function of a multistate series CI network. From the definition of a multistate series system, given in Chapter 2, we conclude that a multistate series network is in the safety-state subset $\{u, u+1, \ldots, z\}$, $u = 1, 2, \ldots, z$, if and only if all its assets are in this subset of safety states.

In the further analysis of multistate networks, we assume that safety functions of subnetworks $\boldsymbol{N}_i$, $i = 1, 2, \ldots, k$, satisfy following condition

$$\boldsymbol{S}^{(i)}(t, u) = 1 \quad \text{for } t < 0, \quad i = 1,2,\ldots,k, \quad u = 1,2,\ldots,z, \tag{5.2.1}$$

and similarly

$$\boldsymbol{S}(t, u) = 1 \quad \text{for} \quad t < 0, \quad u = 1,2,\ldots,z, \tag{5.2.2}$$

and thereby we consider these safety functions in time interval $t \in \langle 0, \infty)$.

Considering a multistate CI network consisting of independent subnetworks $\boldsymbol{N}_i$, $i = 1, 2, \ldots, k$, connected in series, with the safety structure scheme presented in Fig. 5.1, its safety function is given by the vector (Blokus-Roszkowska and Kolowrocki, 2017b,c; Kołowrocki, 2013; Kołowrocki and Soszyńska-Budny, 2011, 2012)

$$\boldsymbol{S}(t, \cdot) = [1, \boldsymbol{S}(t, 1), \ldots, \boldsymbol{S}(t, z)], \quad t \geq 0, \tag{5.2.3}$$

with the coordinates

$$\boldsymbol{S}(t, u) = \prod_{i=1}^{k} \boldsymbol{S}^{(i)}(t, u), \quad t \geq 0, \quad u = 1,2,\ldots,z, \tag{5.2.4}$$

where $\boldsymbol{S}^{(i)}(t, u)$, $i = 1, 2, \ldots, k$, $u = 1, 2, \ldots, z$, denote safety function coordinates of subnetworks $\boldsymbol{N}_i$, $i = 1, 2, \ldots, k$, and can be determined by appropriate formulae, given in Chapter 3, depending on their structure and assumed dependency model.

In a special case, we assume that subnetworks $\boldsymbol{N}_i$, $i = 1, 2, \ldots, k$, have piecewise exponential safety functions

$$\boldsymbol{S}^{(i)}(t, \cdot) = \left[1, \boldsymbol{S}^{(i)}(t, 1), \ldots, \boldsymbol{S}^{(i)}(t, z)\right], \quad t \geq 0, \quad i = 1,2,\ldots,k, \tag{5.2.5}$$

with the coordinates

$$\boldsymbol{S}^{(i)}(t, u) = \exp[-\boldsymbol{\lambda}_i(u)t], \quad t \geq 0, \quad \boldsymbol{\lambda}_i(u) > 0, \quad u = 1,2,\ldots,z. \tag{5.2.6}$$

Fig. 5.1 The scheme of a series network safety structure.

Subsequently, the coordinates of safety function (5.2.3) of a multistate series network consisting of independent subnetworks $\boldsymbol{N}_i$, $i = 1, 2, \ldots, k$, having piecewise exponential safety functions (5.2.5), (5.2.6), are given by

$$\boldsymbol{S}(t, u) = \exp\left[-\sum_{i=1}^{k}\boldsymbol{\lambda}_i(u)t\right], \quad t \geq 0, \quad u = 1, 2, \ldots, z. \tag{5.2.7}$$

If we consider a homogeneous multistate CI network and assume that all its subnetworks $\boldsymbol{N}_i$, $i = 1, 2, \ldots, k$, are independent and have the same safety function with the coordinates

$$\boldsymbol{S}^{(i)}(t, u) = \boldsymbol{S}^{(1)}(t, u), \quad t \geq 0, \quad u = 1, 2, \ldots, z, \quad i = 1, 2, \ldots, k, \tag{5.2.8}$$

then, the CI network safety function (5.2.3), (5.2.4) takes following form

$$\boldsymbol{S}(t, \cdot) = [1, \boldsymbol{S}(t, 1), \ldots, \boldsymbol{S}(t, z)], \quad t \geq 0, \tag{5.2.9}$$

where

$$\boldsymbol{S}(t, u) = \left[\boldsymbol{S}^{(1)}(t, u)\right]^k, \quad t \geq 0, \quad u = 1, 2, \ldots, z. \tag{5.2.10}$$

Next, we consider a multistate series CI network composed of subnetworks $\boldsymbol{N}_i$, $i = 1, 2, \ldots, k$, dependent according to the local load-sharing (LLS) rule, described in Section 3.1 of Chapter 3. In the LLS dependency model the load on a degraded asset and subnetwork is transferred to neighboring assets and subnetworks, and the strength of these dependencies may vary depending on the individual relationship among assets and among subnetworks in the network. Similarly in Zio and Sansavini (2011), the load originated from a failed component is transferred locally to the first neighbors of this component within the network structure it belongs to, and to interdependent components in the other network system that are connected to the failed component.

Here, in the LLS model of dependence among subnetworks, we assume that if the subnetwork $\boldsymbol{N}_j$, $j = 1, 2, \ldots, k$, leaves the safety-state subset $\{u, u+1, \ldots, z\}$, $u = 1, 2, \ldots, z$, then lifetimes of other subnetworks $\boldsymbol{N}_i$, $i = 1, 2, \ldots, k$, $i \neq j$, worsen depending on their exposure to the subnetwork $\boldsymbol{N}_j$ safety-state change. In the adopted model of subnetworks dependency, this means that the subnetwork lifetimes in the safety-state subset $\{v, v + 1, \ldots, z\}$ for $v = u, u - 1, \ldots, 1$ and $u = 1, 2, \ldots, z - 1$, and their mean values decrease according to the formulas

$$\boldsymbol{T}_{i/j}(v) = q(v, \boldsymbol{N}_j, \boldsymbol{N}_i) \cdot \boldsymbol{T}_i(v),$$
$$E[\boldsymbol{T}_{i/j}(v)] = q(v, \boldsymbol{N}_j, \boldsymbol{N}_i) \cdot E[\boldsymbol{T}_i(v)], \quad i = 1, 2, \ldots, k,$$
$$j = 1, 2, \ldots, k, \tag{5.2.11}$$

where $\boldsymbol{T}_{i/j}(v)$ denotes the lifetime in safety-state subset $\{v, v+1, \ldots, z\}$, $v = u, u-1, \ldots, 1$, $u = 1, 2, \ldots, z-1$, of subnetwork $\boldsymbol{N}_i$, $i = 1, 2, \ldots, k$, after departure of the subnetwork $\boldsymbol{N}_j$, $j = 1, 2, \ldots, k$, from the safety-state subset $\{u+1, \ldots, z\}$, $u = 1, 2, \ldots, z$, and where $q(v, \boldsymbol{N}_j, \boldsymbol{N}_i)$ are the coefficients of subnetwork $\boldsymbol{N}_j$, $j = 1, 2, \ldots, k$, impact on other subnetworks $\boldsymbol{N}_i$, $i = 1, 2, \ldots, k$, $i \neq j$, lifetimes in the safety-state subsets. Further, it is assumed that $0 < q(u, \boldsymbol{N}_j, \boldsymbol{N}_i) \leq 1$, $i = 1, 2, \ldots, k$, $i \neq j$, and $q(u, \boldsymbol{N}_j, \boldsymbol{N}_j) = 1$, $j = 1, 2, \ldots, k$.

With such assumptions, the safety function of a multistate series CI network following the LLS dependency rule, using results presented in Section 3.1 and in Blokus-Roszkowska and Kolowrocki (2017b,c), is given by

$$\boldsymbol{S}_{LLS}(t, \cdot) = [1, \boldsymbol{S}_{LLS}(t, 1), \ldots, \boldsymbol{S}_{LLS}(t, z)], \quad t \geq 0, \tag{5.2.12}$$

where

$$\boldsymbol{S}_{LLS}(t, u) = \prod_{i=1}^{k} \boldsymbol{S}^{(i)}(t, u+1)$$
$$+ \int_0^t \sum_{j=1}^{k} \left[\tilde{\boldsymbol{f}}^{(j)}(a, u+1) \cdot \prod_{\substack{i=1 \\ i \neq j}}^{k} \boldsymbol{S}^{(i)}(a, u+1) \cdot \boldsymbol{S}^{(j)}(a, u) \cdot \prod_{i=1}^{k} \boldsymbol{S}^{(i/j)}(t-a, u) \right] da,$$
$$t \geq 0, \quad u = 1, 2, \ldots, z-1, \tag{5.2.13}$$

$$\boldsymbol{S}_{LLS}(t, z) = \prod_{i=1}^{k} \boldsymbol{S}^{(i)}(t, z), \quad t \geq 0, \tag{5.2.14}$$

where

$\boldsymbol{S}^{(i)}(t, u+1)$—the safety function coordinate of a subnetwork $\boldsymbol{N}_i$, $i = 1, 2, \ldots, k$,

$\tilde{\boldsymbol{f}}^{(j)}(t, u+1)$—the density function coordinate of a subnetwork $\boldsymbol{N}_j$, $j = 1, 2, \ldots, k$, corresponding to the distribution function $\tilde{\boldsymbol{F}}^{(j)}(t, u+1)$ given by

$$\widetilde{\boldsymbol{F}}^{(j)}(t, u+1) = 1 - \frac{\boldsymbol{S}^{(j)}(t, u+1)}{\boldsymbol{S}^{(j)}(t, u)}, \quad u = 1, 2, \ldots, z-1, \quad t \geq 0, \tag{5.2.15}$$

$\boldsymbol{S}^{(j)}(t, u)$—the safety function coordinate of a subnetwork $\boldsymbol{N}_j$, $j = 1, 2, \ldots, k$,
$\boldsymbol{S}^{(i/j)}(t, u)$—the safety function coordinate of a subnetwork $\boldsymbol{N}_i$, $i = 1, 2, \ldots, k$, after departure of the subnetwork $\boldsymbol{N}_j$, $j = 1, 2, \ldots, k$, from the safety-state subset $\{u + 1, \ldots, z\}$, $u = 1, 2, \ldots, z - 1$, such that

$$\boldsymbol{S}^{(i/j)}(t-a, u) = \frac{\boldsymbol{S}^{(i/j)}(t, u)}{\boldsymbol{S}^{(i)}(a, u)}, \quad u = 1, 2, \ldots, z-1, \quad 0 < a < t, \quad t \geq 0. \tag{5.2.16}$$

If in an aging series network following the LLS rule, the subnetworks $\boldsymbol{N}_i$, $i = 1, 2, \ldots, k$, have piecewise exponential safety functions (5.2.5), (5.2.6), then the conditional intensities $\boldsymbol{\lambda}_{i/j}(v)$ of the subnetwork $\boldsymbol{N}_i$, $i = 1, 2, \ldots, k$, departure from the subset $\{v, v + 1, \ldots, z\}$, $v = u, u - 1, \ldots, 1$, $u = 1, 2, \ldots, z - 1$, after the departure of the subnetwork $\boldsymbol{N}_j$, $j = 1, 2, \ldots, k$, by (5.2.11), are

$$\boldsymbol{\lambda}_{i/j}(v) = \frac{\boldsymbol{\lambda}_i(v)}{q(v, \boldsymbol{N}_j, \boldsymbol{N}_i)}, \quad i = 1, 2, \ldots, k, \quad j = 1, 2, \ldots, k. \tag{5.2.17}$$

In this case, the safety function of a subnetwork $\boldsymbol{N}_i$, $i = 1, 2, \ldots, k$, after departure of the subnetwork $\boldsymbol{N}_j$, $j = 1, 2, \ldots, k$, from the safety-state subset $\{u + 1, \ldots, z\}$, $u = 1, 2, \ldots, z - 1$, is defined by a vector

$$\boldsymbol{S}^{(i/j)}(t, \cdot) = \left[1, \boldsymbol{S}^{(i/j)}(t, 1), \ldots, \boldsymbol{S}^{(i/j)}(t, z)\right], \quad t \geq 0,$$
$$i = 1, 2, \ldots, k, \quad j = 1, 2, \ldots, k, \tag{5.2.18}$$

with the coordinates

$$\boldsymbol{S}^{(i/j)}(t, v) = \exp\left[-\frac{\boldsymbol{\lambda}_i(v)}{q(v, \boldsymbol{N}_j, \boldsymbol{N}_i)} t\right], \quad v = u, u-1, \ldots, 1,$$
$$u = 1, 2, \ldots, z-1, \tag{5.2.19}$$

$$\boldsymbol{S}^{(i/j)}(t, v) = \exp[-\boldsymbol{\lambda}_i(v) t], \quad v = u+1, \ldots, z, \quad u = 1, 2, \ldots, z-1. \tag{5.2.20}$$

If, the subnetworks $\boldsymbol{N}_i$, $i = 1, 2, \ldots, k$, have piecewise exponential safety functions (5.2.5), (5.2.6), then the distribution function of a subnetwork $\boldsymbol{N}_j$, $j = 1, 2, \ldots, k$, defined by (5.2.15), takes the following form

$$\widetilde{F}^{(j)}(t,u+1)=1-\exp\left[-\left(\lambda_j(u+1)-\lambda_j(u)\right)t\right],$$
$$u=1,2,\ldots,z-1,\quad t\geq 0, \tag{5.2.21}$$

and its corresponding density function is

$$\widetilde{f}^{(j)}(t,u+1)=\left(\lambda_j(u+1)-\lambda_j(u)\right)\cdot\exp\left[-\left(\lambda_j(u+1)-\lambda_j(u)\right)t\right],$$
$$u=1,2,\ldots,z-1,\quad t\geq 0. \tag{5.2.22}$$

Considering (5.2.19)–(5.2.22), for a series network following the LLS rule, with subnetworks $\boldsymbol{N}_i$, $i = 1, 2, \ldots, k$, having piecewise exponential safety functions (5.2.5), (5.2.6), using results presented in Section 3.1 and in Blokus-Roszkowska and Kolowrocki (2017b,c), the network safety function (5.2.12)–(5.2.14) takes the following form

$$\boldsymbol{S}_{LLS}(t,\cdot)=[1,\boldsymbol{S}_{LLS}(t,1),\ldots,\boldsymbol{S}_{LLS}(t,z)],\quad t\geq 0, \tag{5.2.23}$$

where

$$\boldsymbol{S}_{LLS}(t,u)=\exp\left[-\sum_{i=1}^{k}\lambda_i(u+1)t\right]+\sum_{j=1}^{k}\frac{\lambda_j(u+1)-\lambda_j(u)}{\sum_{i=1}^{k}(\lambda_i(u+1)-\lambda_i(u))}$$
$$\cdot\left[\exp\left[-\sum_{i=1}^{k}\frac{\lambda_i(u)}{q\left(u,\boldsymbol{N}_j,\boldsymbol{N}_i\right)}t\right]-\exp\left[-\sum_{i=1}^{k}\left(\lambda_i(u+1)-\lambda_i(u)+\frac{\lambda_i(u)}{q\left(u,\boldsymbol{N}_j,\boldsymbol{N}_i\right)}\right)\cdot t\right]\right],$$
$$t\geq 0,\quad u=1,2,\ldots,z-1, \tag{5.2.24}$$

$$\boldsymbol{S}_{LLS}(t,z)=\exp\left[-\sum_{i=1}^{k}\lambda_i(z)t\right], \tag{5.2.25}$$

where $q(u, \boldsymbol{N}_j, \boldsymbol{N}_i)$, $u = 1, 2, \ldots, z - 1$, are the coefficients of network load increase that express the influence of the subnetwork $\boldsymbol{N}_j$, $j = 1, 2, \ldots, k$, departure from the safety-state subset $\{u, u + 1, \ldots, z\}$, $u = 1, 2, \ldots, z$, on other subnetworks $\boldsymbol{N}_i$, $i = 1, 2, \ldots, k$, $i \neq j$, lifetimes in the safety-state subsets.

As a special case, we consider a homogeneous, aging series network with subnetworks having identical piecewise exponential safety functions (5.2.5) with the coordinates

$$\boldsymbol{S}^{(i)}(t,u)=\exp[-\boldsymbol{\lambda}(u)t],\quad t\geq 0,\quad \boldsymbol{\lambda}(u)>0,\quad u=1,2,\ldots,z,$$
$$i=1,2,\ldots,k. \tag{5.2.26}$$

In that case, the safety function of a series network following the LLS rule is

$$\boldsymbol{S}_{LLS}(t,\ \cdot) = [1, \boldsymbol{S}_{LLS}(t, 1), \ldots, \boldsymbol{S}_{LLS}(t, z)],\quad t \geq 0, \tag{5.2.27}$$

where

$$\boldsymbol{S}_{LLS}(t, u) = \exp[-k\boldsymbol{\lambda}(u+1)t] + \sum_{j=1}^{k} \frac{1}{k} \cdot \left[\exp\left[-\boldsymbol{\lambda}(u) \sum_{i=1}^{n} \frac{1}{q(u, \boldsymbol{N}_j, \boldsymbol{N}_i)} t \right] \right.$$

$$\left. - \exp\left[-\left(k(\boldsymbol{\lambda}(u+1) - \boldsymbol{\lambda}(u)) + \boldsymbol{\lambda}(u) \sum_{i=1}^{k} \frac{1}{q(u, \boldsymbol{N}_j, \boldsymbol{N}_i)} \right) t \right] \right],$$

$$t \geq 0,\quad u = 1, 2, \ldots, z-1, \tag{5.2.28}$$

$$\boldsymbol{S}_{LLS}(t, z) = \exp[-k\boldsymbol{\lambda}(z)t],\quad t \geq 0. \tag{5.2.29}$$

Next, a multistate CI network consisting of subnetworks $\boldsymbol{N}_i$, $i = 1, 2, \ldots, k$, connected in "m out of k" structure (Fig. 5.2), is considered. A multistate "m out of k" CI network is in the safety-state subset $\{u, u+1, \ldots, z\}$, $u = 1, 2, \ldots, z$, if and only if at least m out of its k subnetworks are in this state subset. In that case, the safety function of a multistate "m out of k" CI network, composed of k independent subnetworks, is given by the vector

$$\boldsymbol{S}(t,\ \cdot) = [1, \boldsymbol{S}(t, 1), \ldots, \boldsymbol{S}(t, z)],\quad t \geq 0, \tag{5.2.30}$$

either with the coordinates

$$\boldsymbol{S}(t, u) = \sum_{\substack{r_1, r_2, \ldots, r_k = 0 \\ m \leq r_1 + r_2 + \cdots + r_k \leq k}}^{1} \prod_{i=1}^{k} \left[\boldsymbol{S}^{(i)}(t, u) \right]^{r_i} \left[1 - \boldsymbol{S}^{(i)}(t, u) \right]^{1-r_i},$$

$$t \geq 0,\quad u = 1, 2, \ldots, z, \tag{5.2.31}$$

or with the coordinates

$$\boldsymbol{S}(t, u) = 1 - \sum_{\substack{r_1, r_2, \ldots, r_k = 0 \\ 0 \leq r_1 + r_2 + \cdots + r_k \leq m-1}}^{1} \prod_{i=1}^{k} \left[\boldsymbol{S}^{(i)}(t, u) \right]^{r_i} \left[1 - \boldsymbol{S}^{(i)}(t, u) \right]^{1-r_i},$$

$$t \geq 0,\quad u = 1, 2, \ldots, z, \tag{5.2.32}$$

where $\boldsymbol{S}^{(i)}(t, u)$, $i = 1, 2, \ldots, k$, $u = 1, 2, \ldots, z$, denote safety function coordinates of subnetworks $\boldsymbol{N}_i$, $i = 1, 2, \ldots, k$.

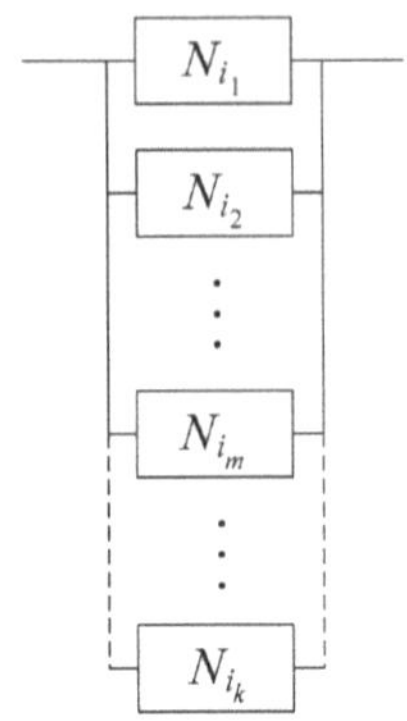

Fig. 5.2 The scheme of a "*m* out of *k*" network safety structure.

If, the subnetworks $\boldsymbol{N}_i$, $i = 1, 2, \ldots, k$, have piecewise exponential safety functions (5.2.5), (5.2.6), then the coordinates of a safety function (5.2.30) of a multistate "*m* out of *k*" CI network, applying (5.2.31), takes either the form

$$\boldsymbol{S}(t, u) = \sum_{\substack{r_1, r_2, \ldots, r_k = 0 \\ m \le r_1 + r_2 + \cdots + r_k \le k}}^{1} \prod_{i=1}^{k} \exp[-r_i \boldsymbol{\lambda}_i(u)t][1 - \exp[-\boldsymbol{\lambda}_i(u)t]]^{1-r_i},$$

$$t \ge 0, \quad u = 1, 2, \ldots, z, \tag{5.2.33}$$

or applying (5.2.32) the following form

$$\boldsymbol{S}(tu) = 1 - \sum_{\substack{r_1, r_2, \ldots, r_k = 0 \\ 0 \le r_1 + r_2 + \cdots + r_k \le m-1}}^{1} \prod_{i=1}^{k} \exp[-r_i \boldsymbol{\lambda}_i(u)t][1 - \exp[-\boldsymbol{\lambda}_i(u)t]]^{1-r_i},$$

$$t \ge 0, \quad u = 1, 2, \ldots, z. \tag{5.2.34}$$

The safety function of a homogeneous multistate "*m* out of *k*" CI network, composed of *k* independent subnetworks $\boldsymbol{N}_i$, $i = 1, 2, \ldots, k$, that have the same safety function with the coordinates (5.2.8), is given by the vector (Kołowrocki, 2013; Kołowrocki and Soszyńska-Budny, 2011)

$$\boldsymbol{S}(t, \cdot) = [1, \boldsymbol{S}(t, 1), \ldots, \boldsymbol{S}(t, z)], \quad t \ge 0, \tag{5.2.35}$$

either with the coordinates

$$\boldsymbol{S}(t,u)=\sum_{i=m}^{k}\binom{k}{i}\left[\boldsymbol{s}^{(1)}(t,u)\right]^{i}\left[1-\boldsymbol{s}^{(1)}(t,u)\right]^{k-i},\quad t\geq 0,$$

$$u=1,2,\ldots,z, \tag{5.2.36}$$

or with the coordinates

$$\boldsymbol{S}(t,u)=1-\sum_{i=0}^{m-1}\binom{k}{i}\left[\boldsymbol{s}^{(1)}(t,u)\right]^{i}\left[1-\boldsymbol{s}^{(1)}(t,u)\right]^{k-i},\quad t\geq 0,$$

$$u=1,2,\ldots,z. \tag{5.2.37}$$

Now, we analyze safety a homogeneous multistate "m out of k" CI network following the equal load-sharing (ELS) dependency rule, described in Section 3.2 of Chapter 3. In this model, the load after deterioration of safety state of a certain number of assets or subnetworks is evenly distributed to other assets or subnetworks and their safety parameters deteriorate equally. A similar analytical model of load distribution is presented in Dobson et al. (2005) for the analysis of large blackouts resulting from cascading failures in electric power transmission systems.

Considering the ELS model of dependency among subnetworks $\boldsymbol{N}_i$, $i = 1, 2, \ldots, k$, we assume that, if ω of subnetworks, $\omega = 0, 1, 2, \ldots, k - m$, leave the safety-state subset $\{u, u + 1, \ldots, z\}$, $u = 1, 2, \ldots, z$, then the lifetimes $\boldsymbol{T}_i'(u)$, $i = 1, 2, \ldots, k$, of other subnetworks decrease depending on the number of subnetworks that had left this subset, and depending on the coefficient of subnetwork stress proportionality correction. Subsequently, the subnetworks lifetimes $\boldsymbol{T}_i'(u)$, $i = 1, 2, \ldots, k$, in safety-state subset $\{u, u + 1, \ldots, z\}$, $u = 1, 2, \ldots, z$, and their mean values decrease according to the following rule

$$\boldsymbol{T}_i'(u)=c(u)\frac{k-\omega}{k}\cdot\boldsymbol{T}_i(u),$$

$$E\left[\boldsymbol{T}_i'(u)\right]=c(u)\frac{k-\omega}{k}\cdot E[\boldsymbol{T}_i(u)],\quad u=1,2,\ldots,z,$$

$$\omega=0,1,\ldots,k-m, \tag{5.2.38}$$

where $\boldsymbol{T}_i'(u)$, $i = 1, \ldots, k$, denotes the lifetime in safety-state subset $\{u, u + 1, \ldots, z\}$, $u = 1, 2, \ldots, z$, of subnetwork $\boldsymbol{N}_i$, $i = 1, 2, \ldots, k$, after the departure of $\omega = 0, 1, 2, \ldots, k - m$, subnetworks from this subset and $c(u)$ is the subnetwork stress proportionality correction coefficient for $u = 1, 2, \ldots, z$.

If we consider a homogeneous "m out of k" network following the ELS dependency rule, composed of subnetworks having identical piecewise exponential safety functions (5.2.5) with the coordinates (5.2.26), then after departure from the safety-state subset of ω subnetworks, $\omega = 0, 1, 2, \ldots, k - m$, the intensities of departure from the subset $\{u, u + 1, \ldots, z\}$, $u = 1, 2, \ldots, z$, of other subnetworks are given by

$$\boldsymbol{\lambda}^{(\omega)}(u) = \frac{k}{k-\omega}\frac{\boldsymbol{\lambda}(u)}{c(u)}, \quad u = 1, 2, \ldots, z, \quad \omega = 0, 1, \ldots, k - m. \qquad (5.2.39)$$

In this case, the safety function of a homogeneous and aging "m out of k" network following the ELS dependency rule, by (5.2.39) and using results presented in Section 3.2 and in Blokus-Roszkowska and Kolowrocki (2017a,c), is given by

$$\boldsymbol{S}_{ELS}(t, \cdot) = [1, \boldsymbol{S}_{ELS}(t, 1), \ldots, \boldsymbol{S}_{ELS}(t, z)], \quad t \geq 0, \qquad (5.2.40)$$

with the coordinates

$$\boldsymbol{S}_{ELS}(t, u) = \sum_{\omega=0}^{k-m} \frac{\left[\frac{k\boldsymbol{\lambda}(u)}{c(u)}t\right]^{\omega}}{\omega!} \exp\left[-\frac{k\boldsymbol{\lambda}(u)}{c(u)}t\right], \quad t \geq 0,$$

$$u = 1, 2, \ldots, z, \qquad (5.2.41)$$

where $c(u)$, $u = 1, 2, \ldots, z$, is the subnetwork stress proportionality correction coefficient.

In particular case when $m = 1$, a multistate "m out of k" CI network is a network with a parallel safety structure (Fig. 5.3).

The safety function of an aging parallel CI network, composed of k independent subnetworks, is given by the vector (Kołowrocki, 2013; Kołowrocki and Soszyńska-Budny, 2011)

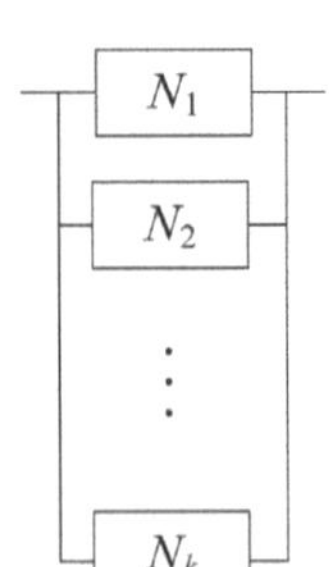

Fig. 5.3 The scheme of a parallel network safety structure.

$$\boldsymbol{S}(t,\ \cdot\ ) = [1, \boldsymbol{S}(t,1), \ldots, \boldsymbol{S}(t,z)], \quad t \geq 0, \tag{5.2.42}$$

where

$$\boldsymbol{S}(t,u) = 1 - \prod_{i=1}^{k}\left[1 - \boldsymbol{S}^{(i)}(t,u)\right], \quad t \geq 0, \quad u = 1,2,\ldots,z, \tag{5.2.43}$$

and $\boldsymbol{S}^{(i)}(t,u)$, $i = 1, 2, \ldots, k$, $u = 1, 2, \ldots, z$, denote safety function coordinates of subnetworks $\boldsymbol{N}_i$, $i = 1, 2, \ldots, k$.

If, the subnetworks $\boldsymbol{N}_i$, $i = 1, 2, \ldots, k$, have piecewise exponential safety functions (5.2.5), (5.2.6), then the coordinates of a safety function (5.2.30) of a multistate parallel CI network, applying (5.2.43), take the following form

$$\boldsymbol{S}(t,u) = 1 - \prod_{i=1}^{k}[1 - \exp[-\boldsymbol{\lambda}_i(u)t]], \quad t \geq 0, \quad u = 1,2,\ldots,z. \tag{5.2.44}$$

The safety function of a homogeneous aging parallel CI network, composed of k independent subnetworks $\boldsymbol{N}_i$, $i = 1, 2, \ldots, k$, that have the same safety function with the coordinates (5.2.8), is given by the vector

$$\boldsymbol{S}(t,\ \cdot\ ) = [1, \boldsymbol{S}(t,1), \ldots, \boldsymbol{S}(t,z)], \quad t \geq 0, \tag{5.2.45}$$

where

$$\boldsymbol{S}(t,u) = 1 - \left[1 - \boldsymbol{S}^{(1)}(t,u)\right]^k, \quad t \geq 0, \quad u = 1,2,\ldots,z. \tag{5.2.46}$$

Next, we consider a homogeneous, aging parallel network with subnetworks dependent according to the ELS rule and having identical piecewise exponential safety functions (5.2.5) with the coordinates (5.2.26). In that case, the network safety function is (Blokus-Roszkowska and Kolowrocki, 2017a,c)

$$\boldsymbol{S}_{ELS}(t,\ \cdot\ ) = [1, \boldsymbol{S}_{ELS}(t,1), \ldots, \boldsymbol{S}_{ELS}(t,z)], \quad t \geq 0, \tag{5.2.47}$$

with the coordinates

$$\boldsymbol{S}_{ELS}(t,u) = \sum_{\omega=0}^{k-1} \frac{\left[\frac{k\boldsymbol{\lambda}(u)}{c(u)}t\right]^{\omega}}{\omega!} \exp\left[-\frac{k\boldsymbol{\lambda}(u)}{c(u)}t\right], \quad t \geq 0,$$

$$u = 1,2,\ldots,z, \tag{5.2.48}$$

where $c(u)$, $u = 1, 2, \ldots, z$, is the subnetwork stress proportionality correction coefficient.

5.2.1 Safety of multistate series network with independent subnetworks and assets

Further, we consider a multistate series network composed of k subnetworks, presented in Fig. 5.4, with exponentially distributed assets. We assume that in the ith subnetwork $\boldsymbol{N}_i$, $i = 1, 2, \ldots, k$, there are l_i assets, denoted by $E_j^{(i)}$ for $i = 1, 2, \ldots, k$ and $j = 1, 2, \ldots, l_i$. Subsequently, $T_j^{(i)}$ are random variables representing lifetimes of assets $E_j^{(i)}$ in the safety-state subset $\{u, u + 1, \ldots, z\}$, $u = 1, 2, \ldots, z$.

If components, i.e., assets, have piecewise exponential safety functions of the form

$$S_j^{(i)}(t, \cdot) = \left[1, S_j^{(i)}(t, 1), \ldots, S_j^{(i)}(t, z)\right], \quad t \geq 0, \quad i = 1, 2, \ldots, k, \\ j = 1, 2, \ldots, l_i, \tag{5.2.49}$$

where

$$S_j^{(i)}(t, u) = \exp\left[-\lambda_j^{(i)}(u)t\right], \quad t \geq 0, \quad \lambda_j^{(i)}(u) > 0, \quad u = 1, 2, \ldots, z, \tag{5.2.50}$$

then the safety function of such series network composed of k multistate subnetworks $\boldsymbol{N}_i$, $i = 1, 2, \ldots, k$, applying (5.2.3), (5.2.4), is given by the vector

$$\boldsymbol{S}(t, \cdot) = [1, \boldsymbol{S}(t, 1), \ldots, \boldsymbol{S}(t, z)], \quad t \geq 0, \tag{5.2.51}$$

with the coordinates

$$\boldsymbol{S}(t, u) = \prod_{i=1}^{k} \boldsymbol{S}^{(i)}(t, u) = \prod_{i=1}^{k}\prod_{j=1}^{l_i} S_j^{(i)}(t, u) \\ = \exp\left[-\sum_{i=1}^{k}\sum_{j=1}^{l_i} \lambda_j^{(i)}(u)t\right], \quad t \geq 0, \quad u = 1, 2, \ldots, z. \tag{5.2.52}$$

Depending on the structure of the CI network as well as its subnetworks that comprise it, we consider different dependency models both among subnetworks and among components in these subnetworks. As an example in next sections, the safety function for a multistate series network with dependent subnetworks and for a multistate series network with dependent assets of its subnetworks is determined.

Fig. 5.4 The scheme of a series network of k subnetworks (Blokus-Roszkowska and Kolowrocki, 2017b).

5.2.2 Safety of multistate series network with dependent series subnetworks

We consider a series network composed of k dependent subnetworks and assume the LLS dependency model among subnetworks, described in Section 3.1 of Chapter 3. In that case, after departure of the subnetwork $\mathbf{N}_g$, $g = 1, 2, \ldots, k$, illustrated in Fig. 5.5, from the safety-state subset $\{u, u+1, \ldots, z\}$, $u = 1, 2, \ldots, z$, the safety parameters of assets $E_j^{(i)}$, $i = 1, 2, \ldots, k$, $i \neq g$, $j = 1, 2, \ldots, l_i$, of remaining subnetworks $\mathbf{N}_i$, $i = 1, 2, \ldots, k$, $i \neq g$, change depending on their exposure to the subnetwork $\mathbf{N}_g$ safety-state change, expressed by the coefficients of network load increase $q(v, \mathbf{N}_g, E_j^{(i)})$, $v = u, u-1, \ldots, 1$, $u = 1, 2, \ldots, z-1$. Within a single subnetwork the assets are independent and linked in series.

We denote by $E[T_j^{(i)}(u)]$ and $E[T_j^{(i/g)}(u)]$ the mean values of lifetimes of assets $E_j^{(i)}$, $i = 1, 2, \ldots, k$, $j = 1, 2, \ldots, l_i$, in the subnetwork $\mathbf{N}_i$, respectively, before and after departure of the subnetwork $\mathbf{N}_g$, $g = 1, 2, \ldots, k$, from the safety-state subset $\{u, u+1, \ldots, z\}$, $u = 1, 2, \ldots, z$. With this notation, in the LLS dependency model used among subnetworks, the mean values of these assets lifetimes in the safety-state subset $\{v, v+1, \ldots, z\}$ for $v = u, u-1, \ldots, 1$ and $u = 1, 2, \ldots, z-1$, decrease according to the formula

$$E\left[T_j^{(i/g)}(v)\right] = q\left(v, \mathbf{N}_g, E_j^{(i)}\right) \cdot E\left[T_j^{(i)}(v)\right], \quad i = 1, 2, \ldots, k,$$
$$g = 1, 2, \ldots, k, \quad j = 1, 2, \ldots, l_i, \tag{5.2.53}$$

where the coefficients of network load increase fulfill conditions $0 < q(v, \mathbf{N}_g, E_j^{(i)}) \leq 1$, $i = 1, 2, \ldots, k$, $g = 1, 2, \ldots, k$, $i \neq g$, $j = 1, 2, \ldots, l_i$ and $q(v, \mathbf{N}_g, E_j^{(g)}) = 1$, $j = 1, 2, \ldots, l_g$.

In particular case, the coefficients of network load increase are concerned with the distance from the subnetwork $\mathbf{N}_g$, $g = 1, 2, \ldots, k$, that has got out of the safety-state subset $\{u, u+1, \ldots, z\}$, $u = 1, 2, \ldots, z$. Subsequently, the coefficients $q(v, \mathbf{N}_g, E_j^{(i)})$ are functions of distance $d_{ig} = |i - g|$ between components $E_j^{(i)}$, $i = 1, 2, \ldots, k$, $i \neq g$, $j = 1, 2, \ldots, l_i$, being in the subnetwork $\mathbf{N}_i$, $i = 1, 2, \ldots, k$, and the subnetwork $\mathbf{N}_g$, $g = 1, 2, \ldots, k$, equivalent to the distance between subnetworks $\mathbf{N}_i$ and $\mathbf{N}_g$. The distance between subnetworks can be interpreted in the metric sense as well as in the sense of

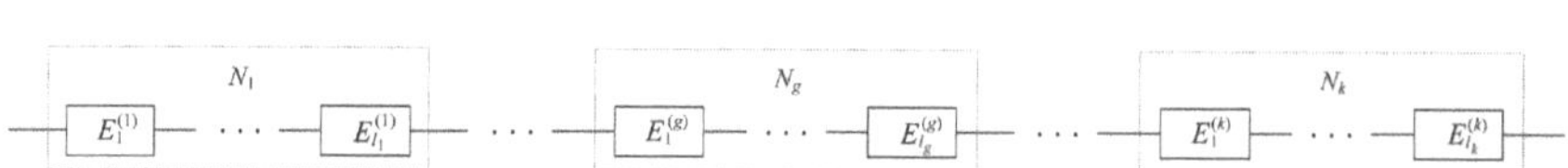

Fig. 5.5 The scheme of a series network of k dependent subnetworks (Blokus-Roszkowska and Kolowrocki, 2017b).

relationships in the network functioning. In this case, the formula (5.2.53) takes form (Blokus-Roszkowska and Kolowrocki, 2017b,c)

$$E\left[T_j^{(i/g)}(v)\right] = q(v, d_{ig}) \cdot E\left[T_j^{(i)}(v)\right], \quad i = 1,2,\ldots,k,$$
$$g = 1,2,\ldots,k, \quad j = 1,2,\ldots,l_i. \tag{5.2.54}$$

If components of CI network have piecewise exponential safety functions given by (5.2.49), (5.2.50), then after departure of the subnetwork $\mathbf{N}_g$, $g = 1, 2, \ldots, k$, from the subset $\{u, u+1, \ldots, z\}$, $u = 1, 2, \ldots, z$, the conditional intensities $\lambda_j^{(i/g)}(v)$ of departure from the safety-state subset $\{v, v+1, \ldots, z\}$, $v = u, u-1, \ldots, 1$, $u = 1, 2, \ldots, z-1$, of assets $E_j^{(i)}$, $i = 1, 2, \ldots, k$, $j = 1, 2, \ldots, l_i$, of other subnetworks $\mathbf{N}_i$, $i = 1, 2, \ldots, k$, $i \neq g$, by (5.2.53), are given by

$$\lambda_j^{(i/g)}(v) = \frac{\lambda_j^{(i)}(v)}{q\left(v, \mathbf{N}_g, E_j^{(i)}\right)}, \quad i = 1,2,\ldots,k, \quad g = 1,2,\ldots,k,$$
$$j = 1,2,\ldots,l_i. \tag{5.2.55}$$

Therefore, the safety function of such series network composed of k-dependent multistate subnetworks $\mathbf{N}_i$, $i = 1, 2, \ldots, k$, applying (5.2.23)–(5.2.25) and using results presented in Blokus-Roszkowska and Kolowrocki (2017b,c), is given by the vector

$$\mathbf{S}_{LLS}(t, \cdot) = [1, \mathbf{S}_{LLS}(t,1), \ldots, \mathbf{S}_{LLS}(t,z)], \quad t \geq 0, \tag{5.2.56}$$

with the coordinates

$$\mathbf{S}_{LLS}(t,u) = \exp\left[-\sum_{i=1}^{k}\sum_{j=1}^{l_i}\lambda_j^{(i)}(u+1)t\right]$$
$$+ \sum_{g=1}^{k} \frac{\sum_{j=1}^{l_i}\left(\lambda_j^{(g)}(u+1) - \lambda_j^{(g)}(u)\right)}{\sum_{i=1}^{k}\sum_{j=1}^{l_i}\left(\lambda_j^{(i)}(u+1) - \lambda_j^{(i)}(u)\right)} \cdot \left[\exp\left[-\sum_{i=1}^{k}\sum_{j=1}^{l_i}\frac{\lambda_j^{(i)}(u)}{q\left(u, \mathbf{N}_g, E_j^{(i)}\right)}t\right]\right.$$
$$\left.-\exp\left[-\sum_{i=1}^{k}\sum_{j=1}^{l_i}\left(\lambda_j^{(i)}(u+1) - \lambda_j^{(i)}(u) + \frac{\lambda_j^{(i)}(u)}{q\left(u, \mathbf{N}_g, E_j^{(i)}\right)}\right)t\right]\right], \quad t \geq 0,$$
$$u = 1,2,\ldots,z-1, \tag{5.2.57}$$

$$\boldsymbol{S}_{LLS}(t,z) = \exp\left[-\sum_{i=1}^{k}\sum_{j=1}^{l_i}\lambda_j^{(i)}(z)t\right], \quad t \geq 0. \tag{5.2.58}$$

5.2.3 Safety of multistate series network with dependent assets of its series subnetworks

In this section, we consider a series network composed of k independent subnetworks, presented in Fig. 5.4. We assume that in the ith series subnetwork $\boldsymbol{N}_i$, $i = 1, 2, \ldots, k$, there are l_i assets dependent according to the LLS rule, denoted by $E_j^{(i)}$, $i = 1, 2, \ldots, k$, $j = 1, 2, \ldots, l_i$. More exactly, we assume that after changing the safety-state subset by one of assets $E_{g_i}^{(i)}$, $g_i = 1, 2, \ldots, l_i$, in the subnetwork $\boldsymbol{N}_i$, $i = 1, 2, \ldots, k$, to the worse safety-state subset, the lifetimes in safety-state subsets of remaining assets in this subnetwork decrease.

Similarly as in previous section, we denote by $E[T_j^{(i)}(u)]$ and $E[T_{j/g_i}^{(i)}(u)]$ the mean values of lifetimes of assets $E_j^{(i)}$, $i = 1, 2, \ldots, k$, $j = 1, 2, \ldots, l_i$, in subnetwork $\boldsymbol{N}_i$, respectively, before and after departure of one fixed component $E_{g_i}^{(i)}$, $g_i = 1, 2, \ldots, l_i$, from the subset $\{u, u+1, \ldots, z\}$, $u = 1, 2, \ldots, z$. The safety parameters of other assets $E_j^{(i)}$, $j = 1, 2, \ldots, l_i$, $j \neq g_i$, in this subnetwork $\boldsymbol{N}_i$ change depending on their exposure to the asset $E_{g_i}^{(i)}$ safety-state change, expressed by the coefficients $q(v, E_{g_i}^{(i)}, E_j^{(i)})$ for $v = u, u-1, \ldots, 1$, $u = 1, 2, \ldots, z-1$. Subsequently, the mean values of these assets' lifetimes in the safety-state subset $\{v, v+1, \ldots, z\}$, $v = u, u-1, \ldots, 1$, $u = 1, 2, \ldots, z-1$, decrease according to the formula

$$E\left[T_{j/g_i}^{(i)}(v)\right] = q\left(v, E_{g_i}^{(i)}, E_j^{(i)}\right) \cdot E\left[T_j^{(i)}(v)\right], \quad i = 1, 2, \ldots, k,$$
$$j = 1, 2, \ldots, l_i, \quad g_i = 1, 2, \ldots, l_i. \tag{5.2.59}$$

If subnetworks are composed of assets with piecewise exponential safety functions given by (5.2.49), (5.2.50), then after departure from the subset $\{u, u+1, \ldots, z\}$, $u = 1, 2, \ldots, z$, of the asset $E_{g_i}^{(i)}$, $g_i = 1, 2, \ldots, l_i$, in the subnetwork $\boldsymbol{N}_i$, $i = 1, 2, \ldots, k$, the intensities $\lambda_{j/g_i}^{(i)}(v)$ of departure from the subset $\{v, v+1, \ldots, z\}$, $v = u, u-1, \ldots, 1$, $u = 1, 2, \ldots, z-1$, of other assets $E_j^{(i)}$, $j = 1, 2, \ldots, l_i$, $j \neq g_i$, in this subnetwork are given by

$$\lambda_{j/g_i}^{(i)}(v) = \frac{\lambda_j^{(i)}(v)}{q\left(v, E_{g_i}^{(i)}, E_j^{(i)}\right)}, \quad i = 1, 2, \ldots, k, \quad j = 1, 2, \ldots, l_i,$$
$$g_i = 1, 2, \ldots, l_i. \tag{5.2.60}$$

The safety function of such series network composed of k multistate subnetworks $\boldsymbol{N}_i$, $i = 1, 2, \ldots, k$, following the LLS rule, applying (5.2.3), (5.2.4), is given by the vector

$$\boldsymbol{S}(t,\ \cdot\ ) = [1, \boldsymbol{S}(t, 1), \ldots, \boldsymbol{S}(t, z)], \quad t \geq 0, \tag{5.2.61}$$

with the coordinates

$$\boldsymbol{S}(t, u) = \prod_{i=1}^{k} \boldsymbol{S}_{LLS}^{(i)}(t, u), \quad t \geq 0, \quad u = 1, 2, \ldots, z, \tag{5.2.62}$$

where $\boldsymbol{S}_{LLS}^{(i)}(t, u)$, $i = 1, 2, \ldots, k$, $u = 1, 2, \ldots, z$, denote safety function coordinates of subnetworks $\boldsymbol{N}_i$, $i = 1, 2, \ldots, k$, with assets dependent according to the LLS rule.

Taking into account (5.2.60), and applying (5.2.24), (5.2.25) for safety function coordinates $\boldsymbol{S}_{LLS}^{(i)}(t, u)$, $i = 1, 2, \ldots, k$, $u = 1, 2, \ldots, z$, the coordinates (5.2.62) of network safety function are given by (Blokus-Roszkowska and Kolowrocki, 2017b)

$$\boldsymbol{S}(t, u) = \prod_{i=1}^{k} \left[\exp\left[-\sum_{j=1}^{l_i} \lambda_j^{(i)}(u+1)t \right] + \sum_{g_i=1}^{l_i} \frac{\lambda_{g_i}^{(i)}(u+1) - \lambda_{g_i}^{(i)}(u)}{\sum_{j=1}^{l_i} \left(\lambda_j^{(i)}(u+1) - \lambda_j^{(i)}(u) \right)} \right.$$

$$\left. \cdot \left[\exp\left[-\sum_{j=1}^{l_i} \frac{\lambda_j^{(i)}(u)}{q\left(u, E_{g_i}^{(i)}, E_j^{(i)}\right)} t \right] - \exp\left[-\sum_{j=1}^{l_i} \left(\lambda_j^{(i)}(u+1) - \lambda_j^{(i)}(u) + \frac{\lambda_j^{(i)}(u)}{q\left(u, E_{g_i}^{(i)}, E_j^{(i)}\right)} \right) t \right] \right] \right],$$

$$t \geq 0, \quad u = 1, 2, \ldots, z-1, \tag{5.2.63}$$

$$\boldsymbol{S}(t, z) = \exp\left[-\sum_{i=1}^{k} \sum_{j=1}^{l_i} \lambda_j^{(i)}(z)t \right], \quad t \geq 0. \tag{5.2.64}$$

5.2.4 Safety of multistate series network with independent "*m* out of *l*" subnetworks

We analyze a multistate network composed of k homogeneous multistate "m_i out of l_i" subnetworks $\boldsymbol{N}_i$, $i = 1, 2, \ldots, k$, connected in series (Fig. 5.6). We assume that in the ith subnetwork $\boldsymbol{N}_i$, $i = 1, 2, \ldots, k$, there are l_i assets, denoted by $E_j^{(i)}$, $i = 1, 2, \ldots, k$ and $j = 1, 2, \ldots, l_i$.

We assume that assets in the ith subnetwork $\boldsymbol{N}_i$, $i = 1, 2, \ldots, k$, have identical piecewise exponential safety functions of the form

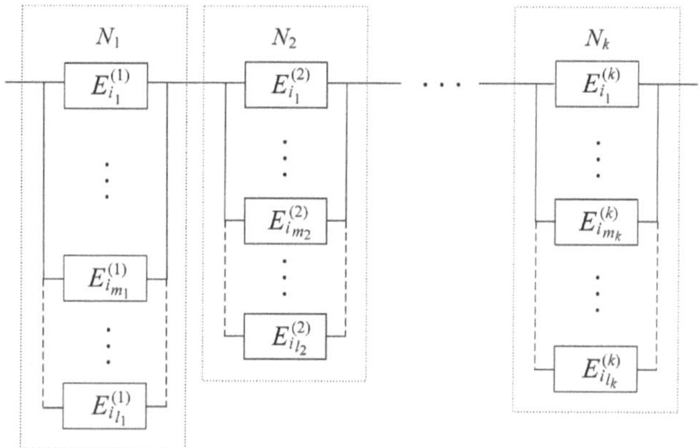

Fig. 5.6 The scheme of a network composed of k homogeneous "m_i out of l_i" subnetworks connected in series.

$$S_j^{(i)}(t,\cdot)=\left[1, S_j^{(i)}(t,1), \ldots, S_j^{(i)}(t,z)\right], \quad t\geq 0, \quad i=1,2,\ldots,k,$$
$$j=1,2,\ldots,l_i, \tag{5.2.65}$$

where

$$S_j^{(i)}(t,u)=\exp\left[-\lambda^{(i)}(u)t\right], \quad t\geq 0, \quad \lambda^{(i)}(u)>0, \quad u=1,2,\ldots,z, \tag{5.2.66}$$

where $\lambda^{(i)}(u)$, $i = 1, 2, \ldots, k$, are intensities of departure from the safety-state subset $\{u, u+1, \ldots, z\}$, $u = 1, 2, \ldots, z$, of assets in subnetwork $\boldsymbol{N}_i$.

Assuming that subnetworks $\boldsymbol{N}_i$, $i = 1, 2, \ldots, k$, and assets in these subnetworks are independent, the safety function of such network composed of k multistate "m_i out of l_i" subnetworks $\boldsymbol{N}_i$, $i = 1, 2, \ldots, k$, connected in series, applying (5.2.3), (5.2.4) and (5.2.36), (5.2.37), is given by the vector

$$\boldsymbol{S}(t,\cdot)=[1, \boldsymbol{S}(t,1), \ldots, \boldsymbol{S}(t,z)], \quad t\geq 0, \tag{5.2.67}$$

with the coordinates

$$\boldsymbol{S}(t,u)=\prod_{i=1}^{k}\boldsymbol{S}^{(i)}(t,u)$$
$$=\prod_{i=1}^{k}\sum_{\omega=m_i}^{l_i}\binom{l_i}{\omega}\exp\left[-\omega\lambda^{(i)}(u)t\right]\cdot\left[1-\exp\left[-\lambda^{(i)}(u)t\right]\right]^{l_i-\omega}, \quad t\geq 0,$$
$$u=1,2,\ldots,z, \tag{5.2.68}$$

or

$$\boldsymbol{S}(t,u)=\prod_{i=1}^{k}\boldsymbol{S}^{(i)}(t,u)$$
$$=\prod_{i=1}^{k}\left[1-\sum_{\omega=0}^{m_i-1}\binom{l_i}{\omega}\exp\left[-\omega\lambda^{(i)}(u)t\right]\cdot\left[1-\exp\left[-\lambda^{(i)}(u)t\right]\right]^{l_i-\omega}\right],$$
$$t\geq 0,\quad u=1,2,\ldots,z. \tag{5.2.69}$$

5.2.5 Safety of multistate series network with dependent assets of its "*m* out of *l*" subnetworks

In this section, we consider a multistate network composed of k homogeneous multistate "m_i out of l_i" subnetworks $\boldsymbol{N}_i$, $i = 1, 2, \ldots, k$, with assets dependent according to the ELS rule. Subnetworks are connected in series and operate independently. We assume that assets in the ith subnetwork $\boldsymbol{N}_i$, $i = 1, 2, \ldots, k$, have identical piecewise exponential safety functions given by (5.2.65), (5.2.66).

In each of multistate "m_i out of l_i" subnetworks $\boldsymbol{N}_i$, $i = 1, 2, \ldots, k$, we considering the ELS model of dependency among their assets. This means that after departure from safety-state subset $\{u, u+1, \ldots, z\}$, $u = 1, 2, \ldots, z$, of ω_i assets, $\omega_i = 0, 1, 2, \ldots, l_i - m_i$, in subnetwork $\boldsymbol{N}_i$, $i = 1, 2, \ldots, k$, the mean values of lifetimes $T'^{(i)}_j(u)$, $i = 1, 2, \ldots, k, j = 1, 2, \ldots, l_i$, in safety-state subset $\{u, u+1, \ldots, z\}$ of this subnetwork remaining assets $E_j^{(i)}$ are given by (Blokus-Roszkowska and Kolowrocki, 2017c,d)

$$E\left[T'^{(i)}_j(u)\right]=c^{(i)}(u)\frac{l_i-{}^{i}\omega_i}{l_i}E\left[T^{(i)}_j(u)\right],\quad \omega_i=0,1,2,\ldots,l_i-m_i,$$
$$u=1,2,\ldots,z, \tag{5.2.70}$$

where $c^{(i)}(u)$, $u = 1, 2, \ldots, z$, is stress proportionality correction coefficient of assets in the subnetwork $\boldsymbol{N}_i$, $i = 1, 2, \ldots, k$.

Assuming that assets in subnetwork $\boldsymbol{N}_i$, $i = 1, 2, \ldots, k$, have identical piecewise exponential safety functions, given by (5.2.65), (5.2.66), and are dependent according to the ELS rule, then after departure from the safety-state subset of ω_i assets, $\omega_i = 0, 1, 2, \ldots, l_i - m_i$, from formula (5.2.70) we conclude that the intensities of departure from the subset $\{u, u+1, \ldots, z\}$, $u = 1, 2, \ldots, z$, of other assets in this subnetwork are given by

$$\tilde{\lambda}^{(i)}(u)=\frac{l_i}{l_i-\omega_i}\frac{\lambda^{(i)}(u)}{c^{(i)}(u)},\quad i=1,2,\ldots,k,\quad \omega_i=0,1,\ldots,l_i-m_i,$$
$$u=1,2,\ldots,z. \tag{5.2.71}$$

The safety function of such network composed of k multistate subnetworks N_i, $i = 1, 2, \ldots, k$, following the ELS rule, applying (5.2.3), (5.2.4), is given by the vector

$$S(t, \cdot) = [1, S(t, 1), \ldots, S(t, z)], \quad t \geq 0, \tag{5.2.72}$$

with the coordinates

$$S(t, u) = \prod_{i=1}^{k} S_{ELS}^{(i)}(t, u), \quad t \geq 0, \quad u = 1, 2, \ldots, z, \tag{5.2.73}$$

where $S_{ELS}^{(i)}(t, u)$, $i = 1, 2, \ldots, k$, $u = 1, 2, \ldots, z$, denote safety function coordinates of subnetworks N_i, $i = 1, 2, \ldots, k$, with assets dependent according to the ELS rule.

Next, taking into account (5.2.71) and applying formula (5.2.41) for safety function coordinates $S_{ELS}^{(i)}(t, u)$, $i = 1, 2, \ldots, k$, $u = 1, 2, \ldots, z$, of homogeneous "m_i out of l_i" subnetworks following the ELS dependency rule, the coordinates (5.2.73) of network safety function are given by

$$S(t, u) = \prod_{i=1}^{k} \left[\sum_{\omega_i=0}^{l_i - m_i} \frac{\left[\frac{l_i \lambda^{(i)}(u)}{c^{(i)}(u)} t \right]^{\omega_i}}{\omega_i!} \exp\left[-\frac{l_i \lambda^{(i)}(u)}{c^{(i)}(u)} t \right] \right], \quad t \geq 0,$$

$$u = 1, 2, \ldots, z, \tag{5.2.74}$$

where $c^{(i)}(u)$, $u = 1, 2, \ldots, z$, is stress proportionality correction coefficient of assets in the subnetwork N_i, $i = 1, 2, \ldots, k$.

5.2.6 Safety of multistate series network with dependent "*m* out of *l*" subnetworks and dependent assets in these subnetworks

We analyze a multistate network composed of k homogeneous multistate "m_i out of l_i" subnetworks N_i, $i = 1, 2, \ldots, k$, illustrated in Fig. 5.7. Subnetworks are connected in series and follow the LLS dependency rule. Assets in these "m_i out of l_i" subnetworks are dependent according to the ELS rule. We assume, as before, that assets $E_j^{(i)}$, $i = 1, 2, \ldots, k$, $j = 1, 2, \ldots, l_i$. in the ith subnetwork N_i have identical piecewise exponential safety functions given by (5.2.65), (5.2.66).

In the ith "m_i out of l_i" subnetwork N_i, $i = 1, 2, \ldots, k$, we assume, similarly as in previous section, dependency of its l_i assets according to

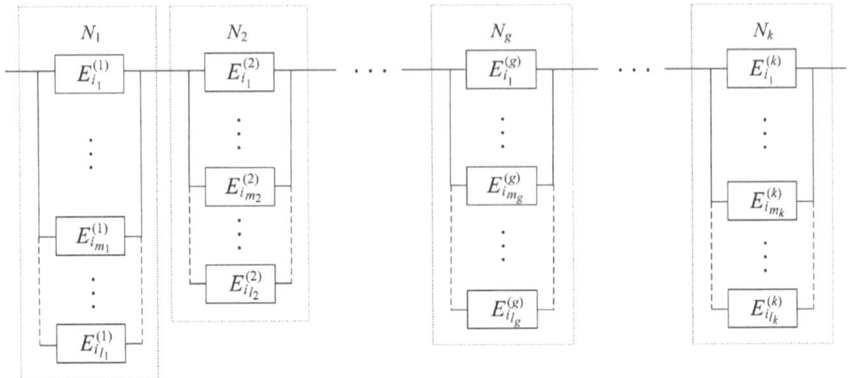

Fig. 5.7 The scheme of a network composed of k dependent homogeneous "m_i out of l_i" subnetworks connected in series.

the ELS rule. In that case, after departure from the safety-state subset of ω_i assets, $\omega_i = 0, 1, 2, \ldots, l_i - m_i$, in subnetwork $\mathbf{N}_i$, the intensities of departure from the subset $\{u, u+1, \ldots, z\}$, $u = 1, 2, \ldots, z$, of other assets in $\mathbf{N}_i$ are given by (5.2.71).

Moreover, we assume that subnetworks $\mathbf{N}_i$, $i = 1, 2, \ldots, k$, connected in series, are dependent according to the LLS rule. In the LLS dependency model, similarly as in Section 5.2.2, we assume that after departure from safety-state subset $\{u, u+1, \ldots, z\}$, $u = 1, 2, \ldots, z$, of the subnetwork $\mathbf{N}_g$, $g = 1, 2, \ldots, k$, illustrated in Fig. 5.7, the safety parameters of other subnetworks $\mathbf{N}_i$, $i = 1, 2, \ldots, k$, $i \neq g$, worsen depending on their exposure to the subnetwork $\mathbf{N}_g$ safety-state change, expressed by the coefficients of network load increase $q(\upsilon, \mathbf{N}_g, \mathbf{N}_i)$, $\upsilon = u, u-1, \ldots, 1$, $u = 1, 2, \ldots, z-1$. Subsequently, the mean values of these subnetworks lifetimes in safety-state subset $\{\upsilon, \upsilon+1, \ldots, z\}$, $\upsilon = u, u-1, \ldots, 1$, $u = 1, 2, \ldots, z-1$, according to (5.2.11), decrease according to the formula

$$E\left[\mathbf{T}_{i/g}(\upsilon)\right] = q\left(\upsilon, \mathbf{N}_g, \mathbf{N}_i\right) \cdot E[\mathbf{T}_i(\upsilon)], \quad i = 1, 2, \ldots, k,$$
$$g = 1, 2, \ldots, k, \tag{5.2.75}$$

where $\mathbf{T}_{i/g}(\upsilon)$ denotes the lifetime in safety-state subset $\{\upsilon, \upsilon+1, \ldots, z\}$, $\upsilon = u, u-1, \ldots, 1$, $u = 1, 2, \ldots, z-1$, of subnetwork $\mathbf{N}_i$, $i = 1, 2, \ldots, k$, after departure of the subnetwork $\mathbf{N}_g$, $g = 1, 2, \ldots, k$, from the safety-state subset $\{u + 1, \ldots, z\}$, $u = 1, 2, \ldots, z$, and where the coefficients of network load increase fulfill the conditions $0 < q(u, \boldsymbol{N}_g, \boldsymbol{N}_i) \leq 1$, $i, g = 1, 2, \ldots, k$, $i \neq g$, and $q(u, \boldsymbol{N}_g, \boldsymbol{N}_g) = 1$, $g = 1, 2, \ldots, k$.

Using the results presented in this chapter and Chapter 3, we obtain the safety function of network composed of dependent according to the LLS rule multistate subnetworks $\boldsymbol{N}_i$, $i = 1, 2, \ldots, k$, and assuming that each of these "m_i out of l_i" subnetworks follows the ELS rule. Applying (5.2.12)–(5.2.14), the network safety function is given by the vector

$$\boldsymbol{S}_{LLS}(t, \cdot) = [1, \boldsymbol{S}_{LLS}(t, 1), \ldots, \boldsymbol{S}_{LLS}(t, z)], \quad t \geq 0, \tag{5.2.76}$$

with the coordinates

$$\boldsymbol{S}_{LLS}(t, u) = \prod_{i=1}^{k} \boldsymbol{S}_{ELS}^{(i)}(t, u+1)$$
$$+ \int_0^t \sum_{g=1}^{k} \left[\tilde{\boldsymbol{f}}_{ELS}^{(g)}(a, u+1) \cdot \prod_{\substack{i=1 \\ i \neq g}}^{k} \boldsymbol{S}_{ELS}^{(i)}(a, u+1) \cdot \boldsymbol{S}_{ELS}^{(g)}(a, u) \cdot \prod_{i=1}^{k} \boldsymbol{S}_{ELS}^{(i/g)}(t-a, u) \right] da,$$
$$t \geq 0, \quad u = 1, 2, \ldots, z-1, \tag{5.2.77}$$

$$\boldsymbol{S}_{LLS}(t, z) = \prod_{i=1}^{k} \boldsymbol{S}_{ELS}^{(i)}(t, z), \quad t \geq 0, \tag{5.2.78}$$

where

$\boldsymbol{S}_{ELS}^{(i)}(t, u)$, $i = 1, 2, \ldots, k$, $u = 1, 2, \ldots, z$, denote the safety function coordinates of subnetwork $\boldsymbol{N}_i$, $i = 1, 2, \ldots, k$, with assets dependent according to the ELS rule,

$\boldsymbol{S}_{ELS}^{(i/g)}(t, u)$, $i, g = 1, 2, \ldots, k$, $u = 1, 2, \ldots, z$, denote the safety function coordinates of subnetwork $\boldsymbol{N}_i$, $i = 1, 2, \ldots, k$, after departure of $\boldsymbol{N}_g$, $g = 1, 2, \ldots, k$, from the safety-state subset $\{u + 1, \ldots, z\}$, $u = 1, 2, \ldots, z - 1$, assuming that assets in subnetworks are dependent according to the ELS rule,

$\tilde{\boldsymbol{f}}_{ELS}^{(g)}(t, u+1)$, $g = 1, 2, \ldots, k$, $u = 1, 2, \ldots, z - 1$, is the density function coordinate of subnetwork $\boldsymbol{N}_g$, $g = 1, 2, \ldots, k$, following the ELS rule.

Further, taking into account formula (5.2.41) for safety function coordinates of homogeneous "m_i out of l_i" subnetworks following the ELS dependency rule, and using results presented in Chapter 3, the coordinates (5.2.77), (5.2.78) of network safety function, assuming that assets in subnetwork $\mathbf{N}_i$, $i = 1, 2, \ldots, k$, have identical piecewise exponential safety functions (5.2.65), (5.2.66), take the following form

$$\boldsymbol{S}_{LLS}(t,u) = \prod_{i=1}^{k}\left[\sum_{\omega_i=0}^{l_i-m_i}\frac{\left[\frac{l_i\lambda^{(i)}(u+1)}{c^{(i)}(u+1)}t\right]^{\omega_i}}{\omega_i!}\right]\cdot\exp\left[-\frac{l_i\lambda^{(i)}(u+1)}{c^{(i)}(u+1)}t\right]$$

$$+\sum_{g=1}^{k}\int_0^t \widetilde{\boldsymbol{f}}^{(g)}_{ELS}(a,u+1)\cdot\prod_{\substack{i=1\\ i\neq g}}^{k}\left[\sum_{\omega_i=0}^{l_i-m_i}\frac{\left[\frac{l_i\lambda^{(i)}(u+1)}{c^{(i)}(u+1)}a\right]^{\omega_i}}{\omega_i!}\right]$$

$$\cdot\sum_{\omega_g=0}^{l_g-m_g}\frac{\left[\frac{l_g\lambda^{(g)}(u)}{c^{(g)}(u)}a\right]^{\omega_g}}{\omega_g!}\cdot\exp\left[-\left[\sum_{\substack{i=1\\ i\neq g}}^{k}\frac{l_i\lambda^{(i)}(u+1)}{c^{(i)}(u+1)}+\frac{l_g\lambda^{(g)}(u)}{c^{(g)}(u)}\right]a\right]$$

$$\cdot\left[\prod_{i=1}^{k}\frac{\sum_{\omega_i=0}^{l_i-m_i}\frac{\left[\frac{l_i\lambda^{(i)}(u)}{c^{(i)}(u)q(u,\mathbf{N}_g,\mathbf{N}_i)}t\right]^{\omega_i}}{\omega_i!}}{\sum_{\omega_i=0}^{l_i-m_i}\frac{\left[\frac{l_i\lambda^{(i)}(u)}{c^{(i)}(u)}a\right]^{\omega_i}}{\omega_i!}}\cdot\exp\left[-\frac{l_i\lambda^{(i)}(u)}{c^{(i)}(u)q(u,\mathbf{N}_g,\mathbf{N}_i)}t+\frac{l_i\lambda^{(i)}(u)}{c^{(i)}(u)}a\right]\right]da,$$

$$t\geq 0,\quad u=1,2,\ldots,z-1, \tag{5.2.79}$$

where $\widetilde{f}^{(g)}_{ELS}(a, u+1)$ is given by

$$\widetilde{f}^{(g)}_{ELS}(a, u+1) = \left[\frac{\sum\limits_{\omega_g=0}^{l_g-m_g} \frac{\left[\frac{l_g\lambda^{(g)}(u+1)}{c^{(g)}(u+1)}a\right]^{\omega_g}}{\omega_g!} \cdot \sum\limits_{\omega_g=0}^{l_g-m_g-1} \frac{\left[\frac{l_g\lambda^{(g)}(u)}{c^{(g)}(u)}\right]^{\omega_g+1} a^{\omega}}{\omega_g!} - \sum\limits_{\omega_g=0}^{l_g-m_g-1} \frac{\left[\frac{l_g\lambda^{(g)}(u+1)}{c^{(g)}(u+1)}\right]^{\omega_g+1} a^{\omega}}{\omega_g!} \cdot \sum\limits_{\omega_g=0}^{l_g-m_g} \frac{\left[\frac{l_g\lambda^{(g)}(u)}{c^{(g)}(u)}a\right]^{\omega_g}}{\omega_g!}}{\left[\sum\limits_{\omega_g=0}^{l_g-m_g} \frac{\left[\frac{l_g\lambda^{(g)}(u)}{c^{(g)}(u)}a\right]^{\omega_g}}{\omega_g!}\right]^2} \right.$$

$$\left. + \frac{\sum\limits_{\omega_g=0}^{l_g-m_g} \frac{\left[\frac{l_g\lambda^{(g)}(u+1)}{c^{(g)}(u+1)}a\right]^{\omega_g}}{\omega_g!} \left(\frac{l_g\lambda^{(g)}(u+1)}{c^{(g)}(u+1)} - \frac{l_g\lambda^{(g)}(u)}{c^{(g)}(u)}\right)}{\sum\limits_{\omega_g=0}^{l_g-m_g} \frac{\left[\frac{l_g\lambda^{(g)}(u)}{c^{(g)}(u)}a\right]^{\omega_g}}{\omega_g!}} \right] \cdot \exp\left[-\left(\frac{l_g\lambda^{(g)}(u+1)}{c^{(g)}(u+1)} - \frac{l_g\lambda^{(g)}(u)}{c^{(g)}(u)}\right)a\right],$$

$$u = 1, 2, \ldots, z-1, \tag{5.2.80}$$

and

$$S_{LLS}(t,z)=\prod_{i=1}^{k}\sum_{\omega_i=0}^{l_i-m_i}\frac{\left[\frac{l_i\lambda^{(i)}(z)}{c^{(i)}(z)}t\right]^{\omega_i}}{\omega_i!}\cdot\exp\left[-\sum_{i=1}^{k}\frac{l_i\lambda^{(i)}(z)}{c^{(i)}(z)}t\right],\quad t\geq 0.\quad (5.2.81)$$

5.3 Application

Proposed theoretical results are applied to the safety analysis of an exemplary academic network and an exemplary distribution network. In both examples, the safety analysis results for the networks with dependent components are compared with the results when the independence of components is assumed.

5.3.1 Safety analysis of exemplary network

We consider a 3-state ($z=2$) series network composed of three subnetworks $\boldsymbol{N}_1$, $\boldsymbol{N}_2$, $\boldsymbol{N}_3$. Its safety structure is given in Fig. 5.8.

The subnetwork $\boldsymbol{N}_1$ is a homogeneous parallel network consisting of four components (Fig. 5.8) and we assume the ELS dependency model among its components.

The components $E_i^{(1)}$, $i=1,2,3,4$, have identical piecewise exponential safety functions

$$S_i^{(1)}(t,\cdot)=\left[1,\exp\left(-\lambda^{(1)}(1)t\right),\exp\left(-\lambda^{(1)}(2)t\right)\right],\quad t\geq 0,$$

$$i=1,2,3,4. \quad (5.3.1)$$

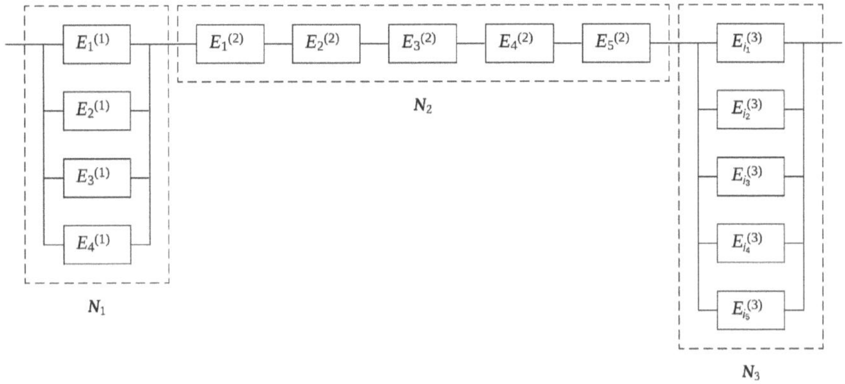

Fig. 5.8 The scheme of the exemplary network $\boldsymbol{N}$ safety structure.

with intensities of departures from safety-states subsets {1,2} and {2}, respectively, given by

$$\lambda^{(1)}(1)=0.02,\ \ \lambda^{(1)}(2)=0.04. \tag{5.3.2}$$

In that case, the safety function of subnetwork $\boldsymbol{N}_1$ following the ELS dependency rule, according to (5.2.47), (5.2.48), is

$$\boldsymbol{S}_{ELS}^{(1)}(t,\ \cdot)=\left[1, \boldsymbol{S}_{ELS}^{(1)}(t,1), \boldsymbol{S}_{ELS}^{(1)}(t,2)\right],\ \ t\geq 0, \tag{5.3.3}$$

where its coordinates are

$$\boldsymbol{S}_{ELS}^{(1)}(t,u)=\sum_{\omega=0}^{3}\frac{\left[\dfrac{4\lambda^{(1)}(u)}{c^{(1)}(u)}t\right]^{\omega}}{\omega!}\cdot\exp\left[-\frac{4\lambda^{(1)}(u)}{c^{(1)}(u)}t\right],\ \ u=1,2. \tag{5.3.4}$$

And next, substituting in (5.3.4) the intensities of departures from the safety-state subsets (5.3.2), the coordinates of $\boldsymbol{N}_1$ safety function are

$$\boldsymbol{S}_{ELS}^{(1)}(t,1)=\left[1+\frac{0.08}{c^{(1)}(1)}t+\frac{0.0032}{\left(c^{(1)}(1)\right)^2}t^2+\frac{0.000085}{\left(c^{(1)}(1)\right)^3}t^3\right]$$
$$\cdot\exp\left[-\frac{0.08}{c^{(1)}(1)}t\right],\ \ t\geq 0, \tag{5.3.5}$$

$$\boldsymbol{S}_{ELS}^{(1)}(t,2)=\left[1+\frac{0.16}{c^{(1)}(2)}t+\frac{0.0128}{\left(c^{(1)}(2)\right)^2}t^2+\frac{0.000683}{\left(c^{(1)}(2)\right)^3}t^3\right]$$
$$\cdot\exp\left[-\frac{0.16}{c^{(1)}(2)}t\right],\ \ t\geq 0. \tag{5.3.6}$$

The graphs of safety function coordinates of subnetwork $\boldsymbol{N}_1$, given by (5.3.5), (5.3.6), assuming the ELS dependency model with coefficient of component stress proportionality correction ranging from 0.5 to 1.0, are presented in Figs. 5.9 and 5.10.

The subnetwork $\boldsymbol{N}_2$ is composed of five components connected with series safety structure (Fig. 5.8) and dependent according to the LLS rule.

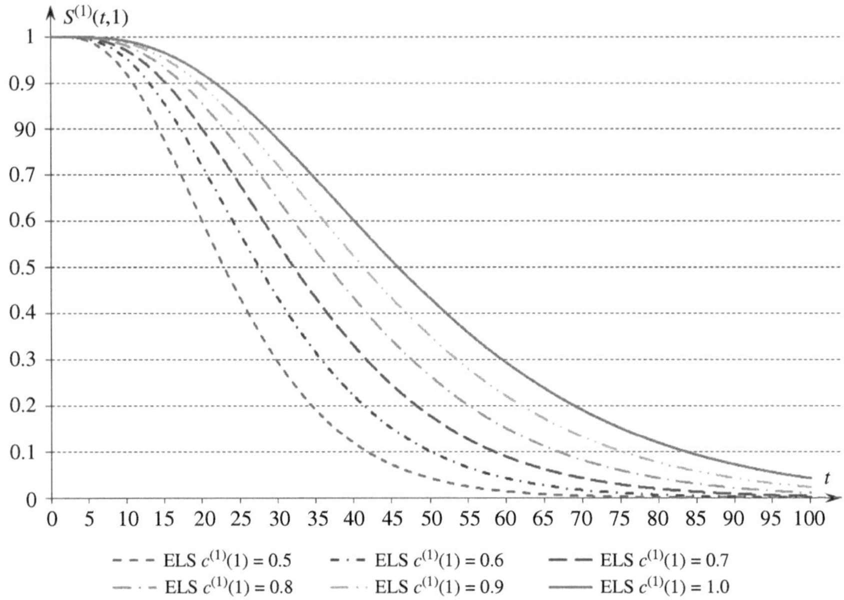

Fig. 5.9 The graphs of $\boldsymbol{N}_1$ safety function coordinate $\boldsymbol{S}^{(1)}(t,1)$ assuming the ELS dependency model with coefficient $c^{(1)}(1)$ ranging from 0.5 to 1.0.

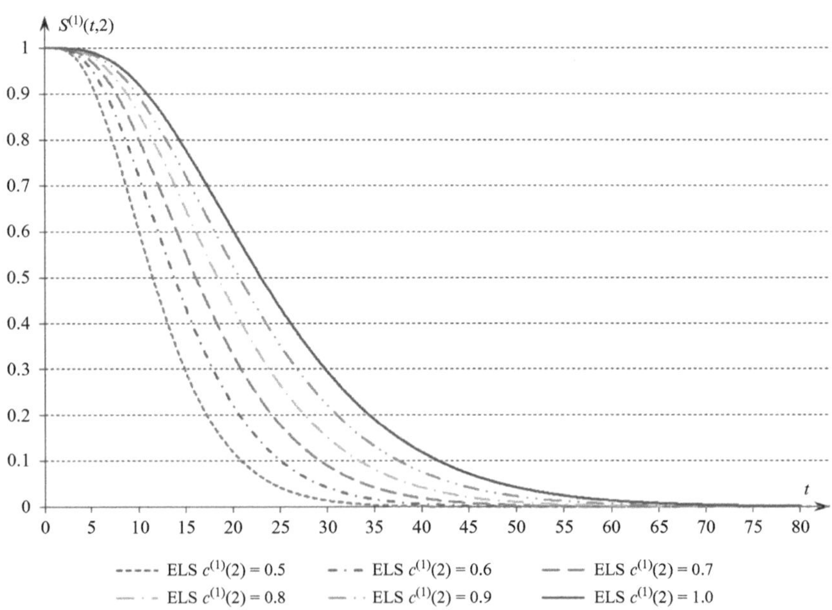

Fig. 5.10 The graphs of $\boldsymbol{N}_1$ safety function coordinate $\boldsymbol{S}^{(1)}(t,2)$ assuming the ELS dependency model with coefficient $c^{(1)}(2)$ ranging from 0.5 to 1.0.

The components $E_i^{(2)}$, $i = 1, 2, 3, 4, 5$, by the assumption, have piecewise exponential safety functions:

$$S_1^{(2)}(t,\ \cdot) = \left[1,\ \exp\left(-\lambda_1^{(2)}(1)t\right),\ \exp\left(-\lambda_1^{(2)}(2)t\right)\right],\quad t \geq 0, \qquad (5.3.7)$$

$$S_i^{(2)}(t,\ \cdot) = \left[1,\ \exp\left(-\lambda_i^{(2)}(1)t\right),\ \exp\left(-\lambda_i^{(2)}(2)t\right)\right],\quad t \geq 0,\quad i = 2,3, \qquad (5.3.8)$$

$$S_i^{(2)}(t,\ \cdot) = \left[1,\ \exp\left(-\lambda_i^{(2)}(1)t\right),\ \exp\left(-\lambda_i^{(2)}(2)t\right)\right],\quad t \geq 0,\quad i = 4,5, \qquad (5.3.9)$$

with intensities, respectively, given by:

$$\lambda_1^{(2)}(1) = 0.005,\quad \lambda_1^{(2)}(2) = 0.015, \qquad (5.3.10)$$

$$\lambda_i^{(2)}(1) = 0.006,\quad \lambda_i^{(2)}(2) = 0.008,\quad i = 2,3, \qquad (5.3.11)$$

$$\lambda_i^{(2)}(1) = 0.008,\quad \lambda_i^{(2)}(2) = 0.011,\quad i = 4,5. \qquad (5.3.12)$$

Next, applying (5.2.23)–(5.2.25), the safety function of subnetwork $\boldsymbol{N}_2$ is

$$\boldsymbol{S}_{LLS}^{(2)}(t,\ \cdot) = \left[1,\ \boldsymbol{S}_{LLS}^{(2)}(t,\ 1),\ \boldsymbol{S}_{LLS}^{(2)}(t,\ 2)\right],\quad t \geq 0, \qquad (5.3.13)$$

with the coordinates

$$\boldsymbol{S}_{LLS}^{(2)}(t,1) = \exp\left[-\sum_{i=1}^{5}\lambda_i^{(2)}(2)t\right] + \sum_{j=1}^{5}\frac{\lambda_j^{(2)}(2) - \lambda_j^{(2)}(1)}{\sum_{i=1}^{5}\left(\lambda_i^{(2)}(2) - \lambda_i^{(2)}(1)\right)}$$

$$\cdot\left[\exp\left[-\sum_{i=1}^{5}\frac{\lambda_i^{(2)}(1)}{q\left(1, E_j^{(2)}, E_i^{(2)}\right)}t\right] - \exp\left[-\sum_{i=1}^{5}\left(\lambda_i^{(2)}(2) - \lambda_i^{(2)}(1) + \frac{\lambda_i^{(2)}(1)}{q\left(1, E_j^{(2)}, E_i^{(2)}\right)}\right)\cdot t\right]\right], \qquad (5.3.14)$$

$$\boldsymbol{S}_{LLS}^{(2)}(t,2) = \exp\left[-\sum_{i=1}^{5}\lambda_i^{(2)}(2)t\right]. \qquad (5.3.15)$$

Substituting in (5.3.14), (5.3.15) the intensities of departures from the safety-state subsets (5.3.10)–(5.3.12), the coordinates of N_2 safety function are

$$
\begin{aligned}
\boldsymbol{S}_{LLS}^{(2)}(t,1) = \exp[-0.053t] \\
&+ 0.5 \cdot \left\{ \exp\left[-\left(\frac{0.005}{q\left(1, E_1^{(2)}, E_1^{(2)}\right)} + \frac{0.006}{q\left(1, E_1^{(2)}, E_2^{(2)}\right)} + \frac{0.006}{q\left(1, E_1^{(2)}, E_3^{(2)}\right)} + \frac{0.008}{q\left(1, E_1^{(2)}, E_4^{(2)}\right)} + \frac{0.008}{q\left(1, E_1^{(2)}, E_5^{(2)}\right)} \right) t \right] \right. \\
&\left. - \exp\left[-\left(0.02 + \frac{0.005}{q\left(1, E_1^{(2)}, E_1^{(2)}\right)} + \frac{0.006}{q\left(1, E_1^{(2)}, E_2^{(2)}\right)} + \frac{0.006}{q\left(1, E_1^{(2)}, E_3^{(2)}\right)} + \frac{0.008}{q\left(1, E_1^{(2)}, E_4^{(2)}\right)} + \frac{0.008}{q\left(1, E_1^{(2)}, E_5^{(2)}\right)} \right) t \right] \right\} \\
&+ 0.1 \cdot \left\{ \exp\left[-\left(\frac{0.005}{q\left(1, E_2^{(2)}, E_1^{(2)}\right)} + \frac{0.006}{q\left(1, E_2^{(2)}, E_2^{(2)}\right)} + \frac{0.006}{q\left(1, E_2^{(2)}, E_3^{(2)}\right)} + \frac{0.008}{q\left(1, E_2^{(2)}, E_4^{(2)}\right)} + \frac{0.008}{q\left(1, E_2^{(2)}, E_5^{(2)}\right)} \right) t \right] \right. \\
&\left. - \exp\left[-\left(0.02 + \frac{0.005}{q\left(1, E_2^{(2)}, E_1^{(2)}\right)} + \frac{0.006}{q\left(1, E_2^{(2)}, E_2^{(2)}\right)} + \frac{0.006}{q\left(1, E_2^{(2)}, E_3^{(2)}\right)} + \frac{0.008}{q\left(1, E_2^{(2)}, E_4^{(2)}\right)} + \frac{0.008}{q\left(1, E_2^{(2)}, E_5^{(2)}\right)} \right) t \right] \right\} \\
&+ 0.1 \cdot \left\{ \exp\left[-\left(\frac{0.005}{q\left(1, E_3^{(2)}, E_1^{(2)}\right)} + \frac{0.006}{q\left(1, E_3^{(2)}, E_2^{(2)}\right)} + \frac{0.006}{q\left(1, E_3^{(2)}, E_3^{(2)}\right)} + \frac{0.008}{q\left(1, E_3^{(2)}, E_4^{(2)}\right)} + \frac{0.008}{q\left(1, E_3^{(2)}, E_5^{(2)}\right)} \right) t \right] \right.
\end{aligned}
$$

$$-\exp\left[-\left(0.02+\frac{0.005}{q\left(1,E_3^{(2)},E_1^{(2)}\right)}+\frac{0.006}{q\left(1,E_3^{(2)},E_2^{(2)}\right)}+\frac{0.006}{q\left(1,E_3^{(2)},E_3^{(2)}\right)}+\frac{0.008}{q\left(1,E_3^{(2)},E_4^{(2)}\right)}+\frac{0.008}{q\left(1,E_3^{(2)},E_5^{(2)}\right)}\right)t\right]\Bigg\}$$

$$+0.15\cdot\Bigg\{\exp\left[-\left(\frac{0.005}{q\left(1,E_4^{(2)},E_1^{(2)}\right)}+\frac{0.006}{q\left(1,E_4^{(2)},E_2^{(2)}\right)}+\frac{0.006}{q\left(1,E_4^{(2)},E_3^{(2)}\right)}+\frac{0.008}{q\left(1,E_4^{(2)},E_4^{(2)}\right)}+\frac{0.008}{q\left(1,E_4^{(2)},E_5^{(2)}\right)}\right)t\right]$$

$$-\exp\left[-\left(0.02+\frac{0.005}{q\left(1,E_4^{(2)},E_1^{(2)}\right)}+\frac{0.006}{q\left(1,E_4^{(2)},E_2^{(2)}\right)}+\frac{0.006}{q\left(1,E_4^{(2)},E_3^{(2)}\right)}+\frac{0.008}{q\left(1,E_4^{(2)},E_4^{(2)}\right)}+\frac{0.008}{q\left(1,E_4^{(2)},E_5^{(2)}\right)}\right)t\right]\Bigg\}$$

$$+0.15\cdot\Bigg\{\exp\left[-\left(\frac{0.005}{q\left(1,E_5^{(2)},E_1^{(2)}\right)}+\frac{0.006}{q\left(1,E_5^{(2)},E_2^{(2)}\right)}+\frac{0.006}{q\left(1,E_5^{(2)},E_3^{(2)}\right)}+\frac{0.008}{q\left(1,E_5^{(2)},E_4^{(2)}\right)}+\frac{0.008}{q\left(1,E_5^{(2)},E_5^{(2)}\right)}\right)t\right]$$

$$-\exp\left[-\left(0.02+\frac{0.005}{q\left(1,E_5^{(2)},E_1^{(2)}\right)}+\frac{0.006}{q\left(1,E_5^{(2)},E_2^{(2)}\right)}+\frac{0.006}{q\left(1,E_5^{(2)},E_3^{(2)}\right)}+\frac{0.008}{q\left(1,E_5^{(2)},E_4^{(2)}\right)}+\frac{0.008}{q\left(1,E_5^{(2)},E_5^{(2)}\right)}\right)t\right]\Bigg\},$$

$$t\geq 0, \tag{5.3.16}$$

$$\boldsymbol{S}_{LLS}^{(2)}(t,2)=\exp[-0.053t],\quad t\geq 0. \tag{5.3.17}$$

In the subnetwork $\boldsymbol{N}_2$ we assume that the coefficients of load increase $q(1,E_j^{(2)},E_i^{(2)})$ for $i=1,2,3,4,5$ and $j=1,2,3,4,5$, in the LLS dependency model are given by:

$$q\left(1,E_j^{(2)},E_i^{(2)}\right)=\frac{a+|i-j|}{a+5},\quad i=1,2,3,4,5,\quad j=1,2,3,4,5,\quad i\neq j, \tag{5.3.18}$$

and

$$q\left(1,E_j^{(2)},E_j^{(2)}\right)=1,\quad j=1,2,3,4,5, \tag{5.3.19}$$

where a can vary from 0 to 5.

For $a=5$ they take following values:

$$\begin{aligned}q\left(1,E_1^{(2)},E_2^{(2)}\right)&=q\left(1,E_2^{(2)},E_1^{(2)}\right)=q\left(1,E_2^{(2)},E_3\right)\\&=q\left(1,E_3^{(2)},E_2^{(2)}\right)=q\left(1,E_3^{(2)},E_4^{(2)}\right)\\&=q\left(1,E_4^{(2)},E_3^{(2)}\right)=q\left(1,E_4^{(2)},E_5^{(2)}\right)\\&=q\left(1,E_5^{(2)},E_4^{(2)}\right)=0.6,\end{aligned} \tag{5.3.20}$$

$$\begin{aligned}q\left(1,E_1^{(2)},E_3^{(2)}\right)&=q\left(1,E_2^{(2)},E_4^{(2)}\right)=q\left(1,E_3^{(2)},E_1^{(2)}\right)\\&=q\left(1,E_3^{(2)},E_5^{(2)}\right)=q\left(1,E_4^{(2)},E_2^{(2)}\right)\\&=q\left(1,E_5^{(2)},E_3^{(2)}\right)=0.7,\end{aligned} \tag{5.3.21}$$

$$\begin{aligned}q\left(1,E_1^{(2)},E_4^{(2)}\right)&=q\left(1,E_2^{(2)},E_5^{(2)}\right)=q\left(1,E_4^{(2)},E_1^{(2)}\right)\\&=q\left(1,E_5^{(2)},E_2^{(2)}\right)=0.8,\end{aligned} \tag{5.3.22}$$

$$q\left(1,E_1^{(2)},E_5^{(2)}\right)=q\left(1,E_5^{(2)},E_1^{(2)}\right)=0.9, \tag{5.3.23}$$

and

$$\begin{aligned}q\left(1,E_1^{(2)},E_1^{(2)}\right)&=q\left(1,E_2^{(2)},E_2^{(2)}\right)=q\left(1,E_3^{(2)},E_3^{(2)}\right)\\&=q\left(1,E_4^{(2)},E_4^{(2)}\right)=q\left(1,E_5^{(2)},E_5^{(2)}\right)=1.\end{aligned} \tag{5.3.24}$$

Substituting in (5.3.16), (5.3.17) the values of coefficients (5.3.20)–(5.3.24), the coordinates of $\boldsymbol{N}_2$ safety function are

$$\boldsymbol{S}^{(2)}_{LLS}(t,1) = \exp[-0.053t] + 0.5\cdot[\exp[-0.04246t] - \exp[-0.06246t]] + 0.1\cdot[\exp[-0.04576t] - \exp[-0.06576t]] + 0.1\cdot[\exp[-0.04790t] - \exp[-0.06790t]] + 0.15\cdot[\exp[-0.04615t] - \exp[-0.6615t]] + 0.15\cdot[\exp[-0.04296t] - \exp[-0.06296t]],\quad t\ge 0, \tag{5.3.25}$$

$$\boldsymbol{S}^{(2)}_{LLS}(t,2) = \exp[-0.053t],\quad t\ge 0. \tag{5.3.26}$$

The graph of safety function coordinate $\boldsymbol{S}^{(2)}(t,1)$ of subnetwork $\boldsymbol{N}_2$, given by (5.3.25) for coefficient of $\boldsymbol{N}_2$ load increase (5.3.20)–(5.3.24) is presented in Fig. 5.11. Fig. 5.11 also shows the graphs of coordinates $\boldsymbol{S}^{(2)}(t,1)$ assuming the LLS dependency model with coefficients given by (5.3.18), (5.3.19) for $a=3$, $a=1$, and $a=0$.

As explained in Chapter 3, in the LLS dependency model among components connected in series, the last coordinate of the safety function,

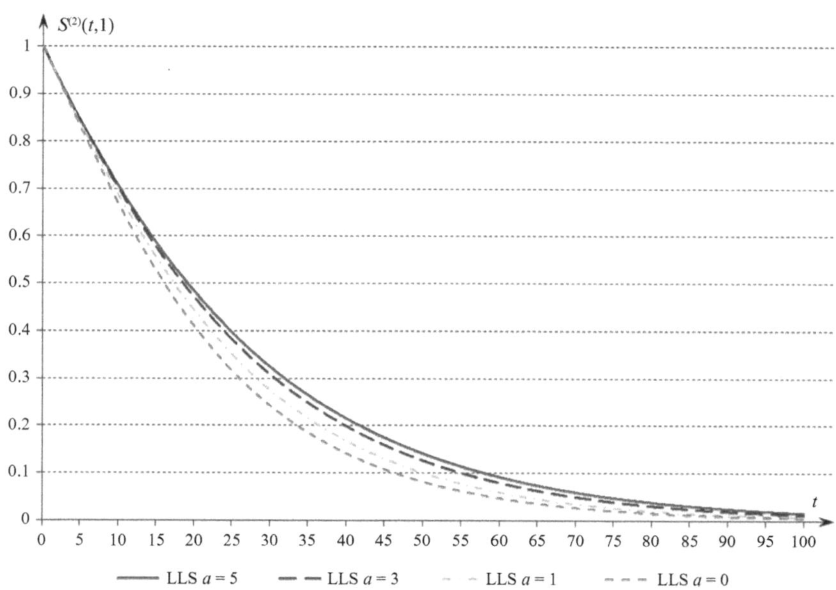

Fig. 5.11 The graphs of $\boldsymbol{N}_2$ safety function coordinate $\boldsymbol{S}^{(2)}(t,1)$ assuming the LLS dependency model with coefficients of load increase (5.3.18), (5.3.19) for a ranging from 5 to 0.

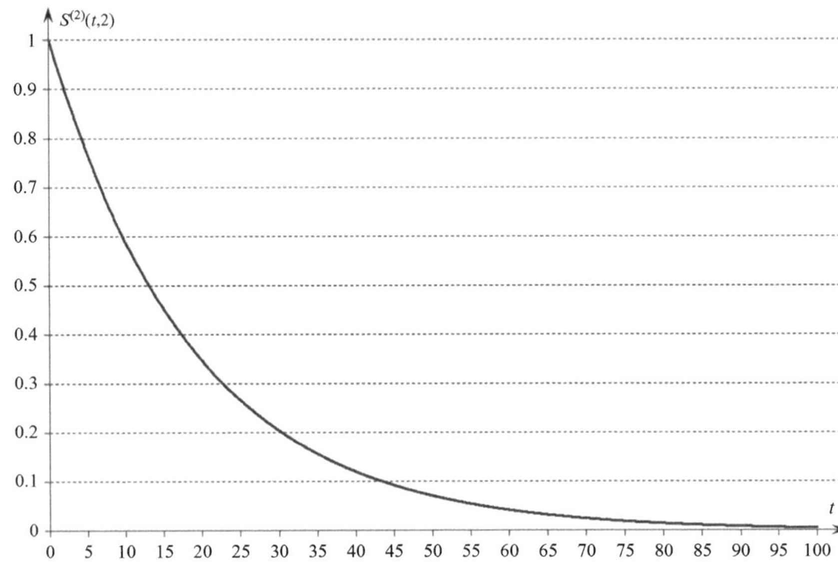

Fig. 5.12 The graphs of N_2 safety function coordinate $\boldsymbol{S}^{(2)}(t,2)$.

i.e., $\boldsymbol{S}^{(2)}(t,2)$ does not depend on the load increase coefficients and is the same as for independent components. This is due to the fact that a series system is in the best state if and only if all its components are in this state. The graphs of $\boldsymbol{N}_2$ safety function coordinate $\boldsymbol{S}^{(2)}(t,2)$ is shown in Fig. 5.12.

The subnetwork $\boldsymbol{N}_3$ is a homogeneous "3 out of 5" network consisting of five components (Fig. 5.8).

The components of subnetwork $\boldsymbol{N}_3$ have the identical, by the assumption, piecewise exponential safety functions

$$S_i^{(3)}(t,\cdot)=\left[1,\exp\left(-\lambda^{(3)}(1)t\right),\exp\left(-\lambda^{(3)}(2)t\right)\right],\quad t\geq 0,$$
$$i=1,2,3,4,5. \tag{5.3.27}$$

with intensities of departures from safety-states subsets $\{1,2\}$ and $\{2\}$, respectively, given by

$$\lambda^{(3)}(1)=0.035,\quad \lambda^{(3)}(2)=0.045. \tag{5.3.28}$$

The safety function of subnetwork $\boldsymbol{N}_3$ following the ELS rule, according to (5.2.40), (5.2.41), is given by

$$\boldsymbol{S}_{ELS}^{(3)}(t,\cdot)=\left[1,\boldsymbol{S}_{ELS}^{(3)}(t,1),\boldsymbol{S}_{ELS}^{(3)}(t,2)\right],\quad t\geq 0, \tag{5.3.29}$$

where its coordinates are

$$\boldsymbol{S}_{ELS}^{(3)}(t,u) = \sum_{\omega=0}^{2} \frac{\left[\frac{5\lambda^{(3)}(u)}{c^{(3)}(u)}t\right]^{\omega}}{\omega!} \cdot \exp\left[-\frac{5\lambda^{(3)}(u)}{c^{(3)}(u)}t\right], \quad u = 1,2. \qquad (5.3.30)$$

And next, substituting in (5.3.30) the intensities of departures from the safety-state subsets (5.3.28), the coordinates of $\boldsymbol{N}_3$ safety function are

$$\boldsymbol{S}_{ELS}^{(3)}(t,1) = \left[1 + \frac{0.175}{c^{(3)}(1)}t + \frac{0.0153125}{\left(c^{(3)}(1)\right)^2}t^2\right] \cdot \exp\left[-\frac{0.175}{c^{(3)}(1)}t\right], \quad t \geq 0, \quad (5.3.31)$$

$$\boldsymbol{S}_{ELS}^{(3)}(t,2) = \left[1 + \frac{0.225}{c^{(3)}(2)}t + \frac{0.0253125}{\left(c^{(3)}(2)\right)^2}t^2\right] \cdot \exp\left[-\frac{0.225}{c^{(3)}(2)}t\right], \quad t \geq 0. \quad (5.3.32)$$

The graphs of safety function coordinates of subnetwork $\boldsymbol{N}_3$, given by (5.3.31), (5.3.32), assuming the ELS dependency model with coefficient of component stress proportionality correction ranging from 0.5 to 1.0, are presented in Figs. 5.13 and 5.14.

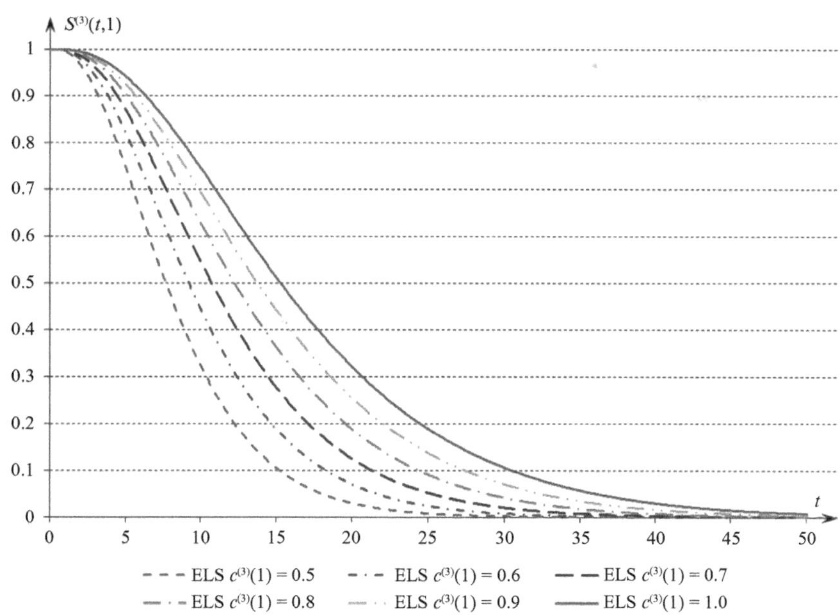

Fig. 5.13 The graphs of $\boldsymbol{N}_3$ safety function coordinate $\boldsymbol{S}^{(3)}(t,1)$ assuming the ELS dependency model with coefficient $c^{(3)}(1)$ ranging from 0.5 to 1.0.

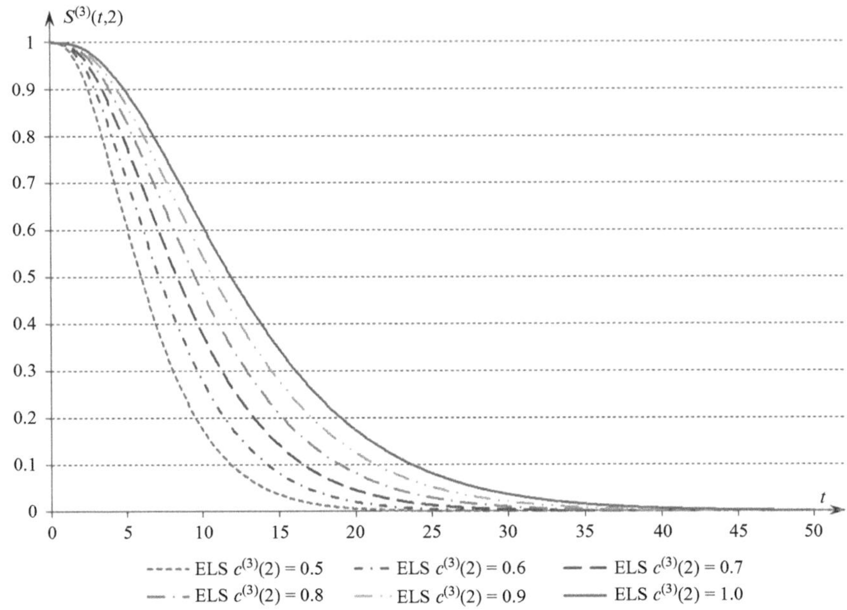

Fig. 5.14 The graphs of $\boldsymbol{N}_3$ safety function coordinate $\boldsymbol{S}^{(3)}(t,2)$ assuming the ELS dependency model with coefficient $c^{(3)}(2)$ ranging from 0.5 to 1.0.

Taking into account the safety structure of network $\boldsymbol{N}$, presented in Fig. 5.8, and applying (5.2.3), (5.2.4), the safety function of network $\boldsymbol{N}$ is given by

$$\boldsymbol{S}_{dep}(t,\ \cdot) = \left[1, \boldsymbol{S}_{dep}(t,1), \boldsymbol{S}_{dep}(t,2)\right],\quad t \geq 0, \tag{5.3.33}$$

where

$$\boldsymbol{S}_{dep}(t,1) = \boldsymbol{S}^{(1)}_{ELS}(t,1) \cdot \boldsymbol{S}^{(2)}_{LLS}(t,1) \cdot \boldsymbol{S}^{(3)}_{ELS}(t,1), \tag{5.3.34}$$

$$\boldsymbol{S}_{dep}(t,2) = \boldsymbol{S}^{(1)}_{ELS}(t,2) \cdot \boldsymbol{S}^{(2)}_{LLS}(t,2) \cdot \boldsymbol{S}^{(3)}_{ELS}(t,2), \tag{5.3.35}$$

and the coordinates of $\boldsymbol{N}_1$ safety function are given by (5.3.5), (5.3.6), the coordinates of $\boldsymbol{N}_2$ safety function are given by (5.3.16), (5.3.17), and the coordinates of $\boldsymbol{N}_3$ safety function are determined in (5.3.31), (5.3.32).

In case the coordinates of $\boldsymbol{N}_2$ safety function are determined by (5.3.25), (5.3.26) for coefficients of load increase (5.3.20)–(5.3.24) and $a = 5$, the safety function of network $\boldsymbol{N}$ is given by the formula

$$\boldsymbol{S}_{dep}(t,\ \cdot) = \left[1, \boldsymbol{S}_{dep}(t,1), \boldsymbol{S}_{dep}(t,2)\right],\quad t \geq 0, \tag{5.3.36}$$

where

$$S_{dep}(t,1) = \left[1 + \frac{0.08}{c^{(1)}(1)}t + \frac{0.0032}{\left(c^{(1)}(1)\right)^2}t^2 + \frac{0.000085}{\left(c^{(1)}(1)\right)^3}t^3\right]$$
$$\cdot \left[1 + \frac{0.175}{c^{(3)}(1)}t + \frac{0.0153125}{\left(c^{(3)}(1)\right)^2}t^2\right] \cdot \exp\left[-\left(\frac{0.08}{c^{(1)}(1)} + \frac{0.175}{c^{(3)}(1)}\right)t\right]$$
$$\cdot \{\exp[-0.053t] + 0.5 \cdot [\exp[-0.04246t] - \exp[-0.06246t]]$$
$$+ 0.1 \cdot [\exp[-0.04576t] - \exp[-0.06576t] + \exp[-0.04790t]$$
$$- \exp[-0.06790t]] + 0.15 \cdot [\exp[-0.04615t] - \exp[-0.6615t]$$
$$+ \exp[-0.04296t] - \exp[-0.06296t]]\}, \quad t \geq 0, \tag{5.3.37}$$

$$S_{dep}(t,2) = \left[1 + \frac{0.16}{c^{(1)}(2)}t + \frac{0.0128}{\left(c^{(1)}(2)\right)^2}t^2 + \frac{0.000683}{\left(c^{(1)}(2)\right)^3}t^3\right]$$
$$\cdot \left[1 + \frac{0.225}{c^{(3)}(2)}t + \frac{0.0253125}{\left(c^{(3)}(2)\right)^2}t^2\right] \cdot \exp\left[-\left(\frac{0.16}{c^{(1)}(2)} + \frac{0.225}{c^{(3)}(2)} + 0.053\right)t\right],$$
$$t \geq 0. \tag{5.3.38}$$

The safety function coordinates of the exemplary network $\boldsymbol{N}$ are illustrated in Figs. 5.15–5.18. Fig. 5.15 presents the graphs of the coordinate $S_{dep}(t,1)$, given by (5.3.37), assuming the LLS dependency model in subnetwork N_2 with coefficients of load increase given by (5.3.18), (5.3.19) for $a = 5$, and the ELS dependency model in subnetworks N_1 and N_3 with coefficients of component stress proportionality correction ranging from 0.5 to 1.0.

Figs. 5.16 and 5.17 present the graphs of the coordinate $S_{dep}(t,1)$, assuming the LLS dependency model in subnetwork N_2 for other values of coefficients of load increase determined for $a = 3$ and $a = 0$.

Fig. 5.18 presents the graphs of the coordinate $S_{dep}(t,2)$, given by (5.3.38), assuming the ELS dependency model in subnetworks N_1 and N_3 with coefficients of component stress proportionality correction ranging from 0.5 to 1.0.

The mean lifetimes of exemplary network $\boldsymbol{N}$ in the safety-state subsets $\{1,2\}$, $\{2\}$, according to (2.1.16), are determined from

$$\mu_{dep}(1) = \int_0^\infty S_{dep}(t,1)dt, \quad \mu_{dep}(2) = \int_0^\infty S_{dep}(t,2)dt, \tag{5.3.39}$$

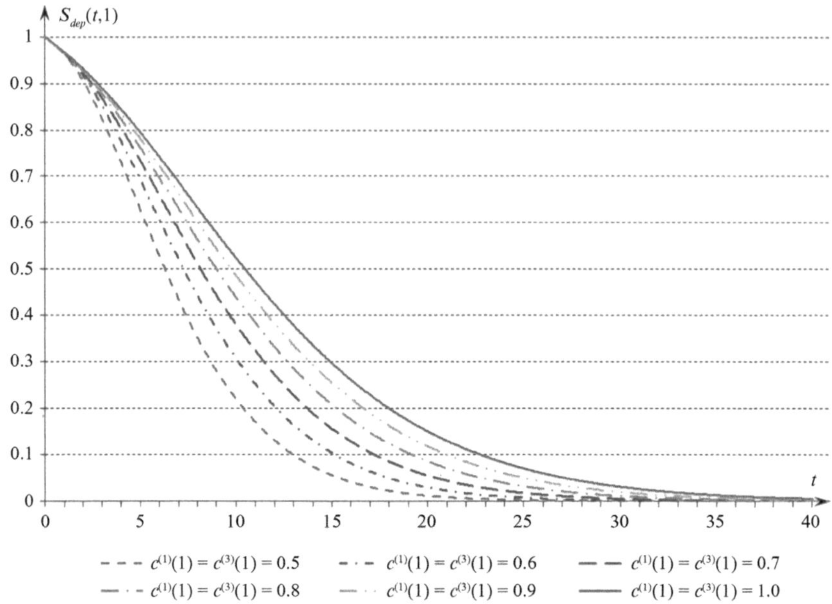

Fig. 5.15 The graphs of the network ***N*** safety function coordinate $\boldsymbol{S}_{dep}(t,1)$ assuming the LLS dependency model in $\boldsymbol{N}_2$ for $a = 5$, and the ELS dependency model in $\boldsymbol{N}_1$ and $\boldsymbol{N}_3$ with coefficients ranging from 0.5 to 1.0.

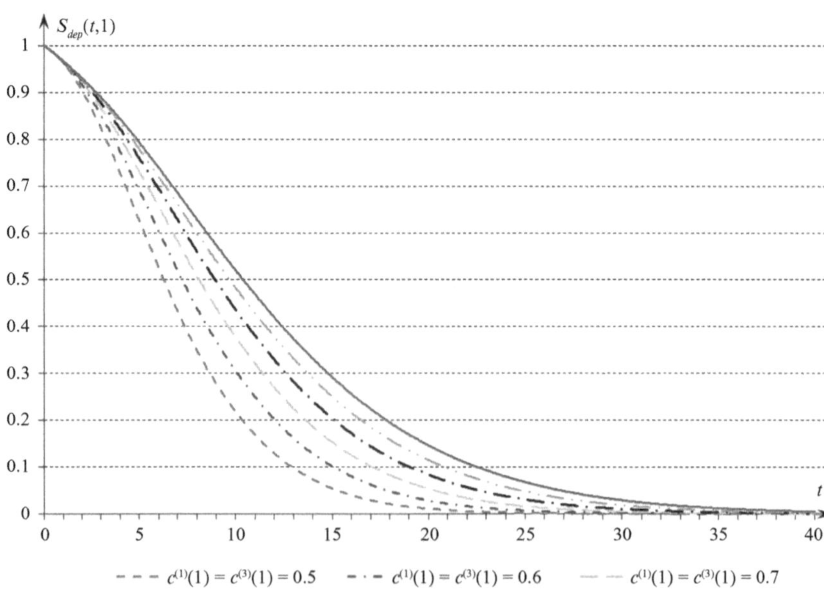

Fig. 5.16 The graphs of the network ***N*** safety function coordinate $\boldsymbol{S}_{dep}(t,1)$ assuming the LLS dependency model in $\boldsymbol{N}_2$ for $a = 3$, and the ELS dependency model in $\boldsymbol{N}_1$ and $\boldsymbol{N}_3$ with coefficients ranging from 0.5 to 1.0.

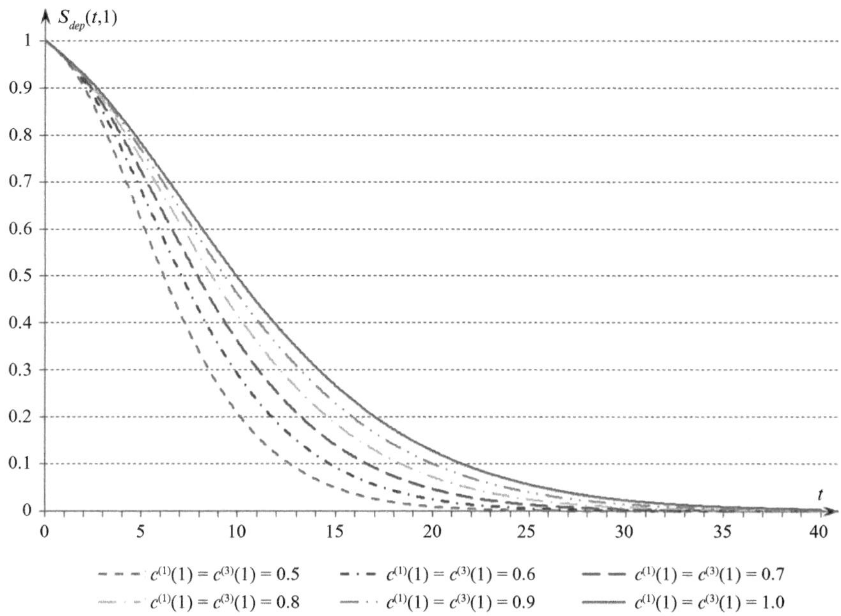

Fig. 5.17 The graphs of the network ***N*** safety function coordinate $\boldsymbol{S}_{dep}(t,1)$ assuming the LLS dependency model in $\boldsymbol{N}_2$ for $a = 0$, and the ELS dependency model in $\boldsymbol{N}_1$ and $\boldsymbol{N}_3$ with coefficients ranging from 0.5 to 1.0.

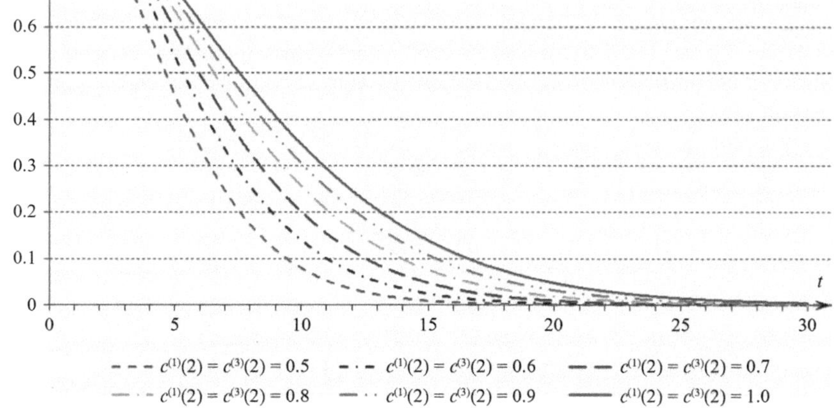

Fig. 5.18 The graphs of the network ***N*** safety function coordinate $\boldsymbol{S}_{dep}(t,2)$ assuming the ELS dependency model in $\boldsymbol{N}_1$ and $\boldsymbol{N}_3$ with coefficients ranging from 0.5 to 1.0.

and their standard deviations, by (2.1.17), (2.1.18), are given by formulas

$$\sigma_{dep}(1) = \sqrt{2\int_0^\infty t \cdot \boldsymbol{S}_{dep}(t, 1)dt - \left[\mu_{dep}(1)\right]^2},$$

$$\sigma_{dep}(2) = \sqrt{2\int_0^\infty t \cdot \boldsymbol{S}_{dep}(t, 2)dt - \left[\mu_{dep}(2)\right]^2}, \qquad (5.3.40)$$

where the safety function coordinates are given by (5.3.34), (5.3.35).

The expected values and standard deviations of network $\boldsymbol{N}$ lifetimes in the safety-state subsets $\{1,2\}$, $\{2\}$, assuming the LLS dependency model in subnetwork $\boldsymbol{N}_2$ with coefficients of load increase given by (5.3.18), (5.3.19) for $a = 5$, $a = 3$ and $a = 0$, and assuming the ELS dependency model in subnetworks $\boldsymbol{N}_1$ and $\boldsymbol{N}_3$ with coefficients of component stress proportionality correction ranging from 0.5 to 1.0, are given in Table 5.1. Different models of dependencies in subnetworks result from different safety structures of these subnetworks.

In the ELS rule component lifetimes decrease uniformly depending on the number of components that have left the safety-state subset, and additionally these changes are influenced by a component stress proportionality correction coefficient, as expressed in formula (5.2.38). Coefficient of component stress proportionality correction expresses the strength of dependence among components. Wherein $c(u) = 1.0$, $u = 1, 2$ means less impact on shortening the lifetimes of other components, and $c(u) = 0.5$, $u = 1, 2$ means the greatest impact.

Table 5.1 The expected values and standard deviations of network $\boldsymbol{N}$ lifetimes in the safety-state subsets {1,2}, {2} for various coefficients of load increase in $\boldsymbol{N}_2$ and coefficients of component stress proportionality correction in $\boldsymbol{N}_1$ and $\boldsymbol{N}_3$ ranging from 0.5 to 1.0.

$c^{(1)}(u) = c^{(3)}(u)$ for $u = 1, 2$	$\mu_{dep}(1)$ $a = 5$	$\sigma_{dep}(1)$ $a = 5$	$\mu_{dep}(1)$ $a = 3$	$\sigma_{dep}(1)$ $a = 3$	$\mu_{dep}(1)$ $a = 0$	$\sigma_{dep}(1)$ $a = 0$	$\mu_{dep}(2)$	$\sigma_{dep}(2)$
0.5	6.89	4.25	6.87	4.23	6.75	4.15	4.91	2.98
0.6	8.00	5.02	7.97	5.00	7.79	4.88	5.69	3.53
0.7	9.02	5.76	8.98	5.73	8.73	5.57	6.41	4.07
0.8	9.98	6.47	9.92	6.43	9.60	6.22	7.09	4.58
0.9	10.87	7.16	10.79	7.10	10.39	6.83	7.71	5.08
1.0	11.70	7.82	11.60	7.74	11.11	7.42	8.30	5.57

In the LLS rule, the strength of dependence among components and the impact of changes in safety-state of one of them on the lifetimes of other components are expressed by the coefficients of load increase. In subnetwork $\boldsymbol{N}_2$ those coefficients of load increase $q(1, E_j^{(2)}, E_i^{(2)})$ for $i = 1, 2, 3, 4, 5$ and $j = 1, 2, 3, 4, 5$, are given by (5.3.18), (5.3.19) depending on parameter a.

The mean values of the exemplary network lifetimes in the particular safety-states 1,2, by (2.1.20) are determined using formulas:

$$\overline{\mu}_{dep}(1) = \mu_{dep}(1) - \mu_{dep}(2), \quad \overline{\mu}_{dep}(2) = \mu_{dep}(2), \tag{5.3.41}$$

and their values for various coefficients of load increase in $\boldsymbol{N}_2$ and coefficients of component stress proportionality correction in $\boldsymbol{N}_1$ and $\boldsymbol{N}_3$ are given in Table 5.2.

Comparing the expected values of network N lifetimes in particular states 1 and 2, it can be seen that mean lifetimes with the same component stress proportionality correction coefficient $c^{(1)}(u) = c^{(3)}(u)$, $u = 1, 2$, for different values of parameter $a = 5$, $a = 3$, $a = 0$, and thus for different values of coefficients of load increase in $\boldsymbol{N}_2$, are at a similar level. However, larger differences can be seen for the set value of parameter a and changing coefficients of component stress proportionality correction $c^{(1)}(u) = c^{(3)}(u)$, $u = 1, 2$, in $\boldsymbol{N}_1$ and $\boldsymbol{N}_3$. This means that the strength of relationship between components in the $\boldsymbol{N}_1$ and $\boldsymbol{N}_3$ subnetworks has an impact on the lifetime of entire network, while changes in the strength of influence between components in the $\boldsymbol{N}_2$ subnetwork on the lifetime of entire network have a negligible effect.

Table 5.2 The expected values of network ***N*** lifetimes in particular states 1 and 2, for various coefficients of load increase in $\boldsymbol{N}_2$ and coefficients of component stress proportionality correction in $\boldsymbol{N}_1$ and $\boldsymbol{N}_3$ ranging from 0.5 to 1.0.

$c^{(1)}(u) = c^{(3)}(u)$ for $u = 1, 2$	$\overline{\mu}_{dep}(1)$ $a = 5$	$\overline{\mu}_{dep}(1)$ (%) $a = 5$	$\overline{\mu}_{dep}(1)$ $a = 3$	$\overline{\mu}_{dep}(1)$ (%) $a = 3$	$\overline{\mu}_{dep}(1)$ $a = 0$	$\overline{\mu}_{dep}(1)$ (%) $a = 0$	$\overline{\mu}_{dep}(2)$	$\overline{\mu}_{dep}(2)$ (%)
0.5	1.98	58%	1.96	59%	1.84	65%	4.91	59%
0.6	2.31	68%	2.28	69%	2.10	75%	5.69	69%
0.7	2.61	77%	2.57	78%	2.32	83%	6.41	77%
0.8	2.89	85%	2.83	86%	2.51	89%	7.09	85%
0.9	3.16	93%	3.08	93%	2.68	95%	7.71	93%
1.0	3.40	100%	3.30	100%	2.81	100%	8.30	100%

The network $\boldsymbol{N}$ mean lifetimes in safety-states 1 and 2, for coefficient of component stress proportionality correction in $\boldsymbol{N}_1$ and $\boldsymbol{N}_3$ equal to 0.5, in relation to the network mean lifetimes for coefficient of 1.0, are shorter by about 40%. This difference decreases to about 30% for coefficient of component stress proportionality correction in $\boldsymbol{N}_1$ and $\boldsymbol{N}_3$ equal to 0.6. Comparing the expected values of network $\boldsymbol{N}$ lifetimes in states 1 and 2, for coefficients of component stress proportionality correction in $\boldsymbol{N}_1$ and $\boldsymbol{N}_3$ of 0.7 and 1.0, the difference is about 23%, for coefficients 0.8 and 1.0 about 15%, respectively, and for coefficients equal to 0.9 and 1.0 about 7%. These differences differ only slightly for parameter $a = 0$.

Further, we compare obtained results with those for an exemplary network $\boldsymbol{N}$ assuming independence of all components in subnetworks $\boldsymbol{N}_1$, $\boldsymbol{N}_2$, $\boldsymbol{N}_3$. The mean lifetimes of network $\boldsymbol{N}$ with independent components, in the safety-state subsets $\{1,2\}$, $\{2\}$, are:

$$\mu(1) \cong 14.485, \quad \mu(2) \cong 10.164 \tag{5.3.42}$$

and in particular states 1,2, take, respectively, values:

$$\bar{\mu}(1) \cong 4.321, \quad \bar{\mu}(2) \cong 10.164. \tag{5.3.43}$$

Obtained mean lifetimes in safety-state 1 for network $\boldsymbol{N}$ assuming ELS dependency model in $\boldsymbol{N}_1$ and $\boldsymbol{N}_3$, and LLS dependency model in $\boldsymbol{N}_2$ (Table 5.2), compared to mean lifetime in state 1 for network $\boldsymbol{N}$ without considering dependencies among components (5.3. 43), allow drawing the following conclusions. For $a = 5$ and $a = 3$ in subnetwork $\boldsymbol{N}_2$, the results are similar and mean lifetime of $\boldsymbol{N}$, if $c^{(1)}(u) = c^{(3)}(u) = 0.5$, $u = 1$, 2, constitutes 45%–46% of the mean lifetime of network $\boldsymbol{N}$ assuming independence of components. This difference decreases at higher values of the coefficient of component stress proportionality correction in subnetworks $\boldsymbol{N}_1$ and $\boldsymbol{N}_3$, and for $c^{(1)}(u) = c^{(3)}(u) = 1.0$, $u = 1, 2$, the mean lifetime of network $\boldsymbol{N}$ is 76%–79% of the network mean lifetime assuming independence of components. For $a = 0$ these results are slightly different, and so for $c(1) = c(2) = 0.5$ the mean lifetime is 43%, and for $c^{(1)}(u) = c^{(3)}(u) = 1.0$, $u = 1, 2$, respectively, 65% of the mean lifetime of network $\boldsymbol{N}$ with independent components.

Comparing the expected values of lifetimes in safety-state 2 of network $\boldsymbol{N}$ composed of subnetworks with dependent components (Table 5.2) and for network $\boldsymbol{N}$ with independent components of subnetworks (5.3.43), the differences are smaller than for expected values of lifetimes in safety-state 1. For coefficient $c^{(1)}(u) = c^{(3)}(u) = 0.5$, $u = 1, 2$, the expected value of lifetime is 52% shorter than the expected value of network $\boldsymbol{N}$ lifetime assuming independence of components. For a coefficient of 0.6, the difference decreases to 44%, and next up to 18% for a coefficient of 1.0.

Assuming that the critical safety state is $r = 1$, we obtain the risk function of network $\boldsymbol{N}$

$$\boldsymbol{r}_{dep}(t) = 1 - \boldsymbol{S}_{dep}(t, 1), \quad t \geq 0. \tag{5.3.44}$$

The time points τ_{dep} of exceeding by risk function an acceptable level, for instance, $\delta = 0.05$ and $\delta = 0.1$, using formula (2.1.22), for coefficients of load increase in $\boldsymbol{N}_2$ given by (5.3.18), (5.3.19) for $a = 5$, $a = 3$, $a = 0$, and for coefficients of component stress proportionality correction in $\boldsymbol{N}_1$ and $\boldsymbol{N}_3$ ranging from 0.5 to 1.0, are determined in Table 5.3.

The difference between values of these moments of exceeding the level $\delta = 0.05$, for individual values of coefficients of load increase in $\boldsymbol{N}_2$ specified for $a = 5$, $a = 3$, $a = 0$, ranges up to 15% for various values of correction coefficients in $\boldsymbol{N}_1$ and $\boldsymbol{N}_3$ in relation to the results obtained for $c^{(1)}(u) = c^{(3)}(u) = 1.0$, $u = 1, 2$. Comparing the results presented in Table 5.3 for $\delta = 0.1$, for correction coefficients in $\boldsymbol{N}_1$ and $\boldsymbol{N}_3$ ranging from 0.5 to 1.0, the time to exceed the level of $\delta = 0.1$ for $c^{(1)}(u) = c^{(3)}(u) = 0.5$, $u = 1, 2$, is shorter by about 25% in relation to the time determined for $c^{(1)}(u) = c^{(3)}(u) = 1.0$, $u = 1, 2$.

It can be noticed, similarly as for expected values, that the dependency among components connected in series in the $\boldsymbol{N}_2$ subnetwork with coefficients of load increase $q(1, E_j^{(2)}, E_i^{(2)})$ for $i, j = 1, 2, 3, 4, 5$, in the LLS

Table 5.3 The time points τ_{dep} of exceeding by risk function an acceptable level $\delta = 0.05$ and $\delta = 0.1$, for various coefficients of load increase in $\boldsymbol{N}_2$ and coefficients of component stress proportionality correction in $\boldsymbol{N}_1$ and $\boldsymbol{N}_3$ ranging from 0.5 to 1.0.

$c^{(1)}(u) = c^{(3)}(u)$	τ_{dep} for $\delta = 0.05$				τ_{dep} for $\delta = 0.1$			
for $u = 1, 2$	$a = 5$	$a = 3$	$a = 0$	[%]	$a = 5$	$a = 3$	$a = 0$	[%]
0.5	1.25	1.24	1.23	85%–86%	2.05	2.05	2.01	75%–76%
0.6	1.33	1.33	1.30	90%–91%	2.26	2.26	2.21	82%–83%
0.7	1.39	1.38	1.35	94%–95%	2.43	2.42	2.36	88%–89%
0.8	1.43	1.42	1.39	97%	2.56	2.55	2.48	93%–94%
0.9	1.45	1.45	1.42	99%	2.67	2.66	2.57	97%
1.0	1.47	1.47	1.43	100%	2.75	2.74	2.65	100%

dependency model, given by (5.3.18), (5.3.19), has a significantly weaker impact on the results for entire network. The times to exceed the risk level, presented in Table 5.3 for different values of parameter $a = 5$, $a = 3$, $a = 0$, affecting the value of coefficients $q(1, E_j^{(2)}, E_i^{(2)})$, take similar values.

In case the coefficients of $\boldsymbol{N}_2$ load increase are given by (5.3.20)–(5.3.24) for $a = 5$, the safety function coordinate $\boldsymbol{S}_{dep}(t,1)$ of network $\boldsymbol{N}$ is given by the formula (5.3.37). In that case the risk function of network $\boldsymbol{N}$, determined applying formula (5.3.44), is illustrated in Fig. 5.19.

5.3.2 Safety analysis of exemplary energy distribution network

Since components of transmission and distribution networks require constant maintenance and degrading causes their properties deterioration over time, the multistate approach to safety analysis of distribution systems seems to be reasonable. In critical and overload states, the component degradation can proceed much faster. The multistate safety analysis of distribution network is performed regarding its assets and subnetworks interdependencies. The transition between safety states of subnetworks or their assets

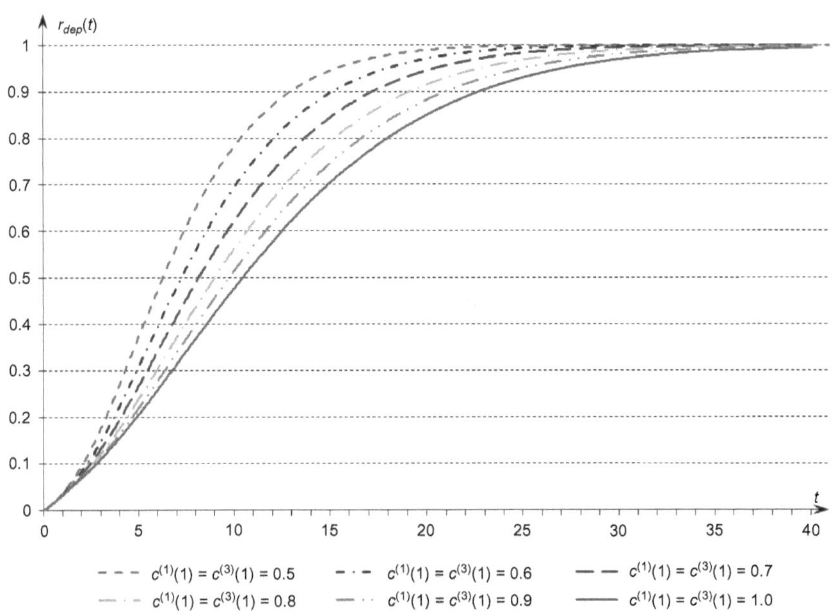

Fig. 5.19 The graphs of the risk function $\boldsymbol{r}_{dep}(t)$ of network $\boldsymbol{N}$ assuming the LLS dependency model in $\boldsymbol{N}_2$ for $a = 5$, and the ELS dependency model in $\boldsymbol{N}_1$ and $\boldsymbol{N}_3$ with coefficients ranging from 0.5 to 1.0.

can have impact on safety characteristics of remaining subnetworks and assets and as a consequence on safety of the whole system. Further, such approach to analysis of interconnections and interdependencies can help to capture the critical points and critical operations that can affect the whole network functioning.

We consider an exemplary distribution network $\boldsymbol{N}$ composed of 40 dependent sections of single-circuit distribution line. We consider line sections as basic component. We assume the LLS dependency model among these sections (assets) connected in series safety structure. Sections form the subnetworks that are also connected in series and dependent according to the LLS rule (Blokus-Roszkowska and Kolowrocki, 2017b).

The network $\boldsymbol{N}$, with the scheme presented in Fig. 5.20, is composed of 15 subnetworks $\boldsymbol{N}_i$, $i = 1, 2, \ldots, 15$. The presentation of network assets in this scheme is related to their location and relationships in the functioning of the entire distribution network (Blokus-Roszkowska and Kolowrocki, 2017b).

For the functioning of the entire network crucial are distribution lines in the main subnetwork $\boldsymbol{N}_1$. The subnetwork $\boldsymbol{N}_1$ does not depend on functioning of other subnetworks and its assets are only dependent on each other. The subnetworks $\boldsymbol{N}_2$, $\boldsymbol{N}_3$, $\boldsymbol{N}_6$, $\boldsymbol{N}_9$, $\boldsymbol{N}_{12}$, and $\boldsymbol{N}_{15}$ are dependent on subnetwork $\boldsymbol{N}_1$. Besides inside interactions among transmission lines in subnetworks may cause the decrease of these lines' lifetimes and have influence on subnetwork's functioning. The subnetworks $\boldsymbol{N}_4$ and $\boldsymbol{N}_5$ depend on subnetworks $\boldsymbol{N}_3$ and $\boldsymbol{N}_1$, the subnetworks $\boldsymbol{N}_7$ and $\boldsymbol{N}_8$ depend on subnetworks $\boldsymbol{N}_6$ and $\boldsymbol{N}_1$, the subnetworks $\boldsymbol{N}_{10}$ and $\boldsymbol{N}_{11}$ depend on subnetworks $\boldsymbol{N}_9$ and $\boldsymbol{N}_1$, the subnetworks $\boldsymbol{N}_{13}$ and $\boldsymbol{N}_{14}$ depend on subnetworks $\boldsymbol{N}_{12}$ and $\boldsymbol{N}_1$.

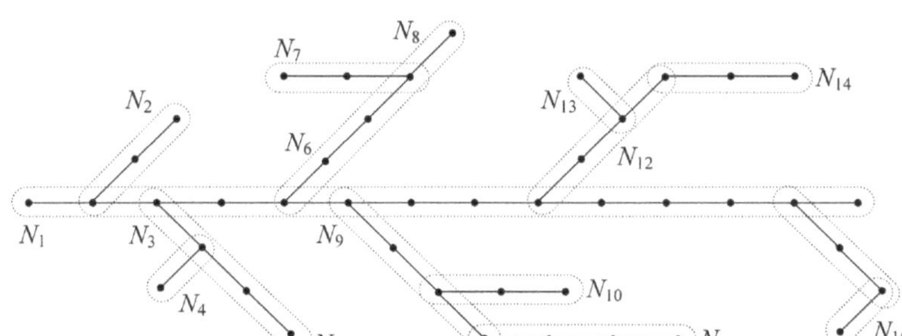

Fig. 5.20 The scheme of the exemplary distribution network N (Blokus-Roszkowska and Kolowrocki, 2017b).

We assume that the network $\boldsymbol{N}$ is a 5-state system ($z = 4$) and we arbitrarily distinguish the following four safety states of the network and its assets (Blokus-Roszkowska and Kolowrocki, 2017b):

- a safety-state 4—the distribution line is new and fully effective, the network operation is fully effective,
- a safety-state 3—the distribution line is not new but fully effective and the network operation is fully effective,
- a safety-state 2—the distribution line is in operation, but it is suitable for further use, the network operation is less effective because of aging,
- a safety-state 1—advanced aging processes in the distribution line and there is a high risk of failure, the network operation is less effective, and does not provide sufficient distribution capacity,
- a safety-state 0—the network or distribution line is destroyed.

The input necessary parameters of the exemplary network safety model are as follows:

– the number of safety states of the network and assets $z = 4$,
– the critical safety state of the network and its assets $r = 2$,
– the network risk permitted level $\delta = 0.2$.

We assume that all sections of distribution lines in the network have identical safety functions

$$S(t,\ \cdot\ ) = [1, S(t,1), S(t,2), S(t,3), S(t,4)],\quad t \geq 0, \tag{5.3.45}$$

with the piecewise exponential coordinates

$$S(t,1) = \exp[-\lambda(1)t],\quad S(t,2) = \exp[-\lambda(2)t],$$
$$S(t,3) = \exp[-\lambda(3)t],\quad S(t,4) = \exp[-\lambda(4)t],\quad t \geq 0. \tag{5.3.46}$$

The lifetimes of distribution lines and their sections, in the safety states, are expressed in years. The intensities of sections' departure from the safety-states subset $\{1,2,3,4\}$, $\{2,3,4\}$, $\{3,4\}$, $\{4\}$ in formula (5.3.46) are as follows (Blokus-Roszkowska and Kolowrocki, 2017b):

$$\lambda(1) = 0.027,\quad \lambda(2) = 0.043,\quad \lambda(3) = 0.100,\quad \lambda(4) = 0.200. \tag{5.3.47}$$

Further, the unknown parameters of a network safety structure, the model of components dependency, and parameters concerned with this model are given.

We assume the LLS dependency model among sections, i.e., assets $E_j^{(i)}$, $i = 1,\ 2, \ldots, 15,\ j = 1,\ 2, \ldots, l_i$, in subnetworks $\boldsymbol{N}_i$, $i = 1,\ 2, \ldots, 15$. Similarly as in Section 5.2.3, the coefficients of network load increase

$q(v, E_g^{(i)}, E_j^{(i)})$, $v = u, u-1, \ldots, 1$, $u = 1, 2, 3$, express the influence of the section $E_g^{(i)}$, $g = 1, 2, \ldots, l_i$, departure from the safety-state subset $\{u, u+1, \ldots, 4\}$, $u = 1, 2, 3, 4$, on lifetimes and safety parameters of other sections $E_j^{(i)}, j = 1, 2, \ldots, l_i, j \neq g$, in the subnetwork $\mathbf{N}_i$, $i = 1, 2, \ldots, 15$. In considered example, these coefficients relate to the distance between sections $E_j^{(i)}$ and $E_g^{(i)}$ in $\mathbf{N}_i$, $i = 1, 2, \ldots, 15$. Subsequently, these coefficients are given in a form $q(v, d_{jg}^{(i)})$, $v = u, u-1, \ldots, 1$, $u = 1, 2, 3$, where $d_{jg}^{(i)}$ denotes the distance between sections $E_j^{(i)}$ and $E_g^{(i)}$ in subnetwork $\mathbf{N}_i$, $i = 1, 2, \ldots, 15$ (Blokus-Roszkowska and Kolowrocki, 2017b).

Under these assumptions, for sections having identical piecewise exponential safety functions (5.3.45), (5.3.46), the formula (5.2.60) for the intensities of departure from the subset $\{v, v+1, \ldots, z\}$, $v = u, u-1, \ldots, 1$, $u = 1, 2, 3$, of other sections $E_j^{(i)}$, $i = 1, 2, \ldots, 15$, $j = 1, 2, \ldots, l_i$, $j \neq g$, after departure of $E_g^{(i)}$, $g = 1, 2, \ldots, l_i$, takes the following form:

$$\lambda_{j/g}^{(i)}(v) = \frac{\lambda(v)}{q\left(v, d_{jg}^{(i)}\right)}, \quad i = 1, 2, \ldots, 15, \quad j = 1, 2, \ldots, l_i,$$

$$g = 1, 2, \ldots, l_i. \tag{5.3.48}$$

The necessary parameters are:

- a subnetwork $\mathbf{N}_1$:
  - the number of sections $l_1 = 13$, denoted by $E_j^{(1)}, j = 1, 2, \ldots, 13$,
  - the coefficients of the subnetwork $\mathbf{N}_1$ load increase in the LLS dependency rule

$$q\left(v, d_{jg}^{(1)}\right) = 1 - \exp\left[-d_{jg}^{(1)}\right], \quad j = 1, 2, \ldots, 13,$$

$$g = 1, 2, \ldots, 13, \quad j \neq g, \quad v = u, u-1, \ldots, 1, \quad u = 1, 2, 3, \tag{5.3.49}$$

where $d_{jg}^{(1)} = |j - g|, j = 1, \ldots, 13, g = 1, \ldots, 13$, denotes the distance between sections $E_j^{(1)}$ and $E_g^{(1)}$, and

$$q(v, 0) = 1, \quad v = u, u-1, \ldots, 1, \quad u = 1, 2, 3; \tag{5.3.50}$$

- a subnetwork $\mathbf{N}_2$:
  - the number of sections $l_2 = 2$, denoted by $E_1^{(2)}$ and $E_2^{(2)}$,
  - the coefficients of the subnetwork $\mathbf{N}_2$ load increase in the LLS dependency rule:

    section $E_2^{(2)}$ depends on functioning of $E_1^{(2)}$

$$q\left(v, d_{21}^{(2)}\right) = 1 - \exp[-1], \quad v = u, u-1, \ldots, 1, \quad u = 1,2,3, \tag{5.3.51}$$

and section $E_2^{(2)}$ has no influence on functioning of $E_1^{(2)}$

$$q\left(v, d_{12}^{(2)}\right) = q(v, 0) = 1, \quad v = u, u-1, \ldots, 1, \quad u = 1,2,3; \tag{5.3.52}$$

- a subnetwork $\boldsymbol{N}_3$:
  – the number of sections $l_3 = 1$,
- a subnetwork $\boldsymbol{N}_4$:
  – the number of sections $l_4 = 1$,
- a subnetwork $\boldsymbol{N}_5$:
  – the number of sections $l_5 = 2$,
  – the coefficients of the subnetwork $\boldsymbol{N}_5$ load increase in the LLS rule:
  section $E_2^{(5)}$ depends on functioning of $E_1^{(5)}$

$$q\left(v, d_{21}^{(5)}\right) = 1 - \exp[-1], \quad v = u, u-1, \ldots, 1, \quad u = 1,2,3, \tag{5.3.53}$$

  and section $E_2^{(5)}$ has no influence on functioning of $E_2^{(5)}$

$$q\left(v, d_{12}^{(5)}\right) = q(v, 0) = 1, \quad v = u, u-1, \ldots, 1, \quad u = 1,2,3; \tag{5.3.54}$$

- a subnetwork $\boldsymbol{N}_6$:
  – the number of sections $l_6 = 3$,
  – the coefficients of the subnetwork $\boldsymbol{N}_6$ load increase in the LLS dependency rule:
  section $E_2^{(6)}$ depends on functioning of $E_1^{(6)}$ with the coefficients

$$q\left(v, d_{21}^{(6)}\right) = 1 - \exp[-1], \quad v = u, u-1, \ldots, 1, \quad u = 1,2,3, \tag{5.3.55}$$

  section $E_3^{(6)}$ depends on functioning of $E_g^{(6)}$, $g = 1, 2$, with the coefficients

$$q\left(v, d_{3g}^{(6)}\right) = 1 - \exp[-(3-g)], \quad g = 1,2, \quad v = u, u-1, \ldots, 1, \quad u = 1,2,3, \tag{5.3.56}$$

  and there are no other dependencies among the sections

$$q\left(v, d_{12}^{(6)}\right) = q\left(v, d_{13}^{(6)}\right) = q\left(v, d_{23}^{(6)}\right) = q(v, 0) = 1, \quad v = u, u-1, \ldots, 1, \quad u = 1,2,3; \tag{5.3.57}$$

- a subnetwork $\boldsymbol{N}_7$:
  - the number of sections $l_7 = 2$,
  - the coefficients of the subnetwork $\boldsymbol{N}_7$ load increase in the LLS dependency rule:

    section $E_2^{(7)}$ depends on functioning of $E_1^{(7)}$

$$q\left(v, d_{21}^{(7)}\right) = 1 - \exp[-1], \quad v = u, u-1, \ldots, 1, \quad u = 1,2,3, \tag{5.3.58}$$

and section $E_2^{(7)}$ has no influence on functioning of $E_1^{(7)}$

$$q\left(v, d_{12}^{(7)}\right) = q(v, 0) = 1, \quad v = u, u-1, \ldots, 1, \quad u = 1,2,3; \tag{5.3.59}$$

- a subnetwork $\boldsymbol{N}_8$:
  - the number of sections $l_8 = 1$,
- a subnetwork $\boldsymbol{N}_9$:
  - the number of sections $l_9 = 2$,
  - the coefficients of the subnetwork $\boldsymbol{N}_9$ load increase in the LLS dependency rule:

    section $E_2^{(9)}$ depends on functioning of $E_1^{(9)}$

$$q\left(v, d_{21}^{(9)}\right) = 1 - \exp[-1], \quad v = u, u-1, \ldots, 1, \quad u = 1,2,3, \tag{5.3.60}$$

and section $E_2^{(9)}$ has no influence on functioning of $E_1^{(9)}$

$$q\left(v, d_{12}^{(9)}\right) = q(v, 0) = 1, \quad v = u, u-1, \ldots, 1, \quad u = 1,2,3; \tag{5.3.61}$$

- a subnetwork $\boldsymbol{N}_{10}$:
  - the number of sections $l_{10} = 2$,
  - the coefficients of the subnetwork $\boldsymbol{N}_{10}$ load increase in the LLS dependency rule:

    section $E_2^{(10)}$ depends on functioning of $E_1^{(10)}$

$$q\left(v, d_{21}^{(10)}\right) = 1 - \exp[-1], \quad v = u, u-1, \ldots, 1, \quad u = 1,2,3, \tag{5.3.62}$$

and section $E_2^{(10)}$ has no influence on functioning of $E_1^{(10)}$

$$q\left(v, d_{12}^{(10)}\right) = q(v, 0) = 1, \quad v = u, u-1, \ldots, 1, \quad u = 1,2,3; \tag{5.3.63}$$

- a subnetwork $\boldsymbol{N}_{11}$:
  - the number of sections $l_{11} = 4$,
  - the coefficients of the subnetwork $\boldsymbol{N}_{11}$ load increase in the LLS dependency rule:

    section $E_2^{(11)}$ depends on functioning of $E_1^{(11)}$ with the coefficients

$$q\left(v, d_{21}^{(11)}\right) = 1 - \exp[-1], \quad v = u, u-1, \ldots, 1, \quad u = 1, 2, 3, \tag{5.3.64}$$

section $E_3^{(11)}$ depends on functioning of $E_g^{(11)}$, $g = 1, 2$, with the coefficients

$$q\left(v, d_{3g}^{(11)}\right) = 1 - \exp[-(3-g)], \quad g = 1, 2,$$
$$v = u, u-1, \ldots, 1, \quad u = 1, 2, 3, \tag{5.3.65}$$

section $E_4^{(11)}$ depends on functioning of $E_g^{(11)}$, $g = 1, 2, 3$, with the coefficients

$$q\left(v, d_{4g}^{(11)}\right) = 1 - \exp[-(4-g)],$$
$$g = 1, 2, 3, \quad v = u, u-1, \ldots, 1, \quad u = 1, 2, 3, \tag{5.3.66}$$

and there are no other dependencies among the sections

$$q\left(v, d_{12}^{(11)}\right) = q\left(v, d_{13}^{(11)}\right) = q\left(v, d_{23}^{(11)}\right) = q\left(v, d_{14}^{(11)}\right)$$
$$= q\left(v, d_{24}^{(11)}\right) = q\left(v, d_{34}^{(11)}\right) = q(v, 0) = 1,$$
$$\text{for } v = u, u-1, \ldots, 1, \quad u = 1, 2, 3; \tag{5.3.67}$$

- a subnetwork $\boldsymbol{N}_{12}$:
  - the number of sections $l_{12} = 2$,
  - the coefficients of the subnetwork $\boldsymbol{N}_{12}$ load increase in the LLS dependency rule:

    section $E_2^{(12)}$ depends on functioning of $E_1^{(12)}$

$$q\left(v, d_{21}^{(12)}\right) = 1 - \exp[-1], \quad v = u, u-1, \ldots, 1, \quad u = 1, 2, 3, \tag{5.3.68}$$

and section $E_2^{(12)}$ has no influence on functioning of $E_1^{(12)}$

$$q\left(v, d_{12}^{(10)}\right) = q(v, 0) = 1, \quad v = u, u-1, \ldots, 1, \quad u = 1, 2, 3; \tag{5.3.69}$$

- a subnetwork $\boldsymbol{N}_{13}$:
  - the number of sections $l_{13} = 1$,
- a subnetwork $\boldsymbol{N}_{14}$:
  - the number of sections $l_{14} = 3$,
  - the coefficients of the subnetwork $\boldsymbol{N}_{14}$ load increase in the LLS dependency rule:

    section $E_2^{(14)}$ depends on functioning of $E_1^{(14)}$ with the coefficients

$$q\left(v, d_{21}^{(14)}\right) = 1 - \exp[-1], \quad v = u, u-1, \ldots, 1, \quad u = 1, 2, 3, \tag{5.3.70}$$

section $E_3^{(14)}$ depends on functioning of $E_g^{(14)}$, $g = 1, 2$, with the coefficients

$$q\left(v, d_{3g}^{(14)}\right) = 1 - \exp[-(3-g)], \quad g = 1, 2,$$
$$v = u, u-1, \ldots, 1, \quad u = 1, 2, 3, \tag{5.3.71}$$

and there are no other dependencies among the sections

$$q\left(vd_{12}^{(14)}\right) = q\left(vd_{13}^{(14)}\right) = q\left(vd_{23}^{(14)}\right) = q(v0) = 1,$$
$$v = u, u-1, \ldots, 1, \quad u = 1, 2, 3; \tag{5.3.72}$$

- a subnetwork $\boldsymbol{N}_{15}$:
  - the number of sections $l_{15} = 3$,
  - the coefficients of the subnetwork $\boldsymbol{N}_{14}$ load increase in the LLS dependency rule:

    section $E_2^{(15)}$ depends on functioning of $E_1^{(15)}$ with the coefficients

$$q\left(v, d_{21}^{(15)}\right) = 1 - \exp[-1], \quad v = u, u-1, \ldots, 1, \quad u = 1, 2, 3, \tag{5.3.73}$$

section $E_3^{(15)}$ depends on functioning of $E_g^{(15)}$, $g = 1, 2$, with the coefficients

$$q\left(v, d_{3g}^{(15)}\right) = 1 - \exp[-(3-g)], \quad g = 1, 2,$$
$$v = u, u-1, \ldots, 1, \quad u = 1, 2, 3, \tag{5.3.74}$$

and there are no other dependencies among the sections

$$q\left(vd_{12}^{(15)}\right) = q\left(vd_{13}^{(15)}\right) = q\left(vd_{23}^{(15)}\right) = q(v0) = 1,$$
$$v = u, u-1, \ldots, 1, \quad u = 1, 2, 3. \tag{5.3.75}$$

Considering the safety model parameters, we determine safety functions of particular subnetworks $\boldsymbol{N}_i$, $i = 1, 2, \ldots, 15$, with sections (assets) connected in series and dependent according to the LLS rule.

The safety function of the subnetwork $\boldsymbol{N}_1$, from (5.2.27)–(5.2.29), is given by the vector

$$\boldsymbol{S}^{(1)}(t, \cdot) = \left[1, \boldsymbol{S}^{(1)}(t,1), \boldsymbol{S}^{(1)}(t,2), \boldsymbol{S}^{(1)}(t,3), \boldsymbol{S}^{(1)}(t,4)\right], \quad t \geq 0, \quad (5.3.76)$$

with the coordinates

$$\begin{aligned}\boldsymbol{S}^{(1)}(t1) = {} & \exp[-13\lambda(2)t] \\ & + \frac{1}{13}\sum_{g=1}^{13}\left[\exp\left[-\lambda(1)\sum_{j=1}^{13}\frac{1}{q\left(1 d_{jg}^{(1)}\right)}t\right]\right. \\ & \left. - \exp\left[-\left(13\lambda(2) - 13\lambda(1) + \lambda(1)\sum_{j=1}^{13}\frac{1}{q\left(1 d_{jg}^{(1)}\right)}\right)t\right]\right], \end{aligned} \quad (5.3.77)$$

$$\begin{aligned}\boldsymbol{S}^{(1)}(t,2) = {} & \exp[-13\lambda(3)t] \\ & + \frac{1}{13}\sum_{g=1}^{13}\left[\exp\left[-\lambda(2)\sum_{j=1}^{13}\frac{1}{q\left(2, d_{jg}^{(1)}\right)}t\right]\right. \\ & \left. - \exp\left[-\left(13\lambda(3) - 13\lambda(2) + \lambda(2)\sum_{j=1}^{13}\frac{1}{q\left(2, d_{jg}^{(1)}\right)}\right)t\right]\right], \end{aligned} \quad (5.3.78)$$

$$\begin{aligned}\boldsymbol{S}^{(1)}(t,3) = {} & \exp[-13\lambda(4)t] \\ & + \frac{1}{13}\sum_{g=1}^{13}\left[\exp\left[-\lambda(3)\sum_{j=1}^{13}\frac{1}{q\left(3, d_{jg}^{(1)}\right)}t\right]\right. \\ & \left. - \exp\left[-\left(13\lambda(4) - 13\lambda(3) + \lambda(3)\sum_{j=1}^{13}\frac{1}{q\left(3, d_{jg}^{(1)}\right)}\right)t\right]\right], \end{aligned} \quad (5.3.79)$$

$$\boldsymbol{S}^{(1)}(t,4) = \exp[-13\lambda(4)t], \quad (5.3.80)$$

where the intensities of assets' departure from the safety-states subset {1,2,3,4}, {2,3,4}, {3,4}, {4} are given by (5.3.47) and the coefficients $q(1, d_{jg}^{(1)})$, $q(2, d_{jg}^{(1)})$, $q(3, d_{jg}^{(1)})$, $j = 1, 2, \ldots, 13$, $g = 1, 2, \ldots, 13$, are given by (5.3.49), (5.3.50). Further after substituting these values, the safety function coordinates of the subnetwork N_1 (5.3.77)–(5.3.80) take following form

$$\boldsymbol{s}^{(1)}(t,1) = \exp[-0.559t] + \frac{1}{13}\sum_{g=1}^{13}\left[\exp\left[-0.027\sum_{j=1}^{13}\frac{1}{1-\exp\left[-d_{jg}^{(1)}\right]}t\right] - \exp\left[-\left(0.208+0.027\sum_{j=1}^{13}\frac{1}{1-\exp\left[-d_{jg}^{(1)}\right]}\right)t\right]\right],\quad t\geq 0, \tag{5.3.81}$$

$$\boldsymbol{s}^{(1)}(t,2) = \exp[-1.3t] + \frac{1}{13}\sum_{g=1}^{13}\left[\exp\left[-0.043\sum_{j=1}^{13}\frac{1}{1-\exp\left[-d_{jg}^{(1)}\right]}t\right] - \exp\left[-\left(0.741+0.043\sum_{j=1}^{13}\frac{1}{1-\exp\left[-d_{jg}^{(1)}\right]}\right)t\right]\right],\quad t\geq 0, \tag{5.3.82}$$

$$\boldsymbol{s}^{(1)}(t,3) = \exp[-2.6t] + \frac{1}{13}\sum_{g=1}^{13}\left[\exp\left[-0.1\sum_{j=1}^{13}\frac{1}{1-\exp\left[-d_{jg}^{(1)}\right]}t\right] - \exp\left[-\left(1.3+0.1\sum_{j=1}^{13}\frac{1}{1-\exp\left[-d_{jg}^{(1)}\right]}\right)t\right]\right],\quad t\geq 0, \tag{5.3.83}$$

$$\boldsymbol{s}^{(1)}(t,4) = \exp[-2.6t],\quad t\geq 0. \tag{5.3.84}$$

The safety function coordinates of the subnetwork N_1 are illustrated in Fig. 5.21.

The safety function of subnetworks N_i, $i = 2, 5, 7, 9, 10, 12$, from (5.2.27) to (5.2.29), is given by the vector (Fig. 5.22).

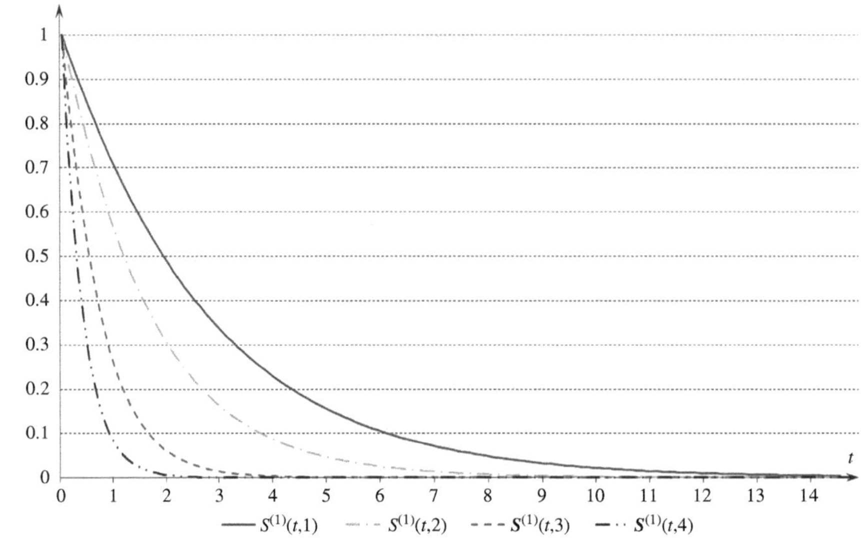

Fig. 5.21 The graphs of the subnetwork $\boldsymbol{N}_1$ safety function coordinates.

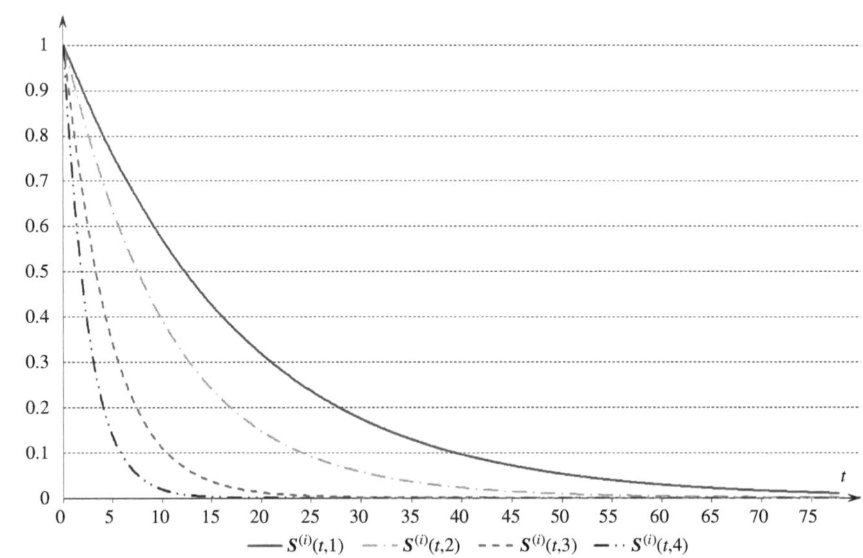

Fig. 5.22 The graphs of the subnetworks $\boldsymbol{N}_i$, $i = 2, 5, 7, 9, 10, 12$, safety function coordinates.

$$\boldsymbol{S}^{(i)}(t,\cdot)=\left[1,\boldsymbol{S}^{(i)}(t,1),\boldsymbol{S}^{(i)}(t,2),\boldsymbol{S}^{(i)}(t,3),\boldsymbol{S}^{(i)}(t,4)\right],\quad t\geq 0,$$
$$i=2,5,7,9,10,12, \tag{5.3.85}$$

with the coordinates

$$\boldsymbol{S}^{(i)}(t,1)=\exp[-2\lambda(2)t]$$
$$+\frac{1}{2}\sum_{g=1}^{2}\left[\exp\left[-\lambda(1)\sum_{j=1}^{2}\frac{1}{q\left(1,d_{jg}^{(i)}\right)}t\right]\right.$$
$$\left.-\exp\left[-\left(2\lambda(2)-2\lambda(1)+\lambda(1)\sum_{j=1}^{2}\frac{1}{q\left(1,d_{jg}^{(i)}\right)}\right)t\right]\right], \tag{5.3.86}$$

$$\boldsymbol{S}^{(i)}(t,2)=\exp[-2\lambda(3)t]$$
$$+\frac{1}{2}\sum_{g=1}^{2}\left[\exp\left[-\lambda(2)\sum_{j=1}^{2}\frac{1}{q\left(2,d_{jg}^{(i)}\right)}t\right]\right.$$
$$\left.-\exp\left[-\left(2\lambda(3)-2\lambda(2)+\lambda(2)\sum_{j=1}^{2}\frac{1}{q\left(2,d_{jg}^{(i)}\right)}\right)t\right]\right], \tag{5.3.87}$$

$$\boldsymbol{S}^{(i)}(t,3)=\exp[-2\lambda(4)t]$$
$$+\frac{1}{2}\sum_{g=1}^{2}\left[\exp\left[-\lambda(3)\sum_{j=1}^{2}\frac{1}{q\left(3,d_{jg}^{(i)}\right)}t\right]\right.$$
$$\left.-\exp\left[-\left(2\lambda(4)-2\lambda(3)+\lambda(3)\sum_{j=1}^{2}\frac{1}{q\left(3,d_{jg}^{(i)}\right)}\right)t\right]\right], \tag{5.3.88}$$

$$\boldsymbol{S}^{(i)}(t,4)=\exp[-2\lambda(4)t]. \tag{5.3.89}$$

Further, for the intensities of assets' departure from the safety-states subset {1,2,3,4}, {2,3,4}, {3,4}, {4} given by (5.3.47) and the coefficients $q(1,d_{jg}^{(i)})$, $q(2,d_{jg}^{(i)})$, $q(3,d_{jg}^{(i)})$, $j=1,2$, $g=1,2$, $i=2,5,7,9,10,12$, given by (5.3.51)–(5.3.54), (5.3.58)–(5.3.63), (5.3.68), (5.3.69), the safety function coordinates of the subnetworks $\boldsymbol{N}_i$, $i=2,5,7,9,10,12$, given by (5.3.86)–(5.3.89), take following form (Fig. 5.22)

$$
\begin{aligned}
\boldsymbol{S}^{(i)}(t,1) = {} & \exp[-0.086t] \\
& + \frac{1}{2}\left[\exp\left[-0.027\left(1+\frac{1}{1-\exp[-1]}\right)t\right]\right. \\
& - \exp\left[-\left(0.032+0.027\left(1+\frac{1}{1-\exp[-1]}\right)\right)t\right]. \\
& \left. + \exp[-0.054t] - \exp[-0.086t]\right] \\
= {} & \exp[-0.086t] + 0.5[\exp[-0.07t] \\
& - \exp[-0.102t] + \exp[-0.054t] \\
& - \exp[-0.086t]], t \geq 0,
\end{aligned} \tag{5.3.90}
$$

$$
\begin{aligned}
\boldsymbol{S}^{(i)}(t,2) = {} & \exp[-0.2t] \\
& + \frac{1}{2}\left[\exp\left[-0.043\left(1+\frac{1}{1-\exp[-1]}\right)t\right]\right. \\
& - \exp\left[-\left(0.114+0.043\left(1+\frac{1}{1-\exp[-1]}\right)\right)t\right] \\
& \left. + \exp[-0.086t] - \exp[-0.2t]\right] \\
= {} & \exp[-0.2t] + 0.5[\exp[-0.111t] - \exp[-0.225t] \\
& + \exp[-0.086t] - \exp[-0.2t]], \quad t \geq 0,
\end{aligned} \tag{5.3.91}
$$

$$
\begin{aligned}
\boldsymbol{S}^{(i)}(t,3) = {} & \exp[-0.4t] \\
& + \frac{1}{2}\left[\exp\left[-0.1\left(1+\frac{1}{1-\exp[-1]}\right)t\right]\right. \\
& - \exp\left[-\left(0.2+0.1\left(1+\frac{1}{1-\exp[-1]}\right)\right)t\right] \\
& \left. + \exp[-0.2t] - \exp[-0.4t]\right] = \exp[-0.4t] \\
& + 0.5[\exp[-0.258t] - \exp[-0.458t] \\
& + \exp[-0.2t] - \exp[-0.4t]], \quad t \geq 0,
\end{aligned} \tag{5.3.92}
$$

$$
\boldsymbol{S}^{(i)}(t,4) = \exp[-0.4t], \quad t \geq 0. \tag{5.3.93}
$$

The safety function of subnetworks N_i, $i = 3, 4, 8, 13$, is given by the vector (Fig. 5.23)

$$S^{(i)}(t,\cdot) = \left[1, S^{(i)}(t, 1), S^{(i)}(t, 2), S^{(i)}(t, 3), S^{(i)}(t, 4)\right], \quad t \geq 0,$$
$$i = 3, 4, 8, 13, \tag{5.3.94}$$

with the coordinates

$$S^{(i)}(t, 1) = \exp[-\lambda(1)t] = \exp[-0.027t], \tag{5.3.95}$$

$$S^{(i)}(t, 2) = \exp[-\lambda(2)t] = \exp[-0.043t], \tag{5.3.96}$$

$$S^{(i)}(t, 3) = \exp[-\lambda(3)t] = \exp[-0.1t], \tag{5.3.97}$$

$$S^{(i)}(t, 4) = \exp[-\lambda(4)t] = \exp[-0.2t]. \tag{5.3.98}$$

The safety function of subnetworks N_i, $i = 6, 14, 15$, from (5.2.27) to (5.2.29), is given by the vector (Fig. 5.24)

$$S^{(i)}(t,\cdot) = \left[1, S^{(i)}(t, 1), S^{(i)}(t, 2), S^{(i)}(t, 3), S^{(i)}(t, 4)\right], \quad t \geq 0,$$
$$i = 6, 14, 15, \tag{5.3.99}$$

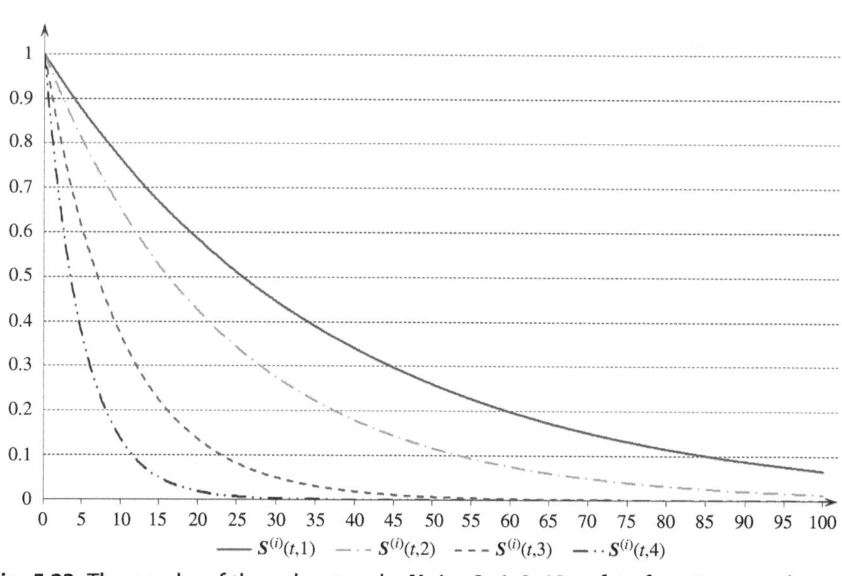

Fig. 5.23 The graphs of the subnetworks N_i, $i = 3, 4, 8, 13$, safety function coordinates.

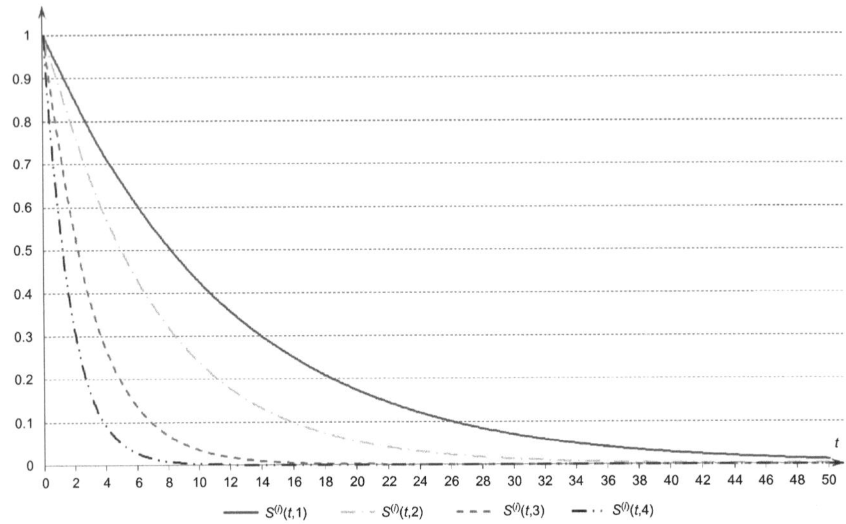

Fig. 5.24 The graphs of the subnetworks $\boldsymbol{N}_i$, $i = 6, 14, 15$, safety function coordinates.

with the coordinates

$$\boldsymbol{s}^{(i)}(t,1) = \exp[-3\lambda(2)t] + \frac{1}{3}\sum_{g=1}^{3}\left[\exp\left[-\lambda(1)\sum_{j=1}^{3}\frac{1}{q\left(1,d_{jg}^{(i)}\right)}t\right] - \exp\left[-\left(3\lambda(2)-3\lambda(1)+\lambda(1)\sum_{j=1}^{3}\frac{1}{q\left(1,d_{jg}^{(i)}\right)}\right)t\right]\right], \quad (5.3.100)$$

$$\boldsymbol{s}^{(i)}(t,2) = \exp[-3\lambda(3)t] + \frac{1}{3}\sum_{g=1}^{3}\left[\exp\left[-\lambda(2)\sum_{j=1}^{3}\frac{1}{q\left(2,d_{jg}^{(i)}\right)}t\right] - \exp\left[-\left(3\lambda(3)-3\lambda(2)+\lambda(2)\sum_{j=1}^{3}\frac{1}{q\left(2,d_{jg}^{(i)}\right)}\right)t\right]\right], \quad (5.3.101)$$

$$\boldsymbol{S}^{(i)}(t,3) = \exp[-3\lambda(4)t]$$
$$+\frac{1}{3}\sum_{g=1}^{3}\left[\exp\left[-\lambda(3)\sum_{j=1}^{3}\frac{1}{q\left(3,d_{jg}^{(i)}\right)}t\right]\right.$$
$$\left.-\exp\left[-\left(3\lambda(4)-3\lambda(3)+\lambda(3)\sum_{j=1}^{3}\frac{1}{q\left(3,d_{jg}^{(i)}\right)}\right)t\right]\right], \quad (5.3.102)$$

$$\boldsymbol{S}^{(i)}(t,4) = \exp[-3\lambda(4)t]. \quad (5.3.103)$$

Further, for the intensities of assets' departure from the safety-states subset $\{1,2,3,4\}$, $\{2,3,4\}$, $\{3,4\}$, $\{4\}$ given by (5.3.47) and the coefficients $q(1, d_{jg}^{(i)})$, $q(2, d_{jg}^{(i)})$, $q(3, d_{jg}^{(i)})$, $j = 1, 2, 3$, $g = 1, 2, 3$, $i = 6, 14, 15$, given by (5.3.55)–(5.3.57), (5.3.70)–(5.3.75), the safety function coordinates of the subnetworks $\boldsymbol{N}_i$, $i = 6, 14, 15$, given by (5.3.100)–(5.3.103), take following form

$$\boldsymbol{S}^{(i)}(t,1) = \exp[-0.129t]$$
$$+\frac{1}{3}\left[\exp\left[-0.027\left(1+\frac{1}{1-\exp[-1]}+\frac{1}{1-\exp[-2]}\right)t\right]\right.$$
$$-\exp\left[-\left(0.048+0.027\left(1+\frac{1}{1-\exp[-1]}+\frac{1}{1-\exp[-2]}\right)\right)t\right]$$
$$+\exp\left[-0.027\left(2+\frac{1}{1-\exp[-1]}\right)t\right]$$
$$-\exp\left[-\left(0.048+0.027\left(2+\frac{1}{1-\exp[-1]}\right)\right)t\right]$$
$$\left.+\exp[-0.081t]-\exp[-0.129t]\right]$$
$$= \exp[-0.129t] + \frac{1}{3}[\exp[-0.101t]$$
$$-\exp[-0.149t]+\exp[-0.097t]-\exp[-0.145t]$$
$$+\exp[-0.081t]-\exp[-0.129t]], \quad t \geq 0, \quad (5.3.104)$$

$$\begin{aligned}
\boldsymbol{S}^{(i)}(t,2) = {} & \exp[-0.3t] \\
& + \frac{1}{3}\Bigg[\exp\left[-0.043\left(1+\frac{1}{1-\exp[-1]}+\frac{1}{1-\exp[-2]}\right)t\right] \\
& - \exp\left[-\left(0.171+0.043\left(1+\frac{1}{1-\exp[-1]}+\frac{1}{1-\exp[-2]}\right)\right)t\right] \\
& + \exp\left[-0.043\left(2+\frac{1}{1-\exp[-1]}\right)t\right] \\
& - \exp\left[-\left(0.171+0.043\left(2+\frac{1}{1-\exp[-1]}\right)\right)t\right] \\
& + \exp[-0.129t] - \exp[-0.3t]\Bigg] \\
= {} & \exp[-0.3t] + \frac{1}{3}[\exp[-0.161t] - \exp[-0.332t] \\
& + \exp[-0.154t] - \exp[-0.325t] \\
& + \exp[-0.129t] - \exp[-0.3t]], \quad t \geq 0, \qquad (5.3.105)
\end{aligned}$$

$$\begin{aligned}
\boldsymbol{S}^{(i)}(t,3) = {} & \exp[-0.6t] \\
& + \frac{1}{3}\Bigg[\exp\left[-0.1\left(1+\frac{1}{1-\exp[-1]}+\frac{1}{1-\exp[-2]}\right)t\right] \\
& - \exp\left[-\left(0.3+0.1\left(1+\frac{1}{1-\exp[-1]}+\frac{1}{1-\exp[-2]}\right)\right)t\right] \\
& + \exp\left[-0.1\left(2+\frac{1}{1-\exp[-1]}\right)t\right] \\
& - \exp\left[-\left(0.3+0.1\left(2+\frac{1}{1-\exp[-1]}\right)\right)t\right] \\
& + \exp[-0.3t] - \exp[-0.6t]\Bigg] \\
= {} & \exp[-0.6t] + \frac{1}{3}[\exp[-0.374t] \\
& - \exp[-0.674t] + \exp[-0.358t] - \exp[-0.658t] \\
& + \exp[-0.3t] - \exp[-0.6t]], \quad t \geq 0, \qquad (5.3.106)
\end{aligned}$$

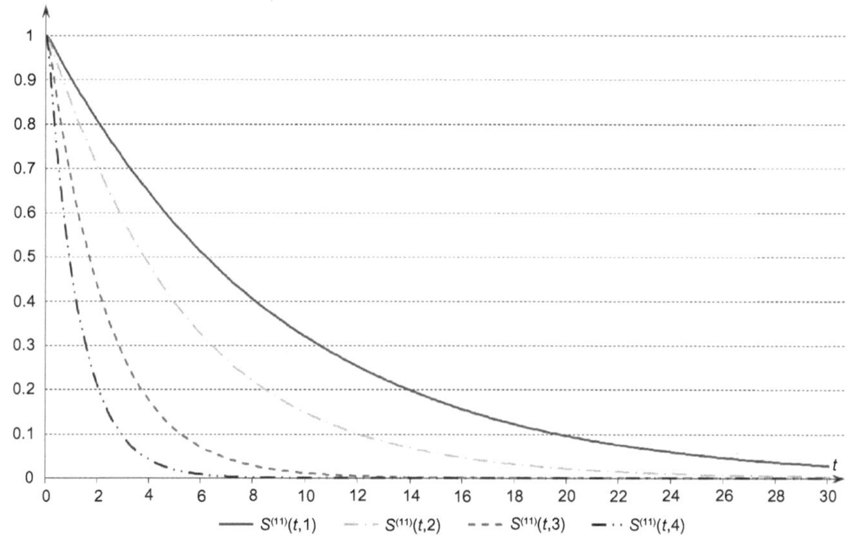

Fig. 5.25 The graphs of the subnetwork $\boldsymbol{N}_{11}$ safety function coordinates.

$$\boldsymbol{s}^{(i)}(t,4) = \exp[-0.6t], \quad t \geq 0. \tag{5.3.107}$$

The safety function of the subnetwork $\boldsymbol{N}_{11}$, from (5.2.27) to (5.2.29), is given by the vector (Fig. 5.25)

$$\boldsymbol{s}^{(11)}(t,\cdot) = \left[1, \boldsymbol{s}^{(11)}(t,1), \boldsymbol{s}^{(11)}(t,2), \boldsymbol{s}^{(11)}(t,3), \boldsymbol{s}^{(11)}(t,4)\right], \quad t \geq 0, \tag{5.3.108}$$

with the coordinates

$$\boldsymbol{s}^{(11)}(t,1) = \exp[-4\lambda(2)t] + \frac{1}{4}\sum_{g=1}^{4}\left[\exp\left[-\lambda(1)\sum_{j=1}^{4}\frac{1}{q\left(1, d_{jg}^{(11)}\right)}t\right] - \exp\left[-\left(4\lambda(2) - 4\lambda(1) + \lambda(1)\sum_{j=1}^{4}\frac{1}{q\left(1, d_{jg}^{(11)}\right)}\right)t\right]\right], \tag{5.3.109}$$

$$\boldsymbol{s}^{(11)}(t,2)=\exp[-4\lambda(3)t]$$
$$+\frac{1}{4}\sum_{g=1}^{4}\left[\exp\left[-\lambda(2)\sum_{j=1}^{4}\frac{1}{q\left(2,d_{jg}^{(11)}\right)}t\right]\right.$$
$$\left.-\exp\left[-\left(4\lambda(3)-4\lambda(2)+\lambda(2)\sum_{j=1}^{4}\frac{1}{q\left(2,d_{jg}^{(11)}\right)}\right)t\right]\right], \tag{5.3.110}$$

$$\boldsymbol{s}^{(11)}(t,3)=\exp[-4\lambda(4)t]$$
$$+\frac{1}{4}\sum_{g=1}^{4}\left[\exp\left[-\lambda(3)\sum_{j=1}^{4}\frac{1}{q\left(3,d_{jg}^{(11)}\right)}t\right]\right.$$
$$\left.-\exp\left[-\left(4\lambda(4)-4\lambda(3)+\lambda(3)\sum_{j=1}^{4}\frac{1}{q\left(3,d_{jg}^{(11)}\right)}\right)t\right]\right], \tag{5.3.111}$$

$$\boldsymbol{s}^{(11)}(t,4)=\exp[-4\lambda(4)t]. \tag{5.3.112}$$

Further, for the intensities of assets' departure from the safety-states subset {1,2,3,4}, {2,3,4}, {3,4}, {4} given by (5.3.47) and the coefficients $q(1,d_{jg}^{(11)})$, $q(2,d_{jg}^{(11)})$, $q(3,d_{jg}^{(11)})$, $j=1, 2, 3, 4$, $g=1, 2, 3, 4$, given by (5.3.64)–(5.3.67), the safety function coordinates of the subnetwork $\boldsymbol{N}_{11}$, given by (5.3.109)–(5.3.112), take following form

$$\boldsymbol{s}^{(11)}(t,1)=\exp[-0.172t]$$
$$+\frac{1}{4}\left[\exp\left[-0.027\left(1+\frac{1}{1-\exp[-1]}+\frac{1}{1-\exp[-2]}+\frac{1}{1-\exp[-3]}\right)t\right]\right.$$
$$-\exp\left[-\left(0.064+0.027\left(1+\frac{1}{1-\exp[-1]}+\frac{1}{1-\exp[-2]}+\frac{1}{1-\exp[-3]}\right)\right)t\right]$$
$$+\exp\left[-0.027\left(2+\frac{1}{1-\exp[-1]}+\frac{1}{1-\exp[-2]}\right)t\right]$$
$$-\exp\left[-\left(0.064+0.027\left(2+\frac{1}{1-\exp[-1]}+\frac{1}{1-\exp[-2]}\right)\right)t\right]$$
$$+\exp\left[-0.027\left(3+\frac{1}{1-\exp[-1]}\right)t\right]-\exp\left[-\left(0.064+0.027\left(3+\frac{1}{1-\exp[-1]}\right)\right)t\right]$$
$$\left.+\exp[-0.108t]-\exp[-0.172t]\right]=\exp[-0.172t]+\frac{1}{4}[\exp[-0.129t]$$
$$-\exp[-0.193t]+\exp[-0.128t]-\exp[-0.192t]+\exp[-0.124t]$$
$$-\exp[-0.188t]+\exp[-0.108t]-\exp[-0.172t]],\quad t\geq 0, \tag{5.3.113}$$

$$
\begin{aligned}
\boldsymbol{S}^{(11)}(t,2) &= \exp[-0.4t] \\
&+\frac{1}{4}\Bigg[\exp\left[-0.043\left(1+\frac{1}{1-\exp[-1]}+\frac{1}{1-\exp[-2]}+\frac{1}{1-\exp[-3]}\right)t\right] \\
&-\exp\left[-\left(0.228+0.043\left(1+\frac{1}{1-\exp[-1]}+\frac{1}{1-\exp[-2]}+\frac{1}{1-\exp[-3]}\right)\right)t\right] \\
&+\exp\left[-0.043\left(2+\frac{1}{1-\exp[-1]}+\frac{1}{1-\exp[-2]}\right)t\right] \\
&-\exp\left[-\left(0.228+0.043\left(2+\frac{1}{1-\exp[-1]}+\frac{1}{1-\exp[-2]}\right)\right)t\right] \\
&+\exp\left[-0.043\left(3+\frac{1}{1-\exp[-1]}\right)t\right]-\exp\left[-\left(0.228+0.043\left(3+\frac{1}{1-\exp[-1]}\right)\right)t\right] \\
&+\exp[-0.172t]-\exp[-0.4t]\Bigg] = \exp[-0.4t]+\frac{1}{4}[\exp[-0.206t] \\
&-\exp[-0.434t]+\exp[-0.204t]-\exp[-0.432t] \\
&+\exp[-0.197t]-\exp[-0.425t]+\exp[-0.172t]-\exp[-0.4t]], \quad t\geq 0, \qquad (5.3.114)
\end{aligned}
$$

$$
\begin{aligned}
\boldsymbol{S}^{(11)}(t,3) &= \exp[-0.8t] \\
&+\frac{1}{4}\Bigg[\exp\left[-0.1\left(1+\frac{1}{1-\exp[-1]}+\frac{1}{1-\exp[-2]}+\frac{1}{1-\exp[-3]}\right)t\right] \\
&-\exp\left[-\left(0.4+0.1\left(1+\frac{1}{1-\exp[-1]}+\frac{1}{1-\exp[-2]}+\frac{1}{1-\exp[-3]}\right)\right)t\right] \\
&+\exp\left[-0.1\left(2+\frac{1}{1-\exp[-1]}+\frac{1}{1-\exp[-2]}\right)t\right] \\
&-\exp\left[-\left(0.4+0.1\left(2+\frac{1}{1-\exp[-1]}+\frac{1}{1-\exp[-2]}\right)\right)t\right] \\
&+\exp\left[-0.1\left(3+\frac{1}{1-\exp[-1]}\right)t\right]-\exp\left[-\left(0.4+0.1\left(3+\frac{1}{1-\exp[-1]}\right)\right)t\right] \\
&+\exp[-0.4t]-\exp[-0.8t]\Bigg] = \exp[-0.8t]+\frac{1}{4}[\exp[-0.479t] \\
&-\exp[-0.879t]+\exp[-0.474t]-\exp[-0.874t]+\exp[-0.458t] \\
&-\exp[-0.858t]+\exp[-0.4t]-\exp[-0.8t]], \quad t\geq 0, \qquad (5.3.115)
\end{aligned}
$$

$$
\boldsymbol{S}^{(11)}(t,4) = \exp[-0.8t], \quad t\geq 0. \qquad (5.3.116)
$$

On the basis of obtained results, we estimate the expected values and standard deviations of lifetimes in the safety-state subsets {1,2,3,4}, {2,3,4}, {3,4}, {4}, for all subnetworks $\boldsymbol{N}_i$, $i = 1, 2, \ldots, 15$. They counted in hours are given in Table 5.4.

Table 5.4 The expected values and standard deviations of lifetimes in the subsets {1,2,3,4}, {2,3,4}, {3,4}, {4}, for subnetworks $\boldsymbol{N}_i$, $i = 1, 2, \ldots, 15$ (Blokus-Roszkowska and Kolowrocki, 2017b).

Subnetwork $\boldsymbol{N}_i$	$\mu^{(i)}(1)$	$\sigma^{(i)}(1)$	$\mu^{(i)}(2)$	$\sigma^{(i)}(2)$	$\mu^{(i)}(3)$	$\sigma^{(i)}(3)$	$\mu^{(i)}(4)$	$\sigma^{(i)}(4)$
$\boldsymbol{N}_1$	2.680	2.601	1.645	1.607	0.713	0.693	0.385	0.385
$\boldsymbol{N}_2$	17.314	16.919	10.596	10.456	4.596	4.514	2.500	2.500
$\boldsymbol{N}_3$	37.037	37.037	23.256	23.256	10.000	10.000	5.000	5.000
$\boldsymbol{N}_4$	37.037	37.037	23.256	23.256	10.000	10.000	5.000	5.000
$\boldsymbol{N}_5$	17.314	16.919	10.596	10.456	4.596	4.514	2.500	2.500
$\boldsymbol{N}_6$	11.489	11.144	7.012	6.869	3.044	2.968	1.667	1.667
$\boldsymbol{N}_7$	17.314	16.919	10.596	10.456	4.596	4.514	2.500	2.500
$\boldsymbol{N}_8$	37.037	37.037	23.256	23.256	10.000	10.000	5.000	5.000
$\boldsymbol{N}_9$	17.314	16.919	10.596	10.456	4.596	4.514	2.500	2.500
$\boldsymbol{N}_{10}$	17.314	16.919	10.596	10.456	4.596	4.514	2.500	2.500
$\boldsymbol{N}_{11}$	8.655	8.402	5.294	5.180	2.296	2.237	1.250	1.250
$\boldsymbol{N}_{12}$	17.314	16.919	10.596	10.456	4.596	4.514	2.500	2.500
$\boldsymbol{N}_{13}$	37.037	37.037	23.256	23.256	10.000	10.000	5.000	5.000
$\boldsymbol{N}_{14}$	11.489	11.144	7.012	6.869	3.044	2.968	1.667	1.667
$\boldsymbol{N}_{15}$	11.489	11.144	7.012	6.869	3.044	2.968	1.667	1.667

To analyze cascading effects in considered distribution network, we take into account also interactions among subnetworks. We assume the LLS dependency model among subnetworks $\boldsymbol{N}_i$, $i = 1, 2, \ldots, 15$, forming a network $\boldsymbol{N}$ with the series safety structure. That means that after departure of one of subnetworks $\boldsymbol{N}_j$, $j = 1, 2, \ldots, 15$, the lifetimes in safety-state subsets and their mean values of other subnetworks $\boldsymbol{N}_i$, $i = 1, 2, \ldots, 15$, $i \neq j$, decrease according to formula (5.2.11). The coefficients $q(v, \boldsymbol{N}_j, \boldsymbol{N}_i)$, existing in (5.2.11), express the influence of state change of subnetwork $\boldsymbol{N}_j$, $j = 1, 2, \ldots, 15$, on lifetimes in the safety-state subset $\{v, v+1, \ldots, 4\}$, $v = u, u-1, \ldots, 1$, $u = 1, 2, 3$, of other subnetworks $\boldsymbol{N}_i$, $i = 1, 2, \ldots, 15$, $i \neq j$. These coefficients of network load increase are as follows:

- the coefficients of load increase of subnetworks $\boldsymbol{N}_i$, $i = 2, \ldots, 15$, after the departure of subnetwork $\boldsymbol{N}_1$ from the safety-state subset $\{u, u+1, \ldots, 4\}$, $u = 1, 2, 3, 4$,

$$q(v, \boldsymbol{N}_1, \boldsymbol{N}_i) = 1 - \exp[-1], \quad i = 2, 3, 6, 9, 12, 15,$$
$$v = u, u-1, \ldots, 1, \quad u = 1, 2, 3, \tag{5.3.117}$$

$$q(v, \boldsymbol{N}_1, \boldsymbol{N}_i) = 1 - \exp[-2], \quad i = 4,5,7,8,10,11,13,14,$$
$$v = u, u-1, \ldots, 1, \quad u = 1,2,3, \qquad (5.3.118)$$

– the coefficients of load increase of subnetworks $\boldsymbol{N}_i$, $i = 4, 5$, after the departure of subnetwork $\boldsymbol{N}_3$ from the safety-state subset $\{u, u+1, \ldots, 4\}$, $u = 1, 2, 3, 4$,

$$q(v, \boldsymbol{N}_3, \boldsymbol{N}_i) = 1 - \exp[-1], \quad i = 4,5,$$
$$v = u, \quad u-1, \ldots, 1, \quad u = 1,2,3, \qquad (5.3.119)$$

– the coefficients of load increase of subnetworks $\boldsymbol{N}_i$, $i = 7, 8$, after the departure of subnetwork $\boldsymbol{N}_6$ from the safety-state subset $\{u, u+1, \ldots, 4\}$, $u = 1, 2, 3, 4$,

$$q(v, \boldsymbol{N}_6, \boldsymbol{N}_i) = 1 - \exp[-1], \quad i = 7,8,$$
$$v = u, u-1, \ldots, 1, \quad u = 1,2,3, \qquad (5.3.120)$$

– the coefficients of load increase of subnetworks $\boldsymbol{N}_i$, $i = 10, 11$, after the departure of subnetwork $\boldsymbol{N}_9$ from the safety-state subset $\{u, u+1, \ldots, 4\}$, $u = 1, 2, 3, 4$,

$$q(v, \boldsymbol{N}_9, \boldsymbol{N}_i) = 1 - \exp[-1], \quad i = 10,11,$$
$$v = u, u-1, \ldots, 1, \quad u = 1,2,3, \qquad (5.3.121)$$

– the coefficients of load increase of subnetworks $\boldsymbol{N}_i$, $i = 13, 14$, after the departure of subnetwork $\boldsymbol{N}_{12}$ from the safety-state subset $\{u, u+1, \ldots, 4\}$, $u = 1, 2, 3, 4$,

$$q(v, \boldsymbol{N}_{12}, \boldsymbol{N}_i) = 1 - \exp[-1], \quad i = 13,14,$$
$$v = u, u-1, \ldots, 1, \quad u = 1,2,3, \qquad (5.3.122)$$

– and other coefficients of load increase of the subnetworks $\boldsymbol{N}_i$, $i = 1, 2, \ldots, 15$, are equal to 1.

Next, we can predict the safety characteristics of exemplary distribution network $\boldsymbol{N}$. Taking into account interactions among subnetworks $\boldsymbol{N}_i$, $i = 1, 2, \ldots, 15$, and assuming the LLS dependency model, applying (5.2.12)–(5.2.14), we determine the safety function of the network $\boldsymbol{N}$

$$\boldsymbol{S}_{LLS}(t, \cdot) = [1, \boldsymbol{S}_{LLS}(t,1), \boldsymbol{S}_{LLS}(t,2), \boldsymbol{S}_{LLS}(t,3), \boldsymbol{S}_{LLS}(t,4)], \quad t \geq 0, \qquad (5.3.123)$$

with the coordinates

$$\boldsymbol{S}_{LLS}(t,u)=\prod_{i=1}^{15}\boldsymbol{S}^{(i)}(t,u+1)$$
$$+\int_0^t\sum_{j=1}^{15}\left[\tilde{f}^{(j)}(a,u+1)\cdot\prod_{\substack{i=1\\i\neq j}}^{15}\boldsymbol{S}^{(i)}(a,u+1)\cdot\boldsymbol{S}^{(j)}(a,u)\cdot\prod_{i=1}^{15}\boldsymbol{S}^{(i/j)}(t-a,u)\right]da,$$
$$u=1,2,3, \tag{5.3.124}$$
$$\boldsymbol{S}_{LLS}(t,4)=\prod_{i=1}^{15}\boldsymbol{S}^{(i)}(t,4), \tag{5.3.125}$$

where $\boldsymbol{S}^{(i)}(t,u)$ are the safety function coordinates of the subnetwork $\boldsymbol{N}_i$, $i=1, 2, \ldots, 15$, for $u=1, 2, 3, 4$, given by (5.3.81)–(5.3.84), (5.3.90)–(5.3.93), (5.3.95)–(5.3.98), (5.3.104)–(5.3.107), and (5.3.113)–(5.3.116).

Further, in order to simplify the calculations, we approximate safety functions of subnetworks $\boldsymbol{N}_i$, $i=1, 2, \ldots, 15$, with piecewise exponential safety functions

$$\boldsymbol{S}^{(i)}(t,\cdot)=\left[1,\boldsymbol{S}^{(i)}(t,1),\boldsymbol{S}^{(i)}(t,2),\boldsymbol{S}^{(i)}(t,3),\boldsymbol{S}^{(i)}(t,4)\right],$$
$$t\geq 0,\quad i=1,2,\ldots,15, \tag{5.3.126}$$

with the coordinates

$$\boldsymbol{S}^{(i)}(t,u)=\exp[-\boldsymbol{\lambda}_i(u)t],\quad u=1,2,3,4,\quad t\geq 0. \tag{5.3.127}$$

The intensities of departure from safety-state subsets of subnetworks $\boldsymbol{N}_i$, $i=1, 2, \ldots, 15$, existing in (5.3.127), are estimated as follows:

$$\boldsymbol{\lambda}_1(1)=0.373,\quad \boldsymbol{\lambda}_1(2)=0.608, \boldsymbol{\lambda}_1(3)=1.403,\quad \boldsymbol{\lambda}_1(4)=2.6, \tag{5.3.128}$$

$$\boldsymbol{\lambda}_i(1)=0.058,\quad \boldsymbol{\lambda}_i(2)=0.094,\quad \boldsymbol{\lambda}_i(3)=0.218,\quad \boldsymbol{\lambda}_i(4)=0.4, i=2,5,7,9,10,12, \tag{5.3.129}$$

$$\boldsymbol{\lambda}_i(1)=0.027,\quad \boldsymbol{\lambda}_i(2)=0.043,\quad \boldsymbol{\lambda}_i(3)=0.1,\quad \boldsymbol{\lambda}_i(4)=0.2,\quad i=3,4,8,13, \tag{5.3.130}$$

$$\lambda_i(1)=0.087,\ \ \lambda_i(2)=0.143,\ \ \lambda_i(3)=0.329,\ \ \lambda_i(4)=0.6,\ \ i=6,14,15,\tag{5.3.131}$$

$$\lambda_{11}(1)=0.116,\ \ \lambda_{11}(2)=0.189,\ \ \lambda_{11}(3)=0.436,\ \ \lambda_{11}(4)=0.8.\tag{5.3.132}$$

In case the subnetworks $\boldsymbol{N}_i$, $i = 1, 2, \ldots, 15$, have piecewise exponential safety functions (5.3.126), (5.3.127), the conditional intensities $\lambda_{i/j}(v)$ of the subnetwork $\boldsymbol{N}_i$, $i = 1, 2, \ldots, 15$, departure from the subset $\{v, v + 1, \ldots, 4\}$, $v = u, u - 1, \ldots, 1$, $u = 1, 2, 3$, after the departure of subnetwork $\boldsymbol{N}_j$, $j = 1, 2, \ldots, 15$, from (5.2.17), are as follows

$$\lambda_{i/j}(v)=\frac{\lambda_i(v)}{q(v,\boldsymbol{N}_j,\boldsymbol{N}_i)},\ \ i=1,2,\ldots,15,\ \ j=1,2,\ldots,15,\tag{5.3.133}$$

where the intensities $\lambda_i(v)$ of $\boldsymbol{N}_i$, $i = 1, 2, \ldots, 15$, are given by (5.3.128)–(5.3.132) and the coefficients $q(v, \boldsymbol{N}_j, \boldsymbol{N}_i)$ are determined by (5.3.117)–(5.3.122).

Next, applying (5.2.23)–(5.2.25), we determine the safety function of the network $\boldsymbol{N}$

$$\boldsymbol{S}_{LLS}(t,\ \cdot\ )=[1,\boldsymbol{S}_{LLS}(t,1),\boldsymbol{S}_{LLS}(t,2),\boldsymbol{S}_{LLS}(t,3),\boldsymbol{S}_{LLS}(t,4)],\quad t\geq 0,\tag{5.3.134}$$

where

$$\begin{aligned}\boldsymbol{S}_{LLS}(t,1) &= \exp\left[-\sum_{i=1}^{15}\lambda_i(2)t\right]\\ &+\sum_{j=1}^{15}\frac{\lambda_j(2)-\lambda_j(1)}{\sum_{i=1}^{15}\lambda_i(2)-\sum_{i=1}^{15}\lambda_i(1)}\cdot\left[\exp\left[-\sum_{i=1}^{15}\frac{\lambda_i(1)}{q(1,\boldsymbol{N}_j,\boldsymbol{N}_i)}t\right]\right.\\ &\left.-\exp\left[-\left(\sum_{i=1}^{15}\lambda_i(2)-\sum_{i=1}^{15}\lambda_i(1)+\sum_{i=1}^{15}\frac{\lambda_i(1)}{q(1,\boldsymbol{N}_j,\boldsymbol{N}_i)}\right)t\right]\right],\end{aligned}\tag{5.3.135}$$

$$\boldsymbol{S}_{LLS}(t,2)=\exp\left[-\sum_{i=1}^{15}\boldsymbol{\lambda}_i(3)t\right]$$

$$+\sum_{j=1}^{15}\frac{\boldsymbol{\lambda}_j(3)-\boldsymbol{\lambda}_j(2)}{\sum_{i=1}^{15}\boldsymbol{\lambda}_i(3)-\sum_{i=1}^{15}\boldsymbol{\lambda}_i(2)}\cdot\left[\exp\left[-\sum_{i=1}^{15}\frac{\boldsymbol{\lambda}_i(2)}{q(2,\boldsymbol{N}_j,\boldsymbol{N}_i)}t\right]\right.$$

$$\left.-\exp\left[-\left(\sum_{i=1}^{15}\boldsymbol{\lambda}_i(3)-\sum_{i=1}^{15}\boldsymbol{\lambda}_i(2)+\sum_{i=1}^{15}\frac{\boldsymbol{\lambda}_i(2)}{q(2,\boldsymbol{N}_j,\boldsymbol{N}_i)}\right)t\right]\right], \quad (5.3.136)$$

$$\boldsymbol{S}_{LLS}(t,3)=\exp\left[-\sum_{i=1}^{15}\boldsymbol{\lambda}_i(4)t\right]$$

$$+\sum_{j=1}^{15}\frac{\boldsymbol{\lambda}_j(4)-\boldsymbol{\lambda}_j(3)}{\sum_{i=1}^{15}\boldsymbol{\lambda}_i(4)-\sum_{i=1}^{15}\boldsymbol{\lambda}_i(3)}\cdot\left[\exp\left[-\sum_{i=1}^{15}\frac{\boldsymbol{\lambda}_i(3)}{q(3,\boldsymbol{N}_j,\boldsymbol{N}_i)}t\right]\right.$$

$$\left.-\exp\left[-\left(\sum_{i=1}^{15}\boldsymbol{\lambda}_i(4)-\sum_{i=1}^{15}\boldsymbol{\lambda}_i(3)+\sum_{i=1}^{15}\frac{\boldsymbol{\lambda}_i(3)}{q(3,\boldsymbol{N}_j,\boldsymbol{N}_i)}\right)t\right]\right], \quad (5.3.137)$$

$$\boldsymbol{S}_{LLS}(t,4)=\exp\left[-\sum_{i=1}^{15}\boldsymbol{\lambda}_i(4)t\right]. \quad (5.3.138)$$

Further, substituting the intensities $\boldsymbol{\lambda}_i(v)$, $v = 1, 2, 3, 4$, of $\boldsymbol{N}_i$, $i = 1, 2, \ldots, 15$, departure from safety-state subsets, given by (5.3.128)–(5.3.132), and the coefficients of network load increase $q(v, \boldsymbol{N}_j, \boldsymbol{N}_i)$, $v = 1, 2, 3, 4$, $i, j = 1, 2, \ldots, 15$, given by (5.3.117)–(5.3.122), the safety function coordinates of the network $\boldsymbol{N}$, given by (5.3.135)–(5.3.138), take following form

$$\boldsymbol{S}_{LLS}(t,1)=\exp[-1.962t]+0.311\cdot[\exp[-1.496t]-\exp[-2.252t]]$$

$$+0.499\cdot[\exp[-1.206t]-\exp[-1.962t]]+0.095\cdot[\exp[-1.255t]$$

$$-\exp[-2.011t]]+0.048\cdot[\exp[-1.307t]-\exp[-2.063t]]$$

$$+0.048\cdot[\exp[-1.272t]-\exp[-2.028t]], \quad t\geq 0, \quad (5.3.139)$$

$$S_{LLS}(t,2) = \exp[-4.534t]$$
$$+0.309 \cdot [\exp[-2.434t] - \exp[-5.006t]] + 0.5 \cdot [\exp[-1.962t] - \exp[-4.534t]]$$
$$+0.094 \cdot [\exp[-2.042t] - \exp[-4.614t]] + 0.048 \cdot [\exp[-2.127t] - \exp[-4.699t]]$$
$$+0.048 \cdot [\exp[-2.070t] - \exp[-4.642t]], \quad t \geq 0, \tag{5.3.140}$$

$$S_{LLS}(t3) = \exp[-8.4t] + 0.310 \cdot [\exp[-5.625t] - \exp[-9.491t]]$$
$$+0.5 \cdot [\exp[-4.534t] - \exp[-8.4t]] + 0.096 \cdot [\exp[-4.719t]$$
$$-\exp[-8.585t]] + 0.047 \cdot [\exp[-4.915t] - \exp[-8.781t]]$$
$$+0.047 \cdot [\exp[-4.784t] - \exp[-8.65t]], \quad t \geq 0, \tag{5.3.141}$$

$$S_{LLS}(t,4) = \exp[-8.4t], \quad t \geq 0. \tag{5.3.142}$$

The safety function coordinates of the exemplary distribution network are illustrated in Fig. 5.26.

The expected values and standard deviations of lifetimes in the safety-state subsets {1,2,3,4}, {2,3,4}, {3,4}, {4}, for the network ***N*** with dependent

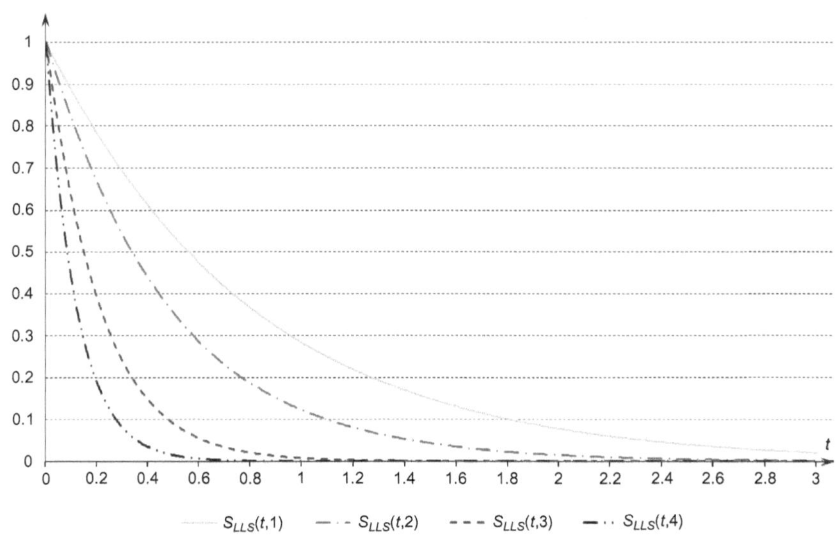

Fig. 5.26 The graphs of the network ***N*** safety function coordinates.

assets, calculated from the results (5.3.139)–(5.3.142) by (2.1.16)–(2.1.18), respectively, in years are:

$$\mu_{LLS}(1) \cong 0.795, \quad \mu_{LLS}(2) \cong 0.481,$$
$$\mu_{LLS}(3) \cong 0.210, \quad \mu_{LLS}(4) \cong 0.119, \tag{5.3.143}$$

$$\sigma_{LLS}(1) \cong 0.783, \quad \sigma_{LLS}(2) \cong 0.477,$$
$$\sigma_{LLS}(3) \cong 0.207, \quad \sigma_{LLS}(4) \cong 0.119, \tag{5.3.144}$$

and further, using (5.3.143), from (2.1.20) it follows that the mean values of the network $\boldsymbol{N}$ lifetimes in particular safety states, counted in years, are:

$$\overline{\mu}_{LLS}(1) \cong 0.314, \quad \overline{\mu}_{LLS}(2) \cong 0.271,$$
$$\overline{\mu}_{LLS}(3) \cong 0.091, \quad \overline{\mu}_{LLS}(4) \cong 0.119. \tag{5.3.145}$$

In order to compare these results for the network $\boldsymbol{N}$ with dependent subnetworks and dependent assets of these subnetworks with results for the network $\boldsymbol{N}$ assuming independence of all assets and subnetworks, the expected values of lifetimes in the safety-state subsets and in the particular safety states are also determined in second of those cases. Subsequently, the expected values of lifetimes in the subsets $\{1,2,3,4\}$, $\{2,3,4\}$, $\{3,4\}$, $\{4\}$, for the network $\boldsymbol{N}$ without dependence, respectively, in years are:

$$\mu(1) \cong 0.926, \quad \mu(2) \cong 0.581, \quad \mu(3) \cong 0.250, \quad \mu(4) \cong 0.125, \tag{5.3.146}$$

and in the particular states 1,2,3,4, respectively, take values:

$$\overline{\mu}(1) \cong 0.345, \quad \overline{\mu}(2) \cong 0.331, \quad \overline{\mu}(3) \cong 0.125, \quad \overline{\mu}(4) \cong 0.125. \tag{5.3.147}$$

As we assumed that the critical safety state is $r = 2$, then the risk function of the network $\boldsymbol{N}$ with dependent assets and subnetworks is given by

$$\begin{aligned}
\boldsymbol{r}_{LLS}(t) &= 1 - \boldsymbol{S}_{LLS}(t, 2) = 1 - \exp[-4.534t] \\
&-0.309 \cdot [\exp[-2.434t] - \exp[-5.006t]] \\
&-0.5 \cdot [\exp[-1.962t] - \exp[-4.534t]] - 0.094 \cdot [\exp[-2.042t] \\
&- \exp[-4.614t]] - 0.048 \cdot [\exp[-2.127t] - \exp[-4.699t] \\
&+ \exp[-2.070t] - \exp[-4.642t]], \quad t \geq 0.
\end{aligned} \tag{5.3.148}$$

Hence, and from (2.1.22), the moment when the risk function of the network $\boldsymbol{N}$ with dependent assets and subnetworks exceeds a permitted level, for instance $\delta = 0.2$, is

$$\tau_{LLS} = \boldsymbol{r}_{LLS}^{-1}(\delta) \cong 0.112 \text{ years} \cong 981 \text{ h}. \tag{5.3.149}$$

The moment when the risk function exceeds a permitted level for the network $\boldsymbol{N}$ without dependence is

$$\tau = \boldsymbol{r}^{-1}(\delta) \cong 0.130 \text{ years} \cong 1139 \text{ h}. \tag{5.3.150}$$

The network $\boldsymbol{N}$ risk function, assuming dependence of its subnetworks and assets according to the LLS rule, is illustrated in Fig. 5.27.

Comparing the results for network $\boldsymbol{N}$ in case of LLS dependency model with the results for network N without dependency, we conclude that the expected values of lifetimes in the safety-state subset $\{1,2,3,4\}$ are shorter by approximately 14%, in the subset $\{2,3,4\}$, by approximately 17% and in the subset $\{3,4\}$ by approximately 5%. The difference is most noticeable for the network lifetime in the state 3 and approximately 27%. The expected values of lifetimes in the state 2 are shorter by approximately 18%, in the state 1 by 9% and in the state 4 by 5%.

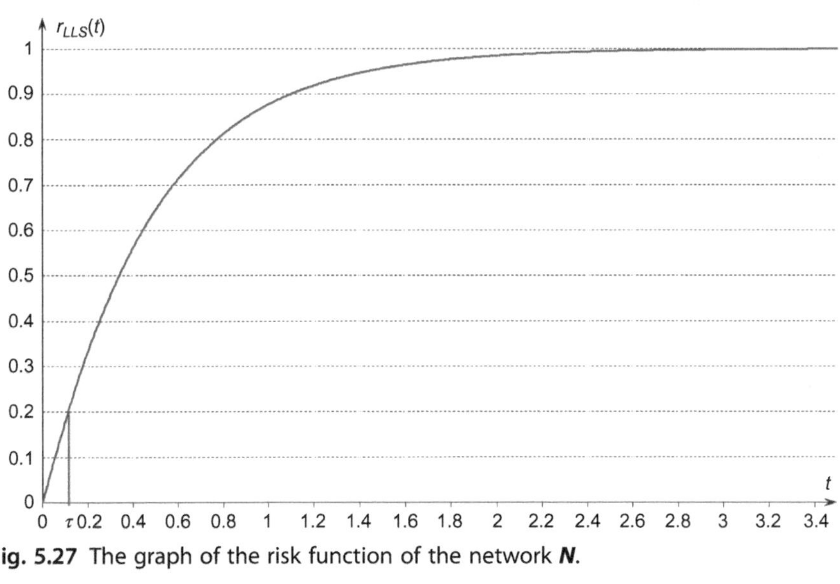

Fig. 5.27 The graph of the risk function of the network ***N***.

References

Allen, S., 1999. Infrastructural urbanism. In: Points + Lines: Diagrams and Projects for the City. Princeton Architectural Press, New York, pp. 46–59.

Almoghathawi, Y., Barker, K., Albert, L.A., 2019. Resilience-driven restoration model for interdependent infrastructure networks. Reliab. Eng. Syst. Saf. 185, 12–23.

Baldick, R., et al., 2008. Initial review of methods for cascading failure analysis in electric power transmission systems IEEE PES CAMS task force on understanding, prediction, mitigation and restoration of cascading failures. In: 2008 IEEE Power and Energy Society General Meeting—Conversion and Delivery of Electrical Energy in the 21st Century, Pittsburgh, PA, pp. 1–8.

Billinton, R., Allan, R.N., 1996. Reliability Evaluation of Power Systems, second ed. Plenum Press, New York.

Blokus-Roszkowska, A., Dziula, P., 2016. An approach to identification of critical infrastructure systems. In: Simos, T., Tsitouras, C. (Eds.), Proceedings of the International Conference of Numerical Analysis and Applied Mathematics (ICNAAM 2015). AIP Conference Proceedings, vol. 1738, AIP Publishing, pp. 440005-1–440005-4, https://doi.org/10.1063/1.4952223.

Blokus-Roszkowska, A., Kolowrocki, K., 2017a. Integrated impact model on interconnected critical infrastructures network cascading related to its operation process. J. Pol. Saf. Reliab. Assoc. Summer Saf. Reliab. Semin. 8 (4), 113–124.

Blokus-Roszkowska, A., Kolowrocki, K., 2017b. Modeling dependencies in critical infrastructure networks. In: Safety and Reliability—Theory and Applications, Proceedings and Monographs in Engineering, European Safety and Reliability Conference (ESREL 2017), pp. 655–663.

Blokus-Roszkowska, A., Kolowrocki, K., 2017c. Modeling safety of interconnected critical infrastructure network cascading. J. Pol. Saf. Reliab. Assoc. Summer Saf. Reliab. Semin. 8 (3), 99–115.

Blokus-Roszkowska, A., Kolowrocki, K., 2017d. Modelling safety of multistate systems with dependent components and subsystems. J. Pol. Saf. Reliab. Assoc. Summer Saf. Reliab. Semin. 8 (3), 23–41.

Blokus-Roszkowska, A., Guze, S., Kołowrocki, K., Soszyńska-Budny, J., Ledóchowski, M., 2016. Methodology for Port Critical Infrastructures safety and resilience to climate change analysis. J. Pol. Saf. Reliab. Assoc. Summer Saf. Reliab. Semin. 7 (2), 139–150.

Blokus-Roszkowska, A., Dziula, P., Kołowrocki, K., 2018a. Safety and resilience of Global Baltic Network of Critical Infrastructure Networks related to cascading effects. J. Pol. Saf. Reliab. Assoc. Summer Saf. Reliab. Semin. 9 (2), 79–86.

Blokus-Roszkowska, A., Guze, S., Kołowrocki, K., 2018b. Safety and resilience of Baltic Port and Shipping Critical Infrastructure Network related to cascading effects and climate-weather change. J. Pol. Saf. Reliab. Assoc. Summer Saf. Reliab. Semin. 9 (2), 87–98.

de Porcellinis, S., Oliva, G., Panzieri, S., Setola, R., 2009. A holistic-reductionistic approach for modeling interdependencies. In: Palmer, C., Shenoi, S. (Eds.), Critical Infrastructure Protection III, 3rd IFIP Int. Conf. on Critical Infrastructure Protection (CIP-2009). Springer, USA, pp. 215–227.

Dobson, I., Carreras, B.A., Newman, D.E., 2005. A loading-dependent model of probabilistic cascading failure. Probab. Eng. Inf. Sci. 19, 15–32.

European Union, European Council, 2008. Council Directive 2008/114/EC of 8 December 2008 on the Identification and Designation of European Critical Infrastructures and the Assessment of the Need to Improve Their Protection. Brussels.

Eusgeld, I., Nan, C., Dietz, S., 2011. System-of systems approach for interdependent critical infrastructures. Reliab. Eng. Syst. Saf. 96, 679–686.

Guze, S., Kołowrocki, K., 2016. Modelling operation process of Baltic Port, Shipping and Ship Traffic and Operation Information Critical Infrastructure Network. TransNav 10 (2), 275–284.

Guze, S., Kołowrocki, K., Soszyńska, J., 2008. Modeling environment and infrastructure influence on reliability and operation processes of port transportation systems. J. Konbin 4 (1), 25–38.

Han, F., Zio, E., 2019. A multi-perspective framework of analysis of critical infrastructures with respect to supply service, controllability and topology. Int. J. Crit. Infrastruct. Prot. 24, 1–13.

Hernandez-Fajardo, I., Dueñas-Osorio, L., 2013. Probabilistic study of cascading failures in complex interdependent lifeline systems. Reliab. Eng. Syst. Saf. 111, 260–272.

Holden, R., Val, D.V., Burkhard, R., Nodwell, S., 2013. A network flow model for interdependent infrastructures at the local scale. Saf. Sci. 53 (3), 51–60.

Huang, C., Liou, J., Chuang, Y., 2014. A method for exploring the interdependencies and importance of critical infrastructures. Knowl.-Based Syst. 55, 66–74.

Johansson, J., Hassel, H., 2010. An approach for modeling interdependent infrastructures in the context of vulnerability analysis. Reliab. Eng. Syst. Saf. 95 (12), 1335–1344.

Kjølle, G.H., Utne, I.B., Gjerde, O., 2012. Risk analysis of critical infrastructures emphasizing electricity supply and interdependencies. Reliab. Eng. Syst. Saf. 105, 80–89.

Kołowrocki, K., 2001. On limit reliability functions of large multi-state systems with ageing components. Appl. Math Comput. 121, 313–361.

Kołowrocki, K., 2008. Reliability and risk analysis of large systems with ageing components. Reliab. Eng. Syst. Saf. 93, 1821–1829.

Kołowrocki, K., 2013. Safety of critical infrastructures. J. Pol. Saf. Reliab. Assoc. Summer Saf. Reliab. Semin. 4 (1), 51–72.

Kołowrocki, K., 2014. Reliability of Large and Complex Systems, second ed. Elsevier, London.

Kołowrocki, K., Soszyńska, J., 2006. Reliability and availability of complex systems. Qual. Reliab. Eng. Int. 22 (1), 79–99.

Kołowrocki, K., Soszyńska-Budny, J., 2011. Reliability and Safety of Complex Technical Systems and Processes: Modeling—Identification—Prediction—Optimization, first ed. Springer-Verlag, London.

Kołowrocki, K., Soszyńska-Budny, J., 2012. Preliminary approach to safety analysis of critical infrastructures. J. Pol. Saf. Reliab. Assoc. Summer Saf. Reliab. Semin. 3 (1), 73–88.

Kołowrocki, K., Soszyńska-Budny, J., 2013. On safety of critical infrastructure modeling with application to port oil transportation system. J. Pol. Saf. Reliab. Assoc. Summer Saf. Reliab. Semin. 4 (2), 189–204.

Kołowrocki, K., Soszyńska-Budny, J., 2017. Integrated impact model on critical infrastructure safety related to its operation process. J. Pol. Saf. Reliab. Assoc. Summer Saf. Reliab. Semin. 8 (4), 1–10.

Kołowrocki, K., Kuligowska, E., Soszyńska-Budny, J., Torbicki, M., 2017a. An approach to safety prediction of critical infrastructure impacted by climate-weather change process. In: The International Conference on Information and Digital Technologies 2017, Žilina, Slovakia, pp. 183–186.

Kołowrocki, K., Kuligowska, E., Soszyńska-Budny, J., Torbicki, M., 2017b. Safety and risk prediction of Baltic Oil Terminal Critical Infrastructure impacted by climate-weather change process. In: The International Conference on Information and Digital Technologies 2017, Žilina, Slovakia, pp. 178–182.

Kołowrocki, K., Soszyńska-Budny, J., Torbicki, M., 2017c. Integrated impact model on critical infrastructure safety related to climate-weather change process including extreme weather hazards. J. Pol. Saf. Reliab. Assoc. Summer Saf. Reliab. Semin. 8 (4), 21–32.

Kołowrocki, K., Soszyńska-Budny, J., Torbicki, M., 2017d. Integrated impact model on critical infrastructure safety related to its operation process and climate-weather change process. J. Pol. Saf. Reliab. Assoc. Summer Saf. Reliab. Semin. 8 (4), 33–48.

Kołowrocki, K., Soszyńska-Budny, J., Torbicki, M., 2017e. Critical infrastructure integrated safety model related to climate-weather change process application to port oil piping transportation system operating at land Baltic seaside area. In: Safety and Reliability, Theory and Applications. Proceedings of European Safety and Reliability Conference—ESREL 2017. Taylor & Francis Ltd./CRC Press/Balkema, The Netherlands, pp. 687–694.

Kotzanikolaou, P., Theoharidou, M., Gritzalis, D., 2013. Assessing n-order dependencies between critical infrastructures. Int. J. Crit. Infrastruct. Prot. 9 (1–2), 93–110.

Lauge, A., Hernantes, J., Sarriegi, J.M., 2015. Critical infrastructure dependencies: a holistic, dynamic and quantitative approach. Int. J. Crit. Infrastruct. Prot. 8, 16–23.

Nieuwenhuijs, A., Luiijf, E., Klaver, M., 2008. Modeling dependencies in critical infrastructures. In: Papa, M., Shenoi, S. (Eds.), Critical Infrastructure Protection II, IFIP Series 253. Springer, Boston, MA, pp. 205–213.

Ouyang, M., 2014. Review on modelling and simulation of interdependent critical infrastructure systems. Reliab. Eng. Syst. Saf. 121, 43–60.

President's Commission on Critical Infrastructure Protection, 1997. The Report of the President's Commission on Critical Infrastructure Protection. Critical Foundation: Protecting America's Infrastructure, USA.

Rehak, D., Senovsky, P., Hromada, M., Lovecek, T., Novotny, P., 2018. Cascading impact assessment in a critical infrastructure system. Int. J. Crit. Infrastruct. Prot. 22, 125–138.

Rinaldi, S., Peerenboom, J., Kelly, T., 2001. Identifying, understanding and analyzing critical infrastructure interdependencies. IEEE Control. Syst. Mag. 21 (6), 11–25.

Saidi, S., Kattan, L., Jayasinghe, P., Hettiaratchi, P., Taron, J., 2018. Integrated infrastructure systems—a review. Sustain. Cities Soc. 36, 1–11.

Svedsen, N., Wolthunsen, S., 2007. Connectivity models of interdependency in mixed-type critical infrastructure networks. Inf. Secur. Tech. Rep. (1), 44–55.

Theoharidou, M., Kotzanikolaou, P., Gritzalis, D., 2011. Risk assessment methodology for interdependent critical infrastructures. Int. J. Risk Assess. Manage. 15 (2/3), 128–148.

Torbicki, M., 2018. Safety of a Critical Network Infrastructure Exposed to Operation and Weather Condition Changes (Ph.D. Thesis). Gdynia Maritime University—Polish Academy of Sciences, System Research Institute, Warsaw.

Utne, I.B., Hokstad, P., Vatn, J., 2011. A method for risk modeling of interdependencies in critical infrastructures. Reliab. Eng. Syst. Saf. 96 (6), 671–678 (ESREL 2009 Special Issue).

Xue, J., 1985. On multi-state system analysis. IEEE Trans. Reliab. 34, 329–337.

Xue, J., Yang, K., 1995a. Dynamic reliability analysis of coherent multi-state systems. IEEE Trans. Reliab. 4 (44), 683–688.

Xue, J., Yang, K., 1995b. Symmetric relations in multi-state systems. IEEE Trans. Reliab. 44 (4), 689–693.

Yu, K., Koren, I., Guo, Y., 1994. Generalised multistate monotone coherent systems. IEEE Trans. Reliab. 43, 242–250.

Zhang, P., Peeta, S., 2011. A generalized modeling framework to analyze interdependencies among infrastructure systems. Transp. Res. B: Methodol. 45 (3), 553–579.

Zio, E., 2016. Challenges in the vulnerability and risk analysis of critical infrastructures. Reliab. Eng. Syst. Saf. 152, 137–150.

Zio, E., Sansavini, G., 2011. Modeling interdependent network systems for identifying cascade-safe operating margins. Trans. Reliab. 60 (1), 94–101.

CHAPTER SIX

Summary

In the book, the approach to reliability analysis of aging multistate systems that considers the subsystems and their components' dependency is presented. The problem of dependence and cascading effect in complex systems and infrastructure networks, presented in literature mainly as a problem of failure dependency among components, is analyzed and developed to multistate aging systems. In the reliability analysis of multistate aging systems, it is assumed that the degradation of reliability state of one or a group of components may cause a degradation of the condition and reliability parameters of other components, and consequently affect the functioning and lifetime of the entire system. From a practical point of view, this assumption is very important as, in practice, often not only the system lifetime depends on its possible shortening caused by changes in the reliability states of its components, but also its lifetime depends on the dependences among its components' aging and changing their reliability state subsets. The proposed approach to degradation of a multistate system with dependent components, also allows us to determine, practically important, the moment of exceeding an acceptable level of system reliability called the permitted risk level, that can be very useful reliability indicator for the system user.

Combining the results of reliability analysis of multistate systems with dependent components and the results of classical renewal theory, the availability analysis of multistate systems under the assumption of imperfect renovation is also performed and the results are included in the book.

Safety of multistate networks with cascading effects, discussed in Chapter 5 of the book, is a very broad topic with a multidimensional problem and many issues still need to be analyzed and resolved. The analysis of interdependencies among infrastructures and dependencies among assets inside infrastructures is very complicated not only because of the variety of processes occurring between infrastructures or their components. This is also due to the problem of determining in real complex networks where and how these relationships occur, what their type is, whether they depend on other external factors, and a number of other practically important

Multistate System Reliability with Dependencies
https://doi.org/10.1016/B978-0-12-821260-8.00006-3

questions. Answering these questions and building one general approach covering them jointly seems to be very challenging.

Referring to cascading effect in infrastructure networks, the issue of initial events that cause degradation and further cascade damage in the network is also raised. The importance of location within the network structure, where the initial damage appeared and the strength of this destruction, should be emphasized. Consequently, very important for critical infrastructure operating environment safety and security, the problems of modeling, identification, prediction, and mitigation of the critical infrastructure accident consequences are arising.

Another important issue regarding dependency analysis in critical networks is how information about interactions among components within the network and among subnetworks be used for critical infrastructure networks design, operation, maintenance, and management and make them more resistant to disturbances, degradation, and failures of other components and subnetworks.

Thus, as a conclusion of this analysis, further research could be focused on reliability and safety analysis of critical infrastructure networks, considering their aging, inside dependencies and outside impacts, and the use of achieved results to improve their safety, strengthen their resilience, and mitigate the effects of their degradation and failures.

Index

Note: Page numbers followed by *f* indicate figures, and *t* indicate tables.

T

FSC
www.fsc.org
MIX
Papier | Fördert
gute Waldnutzung
FSC® C083411

Kolibri 360 GmbH
Ferdinand-Jühlke-Straße 7
99095 Erfurt, Deutschland
produktsicherheit@kolibri360.de